Applied Probability

Valérie Girardin • Nikolaos Limnios

Applied Probability

From Random Sequences to Stochastic Processes

 Springer

Valérie Girardin
Laboratoire de Mathématiques Nicolas
Oresme
Université de Caen Normandie
Caen, France

Nikolaos Limnios
Laboratoire de Mathématiques Appliquées
de Compiègne
Université de Technologie de Compiègne
Compiègne, France

ISBN 978-3-030-07352-7   ISBN 978-3-319-97412-5   (eBook)
https://doi.org/10.1007/978-3-319-97412-5

This Springer imprint is published by the registered company Springer Nature Switzerland AG
The registered company address is: Gewerbestrasse 11, 6330 Cham, Switzerland

Preface

Random elements are defined with reference to a random experiment as functions whose values depend on the result of the experiment. These elements can take values in any measured space. They are called real random variables if the measured space is the real line. A finite family of random variables is a random vector. A denumerable family is a random sequence, taking values in a denumerable power of the real line. Finally, a family of random variables indexed by a continuous set of parameters is a stochastic process, taking values in any general measured space. This volume is dedicated to both random sequences indexed by integers—such as martingales and Markov chains—and random processes indexed by the positive real line—such as Poisson, Markov, and semi-Markov processes. We present all these random elements in order of increasing difficulty, together with notions necessary to their use in applied probability fields. For this purpose, we consider reliability linked to the lifetime of living or industrial systems, entropy linked to information theory, and stationarity and ergodicity linked to signal theory, together with some simulation methods.

The primary target of this book are students studying mathematics at university and PhD students in applied mathematics; in this regard, it can serve as a course at master's and PhD level in applied probability. Further, it is conceived as a support for all researchers and engineers dealing with stochastic modeling issues. Note that this volume is a free translation of the second of our two volumes published in French by Vuibert, Paris, which are in their third edition. We have made it as much self-contained as possible. Still, for avoiding cumbersome definitions and statements, basic notions of a first course of probability, including some integration and measure theory and present in our first volume, are presupposed.

Each chapter is illustrated with a great number of examples and exercises with their solutions. These are intended to serve as solutions of classical problems in probability, complements to the main text, and introduction to more applied fields. Their levels go from mere application of theoretical notions to elaborated problems. A table of notation and a detailed index are given for easy reference. A classified bibliography is presented for further reading, going from theoretical to applied fields.

First, classical notions of probability on random sequences are given. Then, conditioning, martingales, and Markov chains are detailed, together with an introduction to the random processes theory, indexed with either discrete or continuous time. Finally, jump Markov and semi-Markov processes are investigated. Application is proposed in many fields of applied probability, such as reliability, information theory, production, risk, seismic analysis, and queueing theory.

The book is organized as follows:

We present in Chap. 1 basic notions of probability theory, related to studying independent random sequences. We provide many tools necessary to their investigation, from probabilistic tools, such as independence of sequences of events and distributions of sums of random variables, to analytical tools, such as moment transforms of all kinds or entropy. We recall basic definitions and notions on classical distributions through examples. We also present elements of stochastic topology with the different types of convergence of random sequences: almost sure, in mean, in quadratic mean, in probability, and in distribution. We prove different laws of large numbers and central limit theorems.

We present in Chap. 2 the conditional distributions and expectation in order of increasing complexity, together with many examples of practical determination. Then, we investigate the main properties of stopping times and of discrete-time martingales, including the stopping time theorem and many classical inequalities and convergence results. A section is dedicated to the celebrated linear model.

We present in Chap. 3 the discrete Markov chains. We investigate their main properties in great detail, illustrated by numerous typical examples. Then, we apply them especially to reliability and to branching processes theory.

We present in Chap. 4 an introduction to the theory of stochastic processes, mainly through the investigation of their most classical families: jump processes, stationary and ergodic processes, processes with independent increments such as the Brownian motion, point processes with the examples of renewal and Poisson processes, autoregressive time series, etc.

Finally, we present in Chap. 5 jump Markov processes and semi-Markov processes with discrete state spaces. Application to reliability, biology, and queues is realized in particular through exercises with detailed solutions.

Caen, France   Valérie Girardin
Compiègne, France   Nikolaos Limnios
April 2018

Contents

1 Independent Random Sequences 1
  1.1 Denumerable Sequences 1
    1.1.1 Sequences of Events 8
    1.1.2 Independence 9
  1.2 Analytic Tools 12
    1.2.1 Generating Functions 13
    1.2.2 Characteristic Functions 14
    1.2.3 Laplace Transforms 17
    1.2.4 Moment Generating Functions and Cramér Transforms 19
    1.2.5 From Entropy to Entropy Rate 21
  1.3 Sums and Random Sums 25
    1.3.1 Sums of Independent Variables 25
    1.3.2 Random Sums 30
    1.3.3 Random Walks 33
  1.4 Convergence of Random Sequences 34
    1.4.1 Different Types of Convergence 34
    1.4.2 Limit Theorems 37
  1.5 Exercises 44

2 Conditioning and Martingales 59
  2.1 Conditioning 59
    2.1.1 Conditioning with Respect to an Event 60
    2.1.2 Conditional Probabilities 61
    2.1.3 Conditional Distributions 64
    2.1.4 Conditional Expectation 66
    2.1.5 Conditioning and Independence 73
    2.1.6 Practical Determination 75
  2.2 The Linear Model 79
  2.3 Stopping Times 82
    2.3.1 Properties of Stopping Times 83
  2.4 Discrete-Time Martingales 85
    2.4.1 Definitions and Properties 85
    2.4.2 Classical Inequalities 90
    2.4.3 Martingales and Stopping Times 93

	2.4.4	Convergence of Martingales	97
	2.4.5	Square Integrable Martingales	99
2.5		Exercises	102

3 Markov Chains ... 113
3.1		General Properties	113
	3.1.1	Transition Functions with Examples	113
	3.1.2	Martingales and Markov Chains	122
	3.1.3	Stopping Times and Markov Chains	124
3.2		Classification of States	126
3.3		Stationary Distribution and Asymptotic Behavior	132
3.4		Periodic Markov Chains	140
3.5		Finite Markov Chains	145
	3.5.1	Specific Properties	145
	3.5.2	Application to Reliability	151
3.6		Branching Processes	154
3.7		Exercises	158

4 Continuous Time Stochastic Processes ... 175
4.1		General Notions	175
	4.1.1	Properties of Covariance Functions	181
4.2		Stationarity and Ergodicity	183
4.3		Processes with Independent Increments	190
4.4		Point Processes on the Line	193
	4.4.1	Basics on General Point Processes	193
	4.4.2	Renewal Processes	195
	4.4.3	Poisson Processes	200
	4.4.4	Asymptotic Results for Renewal Processes	203
4.5		Exercises	207

5 Markov and Semi-Markov Processes ... 215
5.1		Jump Markov Processes	215
	5.1.1	Markov Processes	215
	5.1.2	Transition Functions	218
	5.1.3	Infinitesimal Generators and Kolmogorov's Equations	221
	5.1.4	Embedded Chains and Classification of States	224
	5.1.5	Stationary Distribution and Asymptotic Behavior	231
5.2		Semi-Markov Processes	236
	5.2.1	Markov Renewal Processes	236
	5.2.2	Classification of States and Asymptotic Behavior	239
5.3		Exercises	242

Further Reading ... 253

Index ... 255

Notation

$\mathbb{N}$	Set of natural numbers $\{0, 1, 2, \ldots\}$
$\mathbb{Z}$	Set of integers $\{\ldots, -2, -1, 0, 1, 2, \ldots\}$
$\mathbb{Q}$	Set of rational numbers
$\mathbb{R}$	Set of real numbers
$[\![a, b]\!]$	Set of natural numbers $a \leq n \leq b$
$]a, b[$	Set of real numbers $a < x < b$
$[a, b]$	Set of real numbers $a \leq x \leq b$
$\overline{\mathbb{R}}$	Extended real line $[-\infty, +\infty]$
$\mathbb{R}_+$	Set of nonnegative numbers $[0, +\infty[$
$\mathbb{C}$	Set of complex numbers
$\mathcal{R}e\, z$	Real part of the complex number z
$\mathcal{I}m\, z$	Imaginary part of the complex number z
i	Square root of -1, i.e., $i^2 = -1$
E^*	Set of all elements of E but 0
∂E	Boundary of the set E
$\overset{\circ}{E}$	Interior of E
$|E|$	Cardinal of E
$\prod_{i=1}^{n} E_i$	Cartesian product of the sets $E_1, \ldots, E_n$
E^n	Order n Cartesian product of E
$E^{\mathbb{N}}$	Set of sequences of elements of E
$n\,!$	Factorial n
$\binom{N}{n}$	Binomial coefficient
$a \vee b$	Maximum of the real numbers a and b
$a \wedge b$	Minimum of the real numbers a and b
δ_{ab}	Kronecker's symbol; equal to 1 if $a = b$, 0 otherwise
$x \to x_0^+$	x tends to x_0, with $x > x_0$
$x \to x_0^-$	x tends to x_0, with $x < x_0$
$[x]$	Integer part of x
Δx_n	Increment of the sequence (x_n), i.e., $\Delta x_n = x_n - x_{n-1}$

θ	Shift operator on sequences,
	i.e., $\theta(x_0, x_1, \ldots) = (x_1, x_2, \ldots)$
$\log$	Neperian logarithm
Id	Identity function
$f \circ g$	Composite of the functions f and g,
$f^{\circ n}$	Order n composite of f
$f(\cdot, y)$	Partial function $x \longrightarrow f(x, y)$ for a fixed y
f^+	Positive part of f, that is $\max(f, 0)$
f^-	Negative part of f, that is $\max(-f, 0)$
$\mathcal{C}^0(I)$	Set of continuous functions on $I \subset \mathbb{R}$
$\mathcal{C}^n(I)$	Set of n times differentiable functions on $I \subset \mathbb{R}$
$o(x)$	Small o of x at $x_0 \in \overline{\mathbb{R}}$, i.e., $\lim_{x \to x_0} o(x)/x = 0$
$O(x)$	Large O of x at $x_0 \in \overline{\mathbb{R}}$,
	i.e., $\lim_{x \to x_0} O(x)/x = c$, with $c \in \mathbb{R}$
$f(x) \approx g(x)$	Equivalent functions at $x_0 \in \overline{\mathbb{R}}$,
	i.e., $\lim_{x \to x_0} f(x)/g(x) = 1$
$f(x) \overset{\sim}{=} c$	$c \in \mathbb{R}$ is an approximated value of $f(x)$
$f(+\infty)$	Limit of the function f at $+\infty$
$\overline{\lim} \, f_n$	Superior limit of the sequence of functions (f_n),
	i.e., $\overline{\lim} \, f_n(x) = \inf_{n \geq 0} \sup_{k \geq n} f_k(x)$
$\underline{\lim} \, f_n$	Inferior limit of (f_n),
	i.e., $\underline{\lim} \, f_n(x) = \sup_{n \geq 0} \inf_{k \geq n} f_k(x)$
$\mathbf{1}_d$	d-Dimensional column vector with components 1
$(x_1, \ldots, x_d)$	Row vector
$(x_1, \ldots, x_d)'$	Column vector
$< v, w >$	Scalar product of $v = (x_1, \ldots, x_d)$ and
	$w = (y_1, \ldots, y_d)$, i.e., $< v, w >= \sum_{i=1}^d x_i y_i$
$\|v\|$	Euclidean norm of $v = (x_1, \ldots, x_d)$,
	i.e., $\|v\| = (\sum_{i=1}^d |x_i|^2)^{1/2}$
$\mathrm{Sp}(v_1, \ldots, v_n)$	Linear subspace spanned by $v_1, \ldots, v_n$
I_d	d-Dimensional identity matrix
$\det M$	Determinant of the matrix M
M'	Transpose of M
$\mathrm{diag}(a_i)$	Diagonal matrix whose i-th coefficient is a_i
$J_\varphi(u)$	Jacobian of the function φ at $u \in \mathbb{R}^d$
Ω	Sample space of a random experiment
ω	Outcome of a random experiment
$\Omega \setminus A = \overline{A}$	Complement of a subset A of Ω
$\mathbb{1}_A$	Indicator function of a subset A of Ω
$\mathcal{F}$	σ-Algebra
$\sigma(\mathcal{C})$	σ-Algebra generated by a family $\mathcal{C}$

$\mathcal{B}(E)$	Borel σ-Algebra of the topological space E
$\mathcal{P}(\Omega)$	Set of all subsets of Ω
μ	Positive measure
δ_ω	Dirac measure at ω
λ	Lebesgue measure on $\mathbb{R}$
λ_d	Lebesgue measure on $\mathbb{R}^d$, for $d > 1$
$\mathbb{P}$	Probability measure
$(\Omega, \mathcal{F}, \mathbb{P})$	Probability space
$\otimes_{i=1}^n \mathcal{F}_i$	Product of the σ-Algebras $\mathcal{F}_1, \ldots, \mathcal{F}_n$
$\mathcal{F}^{\otimes n}$	Order n product of $\mathcal{F}$
$\otimes_{i=1}^n \mathbb{P}_i$	Product of the probabilities $\mathbb{P}_1, \ldots, \mathbb{P}_n$
$\mathbb{P}^{\otimes n}$	Order n product of $\mathbb{P}$
$\sigma(f)$	σ-Algebra generated by the measurable function f
$\sigma(f_i, i \in I)$	σ-Algebra generated by the collection $\{f_i : i \in I\}$
$(f \in B)$	Inverse image of B by f
$(a \le f < b)$	Inverse image of $[a, b[$ by f
$(f \in dx)$	Short for $(x < f \le x + dx)$
$\int f d\mu$	Lebesgue integral of the function f with respect to the measure μ
$\mathcal{L}^p(\Omega, \mathcal{F}, \mu)$	Set of $\mathcal{F}$-measurable functions whose p-th power is μ-integrable
$L^p(\Omega, \mathcal{F}, \mu)$	Set of the equivalence classes for the relation of a.e. equality in $\mathcal{L}^p(\Omega, \mathcal{F}, \mu)$
$\|f\|_p$	L^p-norm
$\mu \ll \nu$	The measure μ is absolutely continuous with respect to the measure ν
a.e.	Almost everywhere
a.s.	Almost surely
X	Random variable
$\mathbb{P}_X$	(Probability) distribution of X
$X \sim P$	The variable X has the distribution P
$X \sim Y$	X and Y have the same distribution
F_X	(Cumulative) distribution function of X
f_X	Density function of the variable X
$\mathbb{E} X$	Expectation (mean) of the variable X
$\mathbb{V}\mathrm{ar}\, X$	Variance of X
$\mathcal{B}(p)$	Bernoulli distribution
$\mathcal{B}(n, p)$	Binomial distribution
$\mathcal{B}_-(r, p)$	Negative binomial distribution
$\mathcal{G}(p)$	Geometric distribution
$\mathcal{P}(\lambda)$	Poisson distribution

$\mathcal{U}(a,b)$	Uniform distribution on the interval $[a,b]$
$\mathcal{N}(\mu,\sigma^2)$	Gaussian distribution with mean $\mu \in \mathbb{R}$ and variance $\sigma^2 \in \mathbb{R}_+$
$\gamma(a,b)$	Gamma distribution
$\chi^2(n)$	Chi square distribution with n degrees of freedom
$\mathcal{E}(n,\lambda)$	Erlang distribution
$\mathcal{E}(\lambda)$	Exponential distribution
$\mathcal{C}(\alpha)$	Cauchy distribution
g_X	Generating function of the variable X
$\widehat{f}$	Fourier transform of the function f
$\widehat{\mu}$	Fourier transform of the measure μ
φ_X	Characteristic function of X
ψ_X	Laplace transform of X
M_X	Moments generating function of X
h_X	Cramér transform of X
R	Survival function or reliability of a system
$\mathcal{S}(P)$	Entropy of P
$\mathcal{S}(X)$	Entropy of the variable X
$\mathbb{S}(X)$	Entropy rate of the sequence $X = (X_n)$
$\mathbb{C}\mathrm{ov}\,(X,Y)$	Covariance of the variables X and Y
$\rho_{X,Y}$	Correlation coefficient of X and Y
$(X_1,\ldots,X_d)$	d-Dimensional row random vector
$Cov(X,Y)$	Covariance matrix of the vectors X and Y
$f * g$	Convolution product of the functions f and g
$\mu * \nu$	Convolution product of the measures μ and ν
$\mathcal{N}_d(M,\Gamma)$	d-Dimensional Gaussian distribution with mean the vector M and covariance the matrix Γ
i.i.d.	Independent and identically distributed
(X_n)	Random sequence indexed by $\mathbb{N}$
$(\Omega,\mathcal{F},\mathbb{P})^{\mathbb{N}}$	Infinite product space of $(\Omega,\mathcal{F},\mathbb{P})$
$\overline{\lim}\,A_n$	Superior limit of the sequence of events (A_n)
$\underline{\lim}\,A_n$	Inferior limit of (A_n)
$\mathcal{CP}(\lambda,P)$	Compound Poisson distribution with parameter $\lambda \in \mathbb{R}_+$
$\xrightarrow{p.s.}$	Almost sure convergence
$\xrightarrow{P}$	Convergence in probability
$\xrightarrow{\mathcal{L}}$	Convergence in distribution
$\xrightarrow{L^p}$	Convergence in L^p-norm
$\mathbb{P}(B \mid A)$	Conditional probability of B given A
$\mathbb{P}(A \mid \mathcal{G})$	Conditional probability of the event A with respect to the sub-σ-algebra $\mathcal{G}$

$\mathbb{E}(X \mid \mathcal{G})$	Conditional expectation of the variable X
$\mathbb{V}\mathrm{ar}(X \mid \mathcal{G})$	Conditional variance of X
$f_{Y\mid X}(\cdot\mid x)$	Density of Y conditional on $(X = x)$
$U_m(a, b)$	Number of passages of a process at levels a and b in $[0, m]$
$(< M >_n)$	Quadratic characteristic of the square integrable martingale (M_n)
P	Transition matrix of a Markov chain
μ	Initial distribution
E	State space
T_i	Return time to the state i
N_i	Total sojourn time in i
N_i^n	Sojourn time of i in $[1, n]$
U^α	α-Potential, for $\alpha \in]0, 1]$
T_i^n	n-th visit time to i
λ	Stationary measure
π	Stationary distribution
E_k	Cyclic class
E_r	Set of the recurrent states
E_t	Set of the transient states
U	Fundamental matrix of a Markov chain
Z	Fundamental matrix of an ergodic chain
$\mathbf{X} = (X_t)$	Stochastic process
$\mathbb{T} \subset \mathbb{R}$	Set of indexes of a process
R_X	Covariance function of $\mathbf{X}$
r_X	Correlation function of $\mathbf{X}$
h_X	Spectral density of $\mathbf{X}$
$\overline{\mathbf{X}}$	Time mean of $\mathbf{X}$
$M_p(\mathbb{R}^d)$	Set of all point measures on $\mathbb{R}^d$
$\mathbf{N} = (N_t)$	Counting process
θ_s	s-Shift operator
$\{P_t : t \geq 0\}$	Transition function of a Markov process
$A = (a_{ij})$	Infinitesimal generator
(J_n)	Embedded Markov chain
(J_n, T_n)	Markov renewal process
Q	Semi-Markov kernel
$\mathbf{Z}$	Semi-Markov process

Independent Random Sequences

1

Random sequences are understood here as real random processes indexed by discrete time. Advanced notions on sequences of real random variables are presented. Apart for their intrinsic value, they are thought as a tool box for the other chapters. They will be completed in Chap. 4 by definitions and properties common to all real random processes, either indexed by continuous or by discrete time.

More precisely, we first give the basic notions of probability linked to the structural probabilistic concept of independence, leading to develop independent random sequences.

We define the functional tools of use for studying random variables, such as generating functions, characteristic functions, Laplace transforms, or entropy. We detail the main properties of classical sums of random variables, leading to define random sums. So doing, we recall the main properties of the classical distributions, through many examples.

We briefly define the main types of convergence of random sequences: almost sure, in mean, in probability, in distribution. We present the weak and strong large numbers laws and central limit theorems, that constitute the most remarkable results of the classical probability theory. They are proven under weak hypothesis, and completed by some basic large deviations results.

For simplifying notation, mainly random variables are considered. Note that most results extend to random vectors, inter alia, by considering them as random sequences indexed by a finite set.

1.1 Denumerable Sequences

A probability space $(\Omega, \mathcal{F}, \mathbb{P})$ is composed of the set Ω of all possible issues of a random experiment, of a σ-algebra $\mathcal{F}$ defined on Ω, and of a probability $\mathbb{P}$ defined on the measurable space $(\Omega, \mathcal{F})$.

© Springer Nature Switzerland AG 2018
V. Girardin, N. Limnios, *Applied Probability*,
https://doi.org/10.1007/978-3-319-97412-5_1

A σ-algebra is a subset of the set $\mathcal{P}(\Omega)$ of all subsets of Ω that is closed under complement and countable union. Its elements—including Ω and $\emptyset$—are called events.

An application $\mathbb{P} \colon \mathcal{F} \longrightarrow \mathbb{R}_+$ is a probability if it is a positive measure with total mass equal to 1. In other words, $\mathbb{P}(\Omega) = 1$, and for any sequence (A_n) of pairwise disjoint events

$$\mathbb{P}\left(\bigcup_{n \geq 0} A_n\right) = \sum_{n \geq 0} \mathbb{P}(A_n).$$

Events with probability zero are said to be null. A property satisfied with probability one is almost sure (a.s.) while it is said to be satisfied almost everywhere (a.e.) for general measures.

Random sequences are rigorously defined through denumerable products of probability spaces.

Definition 1.1 Let $(\Omega, \mathcal{F})$ be a measurable space, $\Omega^{\mathbb{N}}$ the set of sequences of elements of Ω, and π_i the projection on the i-th coordinate, for $i \in \mathbb{N}$.

The infinite product space $(\Omega, \mathcal{F})^{\mathbb{N}}$ is the measurable space $(\Omega^{\mathbb{N}}, \sigma(\{\pi_i^{-1}(\mathcal{F}), i \in \mathbb{N}\})$, where

$$\sigma(\pi_i^{-1}(\mathcal{F}), i \in \mathbb{N}) =$$
$$\sigma(\{A_0 \times \cdots \times A_n \times \Omega \times \Omega \times \cdots : A_i \in \mathcal{F}, i = 0, \ldots, n, \ n \in \mathbb{N}\})$$

is the σ-algebra generated by the infinite cylinders constructed on the rectangles of $\mathcal{F}^{\otimes n}$ for $n \in \mathbb{N}$. $\triangle$

Theorem 1.2 (Kolmogorov Theorem) *For any probability space $(\Omega, \mathcal{F}, \mathbb{P})$, a unique probability $\mathbb{P}^{\otimes \mathbb{N}}$ defined on $(\Omega, \mathcal{F})^{\mathbb{N}}$ exists such that*

$$\mathbb{P}^{\otimes \mathbb{N}}(A_0 \times \cdots \times A_n \times \Omega \times \Omega \times \cdots) = \prod_{i=0}^{n} \mathbb{P}(A_i), \quad (A_0, \ldots, A_n) \in \mathcal{F}^n.$$

The obtained probability space is called the infinite product space and denoted by $(\Omega, \mathcal{F}, \mathbb{P})^{\mathbb{N}}$.

We will denote by $\mathcal{B}(\mathbb{R})$ the Borel σ-algebra of $\mathbb{R}$, that is the σ-algebra generated for example by the intervals $]-\infty, x]$ for $x \in \mathbb{R}$.

A real random variable is a measurable function $X : (\Omega, \mathcal{F}, \mathbb{P}) \longrightarrow (\mathbb{R}, \mathcal{B}(\mathbb{R}))$, such that $X^{-1}(B) \in \mathcal{F}$ for all Borel sets $B \in \mathcal{B}(\mathbb{R})$. Its distribution $\mathbb{P}_X$ is the image probability of $\mathbb{P}$ by X, characterized by the values $\mathbb{P}_X(]-\infty, x])$, that define its distribution function $F_X(x) = \mathbb{P}(X \leq x) = \mathbb{P}_X(]-\infty, x])$ for $x \in \mathbb{R}$. The set $X^{-1}(\mathcal{B}(\mathbb{R}))$ is a σ-algebra on Ω, called the σ-algebra generated by X and denoted by $\sigma(X)$; it is the smallest σ-algebra that makes X measurable.

A random vector $(X_1, \ldots, X_d) : (\Omega, \mathcal{F}, \mathbb{P}) \longrightarrow (\mathbb{R}^d, \mathcal{B}(\mathbb{R}^d))$, with dimension d, is a finite family of d real random variables, or a random variable taking values in $\mathbb{R}^d$. Its distribution is characterized by the values $\mathbb{P}_{(X_1, \ldots, X_d)}(]-\infty, x_1] \times \cdots \times]-\infty, x_d])$ for $(x_1, \ldots, x_d) \in \mathbb{R}^d$. It is also characterized by its marginal distributions $\mathbb{P}_{(X_{i_1}, \ldots, X_{i_n})}$ for all $1 \leq i_1, \ldots, i_n$ and $n \leq d$.

A real random sequence $(X_n) = (X_n)_{n \in \mathbb{N}}$ is a denumerable family of real random variables. It is also a random variable taking values in the set of all sequences of real numbers $\mathbb{R}^{\mathbb{N}} = \{(x_0, x_1, x_2, \ldots) : x_n \in \mathbb{R}, n \in \mathbb{N}\}$. For any fixed $\omega \in \Omega$, the application defined on $\mathbb{N}$ by $n \longrightarrow X_n(\omega)$ is called a trajectory or realization of the sequence.

By definition, the image probability of $\mathbb{P}^{\otimes \mathbb{N}}$ by (X_n) is characterized by its value for infinite cylinders. In other words, the distribution of (X_n) is characterized by its finite marginal distributions $\mathbb{P}_{(X_{i_1}, \ldots, X_{i_n})}$ for all $i_1, \ldots, i_n$ and $n \in \mathbb{N}$. Therefore, two random sequences (X_n) and (Y_n) have the same distribution if and only if $(X_0, \ldots, X_n) \sim (Y_0, \ldots, Y_n)$ for all $n \in \mathbb{N}$.

Let μ and ν be two nonnegative measures on a measure space $(\Omega, \mathcal{F})$. The measure μ is said to be absolutely continuous with respect to ν (or $\mu \ll \nu$) if $\mu(A) = 0$ for all $A \in \mathcal{F}$ such that $\nu(A) = 0$.

Theorem 1.3 (Radon-Nikodym) *Let μ and ν be two σ-finite measures on $(\Omega, \mathcal{F})$. If $\mu \ll \nu$, then a nonnegative Borel function $f : (\Omega, \mathcal{F}, \nu) \longrightarrow (\mathbb{R}_+, \mathcal{B}(\mathbb{R}_+))$ exists such that*

$$\mu(A) = \int_A d\mu = \int_A f \, d\nu, \quad A \in \mathcal{F}.$$

The ν-a.e. defined function f is unique. It is referred to as the density (or Radon-Nikodym derivative) of μ with respect to ν and is typically denoted by $f = d\mu/d\nu$. If h is a μ-integrable function, then $\int_\Omega h \, d\mu = \int_\Omega h f \, d\nu$.

A d-dimensional random vector X is said to be absolutely continuous if its distribution is absolutely continuous with respect to the Lebesgue measure λ on $\mathbb{R}^d$, that is if a nonnegative Borel function $f_X : \mathbb{R}^d \longrightarrow \mathbb{R}_+$ exists such that $d\mathbb{P}_X(x) = f_X(x)dx$, or if

$$\mathbb{E}[h(X)] = \int_{\mathbb{R}^d} h(x) f_X(x) \, dx,$$

for all bounded Borel functions h. The function f_X is called the (probability) density (function) of the random variable X.

▷ *Example 1.4 (Gaussian Variables, Vectors and Sequences)* A random variable X has a Gaussian (or normal) distribution with parameters $m \in \mathbb{R}$ and $\sigma^2 \in \mathbb{R}_+$ if its density on $\mathbb{R}$ is

$$\frac{1}{\sqrt{2\pi}\sigma} e^{-(x-m)^2/2\sigma^2}.$$

We will write $X \sim \mathcal{N}(m, \sigma^2)$. If $m = 0$ ans $\sigma^2 = 1$, it is said to be standard Gaussian. If $\sigma = 0$, it reduces to the Dirac distribution δ_m at m, and said to be degenerated.

A d-dimensional random vector $(X_1, \ldots, X_d)$ is called a Gaussian vector if all the linear combinations of its coordinates are Gaussian variables, in other words if $\sum_{i=1}^d a_i X_i \sim \mathcal{N}(m_a, \sigma_a^2)$, for all $a = (a_1, \ldots, a_d)' \in \mathbb{R}^d$, where (m_a, σ_a^2) depends on a. If $(X_1, \ldots, X_d)$ is a Gaussian vector, then all $X_1, \ldots, X_d$ are Gaussian variables and, for any linear or affine function $f : \mathbb{R}^d \longrightarrow \mathbb{R}^{d'}$, the image $f(X)$ is a d'-dimensional Gaussian vector.

A random sequence (X_n) such that $(X_{i_1}, \ldots, X_{i_d})$ is a Gaussian vector for all $i_1, \ldots, i_d$ and $d \in \mathbb{N}^*$ is called a Gaussian sequence.   ◁

The expectation is the mean value of a random variable. Its rigorous definition is based on the definition of Lebesgue integral with respect to the probability.

Definition 1.5 Let $(\Omega, \mathcal{F}, \mathbb{P})$ be a probability space.

1. Let X be a nonnegative random variable. The expected value of X is defined by

$$\mathbb{E} X = \int_\Omega X(\omega) d\mathbb{P}(\omega)$$

when this quantity is finite, and as $\mathbb{E} X = +\infty$ otherwise.
2. Let X be a random variable. Set $X^+ = \sup(X, 0)$ and $X^- = -\inf(X, 0)$. The expected value (or mean) of X is defined by

$$\mathbb{E} X = \mathbb{E} X^+ - \mathbb{E} X^-$$

when both expected values are finite. If one is finite and the other is not, then $\mathbb{E} X = +\infty$ (or $-\infty$). If none are finite, the expected value of X is not defined.

The transfer theorem allows an integral on Ω to be transformed into an integral on $\mathbb{R}$ by using the image probability, that is the distribution of the variable.

Theorem 1.6 (Transfer) *Let X be a random variable and let $h : \mathbb{R} \longrightarrow \mathbb{R}$ be a Borel function. If $h \circ X$ is integrable, then*

$$\int_\Omega h(X(\omega)) d\mathbb{P}(\omega) = \mathbb{E}[h(X)] = \int_\mathbb{R} h(x) d\mathbb{P}_X(x).$$

The variance of a random variable X is more often of use than its second order moment $\mathbb{E}(X^2)$.

Definition 1.7 The variance of a square integrable random variable X is defined as

$$\mathbb{V}\mathrm{ar}\, X = \mathbb{E}[(X - \mathbb{E} X)^2].$$

The variance is a dispersion parameter: the closer to its mean are the values of X, the smaller is its variance. Note that X is said to be centered if $\mathbb{E}\,X = 0$ and standard if, moreover, $\mathrm{Var}\,X = 1$. The following formula is often of use for computing variances.

Proposition 1.8 (König's Formula) $\mathrm{Var}\,X = \mathbb{E}\,(X^2) - (\mathbb{E}\,X)^2$.

Proof Since expectation is linear,

$$\mathbb{E}\,[(X - \mathbb{E}\,X)^2] = \mathbb{E}\,[X^2 + (\mathbb{E}\,X)^2 - 2X\mathbb{E}\,X]$$
$$= \mathbb{E}\,(X^2) + (\mathbb{E}\,X)^2 - 2(\mathbb{E}\,X)(\mathbb{E}\,X),$$

and the result follows. $\qquad\square$

Let us now state without some of the most classical theorems of integration theory for a measure space $(\Omega, \mathcal{F}, \mu)$.

Theorem 1.9 (Lebesgue Monotone Convergence) *Let* (f_n) *be an increasing sequence of nonnegative Borel functions converging μ-a.e. to a function f. Then* $\int_\Omega f_n d\mu$ *converges to* $\int_\Omega f d\mu$.

Note that the limit f need not be finite.

Theorem 1.10 (Fatou's Lemma) *Let* (f_n) *be a sequence of nonnegative Borel functions. Then*

$$\int_\Omega \underline{\lim}\, f_n d\mu \leq \underline{\lim} \int_\Omega f_n d\mu.$$

Theorem 1.11 (Lebesgue Dominated Convergence) *Let* (f_n) *be a sequence of Borel functions such that $f_n \longrightarrow f$ μ-a.e. and $|f_n| \leq g$ μ-a.e. for all integers n. Then f is integrable and $\int_\Omega f_n d\mu$ converges to $\int_\Omega f d\mu$. Moreover $\int_\Omega |f_n - f| d\mu$ converges to 0.*

This result applies to integrals depending on real parameters as follows.

Proposition 1.12 *Let* $f : I \times \Omega \longrightarrow \mathbb{R}$ *be a function, where I is any interval of $\mathbb{R}$.*

1. **(Continuity)** *If $f(t, \cdot)$ is measurable for all $t \in I$, if $f(\cdot, \omega)$ is continuous at t_0 μ-a.e., and if $|f(t, \omega)| \leq g(\omega)$ μ-a.e. for $t \in I$, where g is some μ-integrable function, then $t \longrightarrow \int_\Omega f(t, \omega) d\mu(\omega)$ is continuous at t_0.*
2. **(Differentiability)** *If $f(t, \cdot)$ is integrable for all $t \in I$, if $f(\cdot, \omega)$ is differentiable on I μ-a.e., and if $\left|\dfrac{\partial f}{\partial t}(t, \omega)\right| \leq g(\omega)$ for $t \in I$ μ-a.e., where g is some*

μ-integrable function, then $\frac{\partial f}{\partial t}(t, \cdot)$ is integrable, $\int_\Omega f(\cdot, \omega)d\mu(\omega)$ is differentiable, and

$$\frac{d}{dt}\int_\Omega f(t, \omega)d\mu(\omega) = \int_\Omega \frac{\partial f}{\partial t}(t, \omega)d\mu(\omega).$$

The following result, easily proven for step functions, then for nonnegative functions and finally for general functions by difference of nonnegative functions, is the basis of the transfer theorem stated above.

Theorem 1.13 *Let* $(\Omega, \mathcal{F}, \mu)$ *be a measure space,* $h: \Omega \longrightarrow \Omega'$ *a measurable function and* $\mu' = \mu \circ h^{-1}$ *the image measure of* μ *by* h*. Let* $f: \Omega' \longrightarrow \mathbb{R}$ *be a Borel function. Then*

$$\int_\Omega (f \circ h)d\mu = \int_{\Omega'} f d\mu'.$$

The following result is one of the most famous concerning multi-dimensional functions.

Theorem 1.14 (Fubini) *Let* $(\Omega, \mathcal{F}, \mu)$ *and* $(\Omega', \mathcal{F}', \nu)$ *be two measure spaces. If* $f: (\Omega \times \Omega', \mathcal{F} \times \mathcal{F}', \mu \otimes \nu) \longrightarrow (\mathbb{R}, \mathcal{B}(\mathbb{R}))$ *is a Borel function, then both partial functions* $f(\cdot, \omega')$ *and* $f(\omega, \cdot)$ *are measurable on their respective spaces.*
If moreover f *is nonnegative or* $\mu \otimes \nu$*-integrable, then*

$$\iint_{\Omega \times \Omega'} f(\omega, x)d(\mu \otimes \nu)(\omega, \omega') = \int_\Omega \left(\int_{\Omega'} f(\omega, \omega')d\nu(\omega') \right)d\mu(\omega)$$

$$= \int_{\Omega'} \left(\int_\Omega f(\omega, \omega')d\mu(\omega) \right)d\nu(\omega').$$

The next change of variables formula is especially convenient for computing distributions of functions of random variables. Let us recall that a $\mathcal{C}^1$-diffeomorphism is a continuously differentiable mapping ψ whose inverse is continuously differentiable too. Its Jacobian is

$$J_\psi(u) = \det \begin{pmatrix} \frac{\partial \psi_1}{\partial x_1}(u) & \cdots & \frac{\partial \psi_1}{\partial x_d}(u) \\ \vdots & & \vdots \\ \frac{\partial \psi_d}{\partial x_1}(u) & \cdots & \frac{\partial \psi_d}{\partial x_d}(u) \end{pmatrix}.$$

Note that $J_{\psi^{-1}}(u) = 1/J_\psi(\psi^{-1}(u))$.

Theorem 1.15 (Change of Variable Formula) *Let* D_1 *and* D_2 *be two open sets of* $\mathbb{R}^d$*. Let* $f: D_2 \longrightarrow \mathbb{R}^d$ *be a Borel function. Let* $\psi: D_1 \longrightarrow D_2$ *be a* $\mathcal{C}^1$-

diffeomorphism with Jacobian $J_\psi(u)$. The function $x \longrightarrow f(x)$ is integrable on D_2 if and only if the function $u \longrightarrow f \circ \psi(u)|J_\psi(u)|$ is integrable on D_1, and we have

$$\int_{D_2} f(x)\,dx = \int_{D_1} f \circ \psi(u)|J_\psi(u)|du. \tag{1.1}$$

For $d = 1$ and $D_1 = [a, b]$, this amounts to the well-known formula

$$\int_{\psi(a)}^{\psi(b)} f(x)\,dx = \int_a^b f \circ \psi(u)\psi'(u)du. \tag{1.2}$$

Let us now present shortly the spaces L^p, that constitute a natural framework for many of the notions investigated in this book.

Let $(\Omega, \mathcal{F}, \mu)$ be any measure space, with μ a positive measure. Let p be a positive integer. The linear space of $\mathcal{F}$-measurable real-valued functions whose p-th power is μ-integrable is typically denoted by $\mathcal{L}^p(\Omega, \mathcal{F}, \mu)$. The quantity

$$\|f\|_p = \left(\int_\Omega |f(\omega)|^p d\mu(\omega) \right)^{1/p}$$

is referred to as the L^p-norm. The space $\mathcal{L}^\infty(\Omega, \mathcal{F}, \mu)$ is defined as the set of all $\mathcal{F}$-measurable functions f whose sup-norm is finite, where

$$\|f\|_\infty = \sup\{x \geq 0 : \mu(|f| \geq x) > 0\}$$
$$= \inf\{x \geq 0 : \mu(|f| \geq x) = 0\} < +\infty. \tag{1.3}$$

If f and g are μ-integrable and if $f = g$ μ-a.e., then $\int_\Omega f d\mu = \int_\Omega g d\mu$, and hence $\|\cdot\|_p$ is a semi-norm on $\mathcal{L}^p(\Omega, \mathcal{F}, \mu)$, for $p \in \overline{\mathbb{N}}^*$. Note that the triangular inequality is given by Minkowski's inequality.

Consider the equivalence relation defined on $\mathcal{L}^p(\Omega, \mathcal{F}, \mu)$ by $f \sim g$ if the two functions are equal almost everywhere (a.e.), that is $\mu(\{\omega \in \Omega : f(\omega) \neq g(\omega)\}) = 0$. The value of $\|f\|_p$ depends only on the class of f. The quotient space is typically denoted by

$$L^p(\Omega, \mathcal{F}, \mu) = \mathcal{L}^p(\Omega, \mathcal{F}, \mu)/\sim,$$

or simply L^p when no ambiguity may arise. Each class is identified to one of its elements, and hence, speaking of μ-a.e. equality remains possible in L^p. We will say that a function is L^p if its p-th power is integrable, that is if $\|f\|_p < +\infty$.

Proposition 1.16 *If $\mu(\Omega) < +\infty$, then $L^{p'} \subset L^p$ for all $1 \leq p \leq p' \leq +\infty$.*

In particular, the above inclusion holds true for any probability measure μ.

All L^p spaces for $p \geq 1$ are Banach spaces. Moreover, $L^2(\Omega, \mathcal{F}, \mu)$, equipped with the scalar product $< f, g >_2 = \int_\Omega fg d\mu$ is a Hilbert space.

Similar definitions for sequences follow.

Definition 1.17 Let (X_n) be any real random sequence. The sequence is:

- integrable (square integrable, L^p for $p \in \mathbb{N}^*$, nonnegative, discrete,...) if each of the random variables X_n is integrable (square integrable, L^p for $p \in \mathbb{N}^*$, nonnegative, discrete,...);
- equi-integrable if $\sup_{n \geq 0} \mathbb{E}[\mathbb{1}_{(|X_n| \geq N)}|X_n|] \longrightarrow 0, \quad N \to +\infty$;
- L^p-bounded if $\sup_{n \geq 0} \|X_n\|_p < +\infty$.

▷ *Example 1.18* Let (X_n) be a random sequence. If $X \in L^1$ is such that $|X_n| \leq X$ almost surely for all $n \geq 0$, then

$$\mathbb{E}[\mathbb{1}_{(|X_n| \geq N)}|X_n|] \leq \mathbb{E}[\mathbb{1}_{(X \geq N)}X], \quad n \geq 0,$$

meaning that (X_n) is equi-integrable.   ◁

1.1.1   Sequences of Events

Inferior and superior limit events constitute a key to the investigation of sequences of events.

Definition 1.19 Let $(\Omega, \mathcal{F}, \mathbb{P})$ be a probability space and let (A_n) be a sequence of events. The set of all $\omega \in \Omega$ belonging to infinitely many A_n, that is

$$\overline{\lim} A_n = \bigcap_{n \geq 0} \bigcup_{k \geq n} A_k$$

$$= \left\{ \omega \in \Omega : \sum_{n \geq 0} \mathbb{1}_{A_n}(\omega) = +\infty \right\} = \{\omega \in \Omega : \forall n, \exists k \geq n, \ \omega \in A_k\},$$

is called the superior limit of (A_n), or limsup.

The set of all the $\omega \in \Omega$ belonging to all A_n but perhaps a finite number, that is

$$\underline{\lim} A_n = \bigcup_{n \geq 0} \bigcap_{k \geq n} A_k$$

$$= \left\{ \omega \in \Omega : \sum_{n \geq 0} \mathbb{1}_{\overline{A_n}}(\omega) < +\infty \right\} = \{\omega \in \Omega : \exists n, \forall k \geq n, \ \omega \in A_k\},$$

is called the inferior limit of (A_n), or liminf.

Properties of Superior and Inferior Limits of Events

1. $\underline{\lim} A_n \subset \overline{\lim} A_n$.
2. A sequence (A_n) is said to converge, if $\underline{\lim} A_n = \overline{\lim} A_n$, then denoted by $\lim A_n$. In particular, if (A_n) is increasing, then $\lim A_n = \cup_{n \geq 0} A_n$. Indeed, for an increasing sequence, ω belongs to $\cup_{n \geq 0} A_n$ if and only if it belongs to some given A_{n_0}; hence it belongs to infinitely many A_n (for all $n \geq n_0$), and so to all but perhaps the first $n_0 - 1$ ones. Symmetrically, if the sequence is decreasing, then $\lim A_n = \cap_{n \geq 0} A_n$.
3. If the A_n are pairwise disjoint, then $\lim A_n = \emptyset$.
4. Let (A_n) and (B_n) be two sequences of events. For union, $\overline{\lim}(A_n \cup B_n) = \overline{\lim} A_n \cup \overline{\lim} B_n$. For intersection, we only have

$$\overline{\lim} A_n \cap \underline{\lim} B_n \subset \overline{\lim}(A_n \cap B_n) \subset \overline{\lim} A_n \cap \overline{\lim} B_n.$$

5. $\underline{\lim} A_n$ and $\overline{\lim} \overline{A_n}$ are complementary in Ω: they are disjoint and their union is Ω. In other words,

$$\overline{\lim} A_n = \bigcup_{n \geq 0} \bigcap_{k \geq n} A_k = \overline{\bigcap_{n \geq 0} \bigcap_{k \geq n} A_k} = \bigcap_{n \geq 0} \bigcup_{k \geq n} \overline{A_k} = \overline{\lim} \, \overline{A_n}.$$

6. $\mathbb{P}(\underline{\lim} A_n) \leq \underline{\lim} \mathbb{P}(A_n) \leq \overline{\lim} \mathbb{P}(A_n) \leq \mathbb{P}(\overline{\lim} A_n)$. The first inequality comes from considering the increasing sequence $C_n = \cap_{k \geq n} A_k$ and the second from considering the decreasing sequence $B_n = \cup_{k \geq n} A_k$.

1.1.2 Independence

The independence of denumerable sequences of events or of random variables is based on the independence of sequences of σ-algebras.

Definition 1.20 Let $(\mathcal{G}_n)$ be a sequence of σ-algebras included in $\mathcal{F}$. The sequence is said to be independent if all finite subcollection is constituted of (mutually) independent σ-algebras, that is if for all $\{\mathcal{G}_{n_k} : k = 1, \ldots, N\} \subset (\mathcal{G}_n)$,

$$\mathbb{P}(B_{n_1} \cap \cdots \cap B_{n_N}) = \mathbb{P}(B_{n_1}) \ldots \mathbb{P}(B_{n_N}), \quad B_{n_1} \in \mathcal{G}_{n_1}, \ldots, B_{n_N} \in \mathcal{G}_{n_N}.$$

The definition applies to finite families of events. If two subsets are independent, then their generated σ-algebras are also independent. For sequences of σ-algebras, this is precisely stated by the associativity principle.

Proposition 1.21 (Associativity Principle) *Let $(\mathcal{G}_n)$ be an independent sequence of σ-algebras included in $\mathcal{F}$. Let I_1 and I_2 be two disjoint subsets of $\mathbb{N}$. Then, the two σ-algebras generated by $(\mathcal{G}_n)_{n \in I_1}$ and by $(\mathcal{G}_n)_{n \in I_2}$ are independent.*

Definition 1.22 Let $(\mathcal{G}_n)$ be a sequence of σ-algebras included in $\mathcal{F}$. The σ-algebra

$$\mathcal{L} = \bigcap_{n \geq 0} \sigma \left(\bigcup_{k \geq n} \mathcal{G}_k \right)$$

is called the asymptotic σ-algebra of $(\mathcal{G}_n)$.

The events of $\mathcal{L}$ are referred to as asymptotic events. If the index sequence is regarded as a time sequence, the asymptotic events depend only on the events posterior to time n, and this for all integers n.

Theorem 1.23 (Kolmogorov's Zero-One Law) *Let* $(\mathcal{G}_n)$ *be an independent sequence of σ-algebras included in* $\mathcal{F}$. *Every asymptotic event of* $(\mathcal{G}_n)$ *is either almost sure (a.s.) or null.*

Proof Let us set $\mathcal{F}_n = \sigma(\cup_{k \geq n} \mathcal{G}_k)$. For any integer n, the σ-algebra $\mathcal{G}_{n-1}$ is independent of $(\mathcal{G}_k)_{k \geq n}$. So, by the associativity principle, it is independent of $\mathcal{F}_n$, and hence also of $\mathcal{L} = \cap_{n \geq 0} \mathcal{F}_n$.

Since $\mathcal{L}$ is independent of $\mathcal{G}_k$ for all integers k, it is also independent of $\mathcal{F}_n = \sigma(\cup_{k \geq n} \mathcal{G}_k)$ for all n, again by the associativity principle. Hence it is independent of $\mathcal{L} = \cap_{n \geq 0} \mathcal{F}_n$. Therefore, every event L of $\mathcal{L}$ is independent of itself, $\mathbb{P}(L) = \mathbb{P}(L \cap L) = \mathbb{P}(L)\mathbb{P}(L)$, that is $\mathbb{P}(L) = 0$ or 1.   □

The independence of finite or denumerable sequences of events is equivalent to the independence of their generated σ-algebras.

Definition 1.24 A sequence of events (A_n) is said to be independent if the sequence of their generated σ-algebras is independent, that is if every finite subsequence is constituted of independent events.

The condition

$$\mathbb{P}(A_1 \cap \cdots \cap A_n) = \prod_{i=1}^{n} \mathbb{P}(A_i), \quad n \in \mathbb{N}.$$

can be proven to be sufficient. Considering the decreasing sequence $B_n = \cap_{k \leq n} A_n$ for $n \in \mathbb{N}$, and taking the limits then yields

$$\mathbb{P} \left(\bigcap_{n \geq 0} A_n \right) = \prod_{n \geq 0} \mathbb{P}(A_n).$$

Borel-Cantelli lemma says that $\overline{\lim} A_n$ is either an almost sure or a null event. This is a particular case of Kolmogorov zero-one law. Due to its paramount importance in applications, we present a specific proof.

Theorem 1.25 (Borel-Cantelli Lemma) *Let (A_n) be a sequence of events.*

1. If $\sum_{n \geq 0} \mathbb{P}(A_n) < +\infty$, then $\mathbb{P}(\overline{\lim} A_n) = 0$.
2. If $\sum_{n \geq 0} \mathbb{P}(A_n) = +\infty$ and if (A_n) is independent, then $\mathbb{P}(\overline{\lim} A_n) = 1$.

Proof

1. By definition, $\overline{\lim} A_n = \cap_{n \geq 0}(\cup_{k \geq n} A_k)$, so

$$0 \leq \mathbb{P}(\overline{\lim} A_n) \leq \mathbb{P}\Big(\bigcup_{k \geq n} A_k\Big) \leq \sum_{k \geq n} \mathbb{P}(A_k), \quad n \in \mathbb{N}.$$

The series whose general term is $\mathbb{P}(A_n)$ converges, so $\sum_{k \geq n} \mathbb{P}(A_k)$ tends to zero when n tends to infinity, and the result follows.
2. We know that $\overline{\overline{\lim} A_n} = \underline{\lim} \overline{A_n} = \cup_{n \geq 0} \cap_{k \geq n} \overline{A_k}$. Set $B_n = \cap_{k \geq n} \overline{A_n}$ for $n \in \mathbb{N}$. Since the sequence (B_n) is increasing, we have $\mathbb{P}(\underline{\lim} \overline{A_n}) = \mathbb{P}(\cup_{n \geq 0} B_n) = \lim_{n \to +\infty} \mathbb{P}(B_n)$. The event B_n can be decomposed into an intersection of independent events, under the form

$$B_n = \Big(\bigcap_{k=n}^{N-1} \overline{A_k}\Big) \cap \Big(\bigcap_{k \geq N} \overline{A_k}\Big), \quad N \geq n+1.$$

So $\mathbb{P}(B_n) = \mathbb{P}(\cap_{n \leq k < N} \overline{A_k}) \mathbb{P}(\cap_{k \geq N} \overline{A_k})$. Since the A_k are independent, we have $\mathbb{P}(\cap_{n \geq k \geq N-1} \overline{A_k}) = \prod_{k=n}^{N-1} \mathbb{P}(\overline{A_k})$ and $\mathbb{P}(\cap_{k \geq N} \overline{A_k}) \leq 1$, so $\mathbb{P}(B_n) \leq \prod_{k=n}^{N-1}[1 - \mathbb{P}(A_k)]$. Since $1 - x \leq e^{-x}$ for all nonnegative x, we have $1 - \mathbb{P}(A_k) \leq \exp[-\mathbb{P}(A_k)]$ for $k \in \mathbb{N}$ from which it follows that

$$0 \leq \mathbb{P}(B_n) \leq \exp[-\sum_{k=n}^{N-1} \mathbb{P}(A_k)].$$

The series whose general term is $\mathbb{P}(A_k)$ diverges, and hence $\sum_{k=n}^{N-1} \mathbb{P}(A_k)$ tends to infinity when N tends to infinity. Therefore $\mathbb{P}(B_n) = 0$, for all n. Finally, $\mathbb{P}(\underline{\lim} \overline{A_n}) = 0 = 1 - \mathbb{P}(\overline{\lim} A_n)$ and the result follows. $\qquad \square$

Independence of random sequences also relies on independence of σ-algebras.

Definition 1.26 A random sequence (X_n) is said to be independent if the corresponding sequence of generated σ-algebras $(\sigma(X_n))$ is independent, that is if every finite subsequence is constituted of independent variables.

Independence of $X_0, \ldots, X_n$ for all n can be proven to be sufficient.

An independent sequence of random variables each of which obeying the same distribution P is referred to as i.i.d. (independent identically distributed) with (common) distribution P. We will say that two sequences are independent if all their finite subsequences are independent, and that a random variable X and an independent random sequence (X_n) are independent if the sequence $(X, X_0, X_1, \ldots)$ is an independent random sequence.

Definition 1.27 Let $(\Omega, \mathcal{F}, \mathbb{P})$ be a probability space. Let $\mathbf{F} = (\mathcal{F}_n)$ be a sequence of σ-algebras $\mathcal{F}_n \subset \mathcal{F}$. If $\mathcal{F}_n \subset \mathcal{F}_{n+1}$ for all n, then $\mathbf{F}$ is called a filtration (or history) of $\mathcal{F}$. We will set $\mathcal{F}_\infty = \sigma(\cup_{n \geq 0} \mathcal{F}_n)$.

In particular, if (X_n) is a random sequence, then the sequence of generated σ-algebras $(\sigma(X_k), 0 \leq k \leq n))$ is called the natural filtration (or internal history) of the sequence, and the σ-algebra $\cap_{n \geq 0} \sigma(X_n, X_{n+1}, \ldots)$ is its tail σ-algebra.

The next result appears as a corollary of Kolmogorov's zero-one law.

Corollary 1.28 *Every event of the tail σ-algebra of an independent random sequence has probability either zero or one.*

▷ *Example 1.29 (Convergence of Series)* Let (X_n) be an independent random sequence. Then the series whose general term is X_n either diverges or converges almost surely. Indeed, the event "$\sum_{n \geq 0} X_n$ converges" belongs to the tail σ-algebra of (X_n).   ◁

Finally, the next definitions will also be of use in the following.

Definition 1.30 Let $\mathbf{F} = (\mathcal{F}_n)$ be a filtration of $\mathcal{F}$ and let (X_n) be a random sequence. The sequence is said to be $\mathbf{F}$-adapted if X_n is $\mathcal{F}_n$-measurable for all n; it is said to be $\mathbf{F}$-predictable if X_n is $\mathcal{F}_{n-1}$-measurable.

1.2 Analytic Tools

We detail in this section the analytic tools useful for investigating random variables, sums and sequences: generating functions for discrete variables, characteristic functions linked to Fourier transform, Laplace transform, Cramér transform. These analytic transforms characterize distributions, and thus are involved in many aspects of the determination of limits of random sequences. Thanks to their property of converting convolution into product, they are also essential in determining distributions of sums.

We also introduce Shannon entropy linked to uncertainty, with the notion of entropy rate especially suited to denumerable random sequences.

1.2.1 Generating Functions

First, the generating function is classically defined for random variables taking integer values.

Definition 1.31 Let X be a random variable taking nonnegative integer values. The generating function of X is defined by

$$g_X(t) = \mathbb{E}(t^X) = \sum_{n \geq 0} \mathbb{P}(X = n)t^n.$$

Properties of Generating Functions

1. $g_X(t)$ is finite at least for $t \in [-1, 1]$, because g_X is an entire series whose convergence radius is at least equal to 1.
2. $g_X(0) = \mathbb{P}(X = 0)$ and $g_X(1) = 1$. Moreover, $g_X(t) \in [0, 1]$ for $t \in [0, 1]$.
3. g_X belongs to $\mathcal{C}^\infty(]-1, 1[)$, with $g_X^{(n)}(0) = n! \, \mathbb{P}(X = n)$ for $n \geq 0$, and hence g_X characterizes the distribution of X.
4. g_X is a convex function—at least on $]-1, 1[$, because $g_X''(s) = \sum_{n \geq 2} n(n-1)\mathbb{P}(X = n)s^{n-2}$ is nonnegative.

$\triangleright$ *Example 1.32 (Generating Functions of Classical Distributions)* The Bernoulli distribution, with parameter p, is the distribution of experiments with two issues. Let $X \sim \mathcal{B}(p)$, such that $\mathbb{P}(X = 1) = p = 1 - \mathbb{P}(X = 0)$. We compute $g_X(t) = pt + 1 - p$ for all $t \in \mathbb{R}$.

The Poisson distribution, with parameter λ, is the distribution of rare events. Let $Y \sim \mathcal{P}(\lambda)$, such that $\mathbb{P}(Y = n) = e^{-\lambda}\lambda^n/n!$ for all $n \in \mathbb{N}$. We compute

$$g_Y(t) = \sum_{n \geq 0} t^n \, \mathbb{P}(Y = n) = e^{-\lambda} \sum_{n \geq 0} \frac{(t\lambda)^n}{n!} = e^{\lambda(t-1)}, \quad t \in \mathbb{R}.$$

The negative binomial distribution with parameters $r \in \mathbb{N}^*$ and $p \in [0, 1]$, is the distribution of the r first successes in the repetition of an experiment with two issues. Let $Z \sim \mathcal{B}_-(r, p)$, such that $\mathbb{P}(X = r + k) = \binom{r+k-1}{r-1}p^r(1-p)^k$, for all $k \geq 0$. We compute

$$g_{Z_r}(t) = \sum_{k \geq 0} \mathbb{P}(Z_r = r + k)t^{r+k} = p^r t^r \sum_{k \geq 0} \binom{r+k-1}{r-1}t^k q^k = \frac{p^r t^r}{(1 - qt)^r},$$

defined for $|t| < 1/q$, where $q = 1 - p$. $\triangleleft$

The so-called generating functions generate moments of random variables as follows.

Proposition 1.33 *Let X be a random variable taking integer values. Let g_X denote its generating function. Then $\mathbb{E}\,X = \lim_{t \to 1^-} g'_X(t)$.*
   If moreover $\mathbb{E}\,X$ is finite, then $\mathbb{V}\mathrm{ar}\,X = g'_X(1)[1 - g'_X(1)] + \lim_{t \to 1^-} g''_X(t)$.

Proof We compute $g'_X(t) = \mathbb{E}\,(Xt^{X-1}) = \sum_{n \geq 1} n\mathbb{P}(X = n)t^{n-1}$, and hence g'_X converges to a (possibly non finite) limit, when $t \to 1^-$; by continuity, $X(\omega)t^{X(\omega)-1}$ converges to $X(\omega)$ for all ω, so the first equality holds.
   In the same way, $g''_X(t) = \mathbb{E}\,[X(X-1)t^{X-2}]$, so

$$\lim_{t \to 1^-} g''_X(s) = \mathbb{E}\,[X(X-1)] = \mathbb{E}\,(X^2) - \mathbb{E}\,X,$$

and the conclusion follows.   □

▷ *Example 1.34 (Moments of Poisson Distributions)* Let $X \sim \mathcal{P}(\lambda)$. We compute

$$g_X(t) = \sum_{n \geq 0} t^n\,\mathbb{P}(X = n) = e^{-\lambda} \sum_{n \geq 0} (t\lambda)^n/n! = e^{\lambda(t-1)}.$$

The generating function is two times differentiable on $\mathbb{R}$, with $g'_X(t) = \lambda g_X(t)$ and $g''_X(t) = \lambda^2 g(t)$. Hence, due to Proposition 1.33, $\mathbb{E}\,X = g'_X(1) = \lambda g_X(1) = \lambda$, and $\mathbb{V}\mathrm{ar}\,X = g_X{''}_X(1) + g'_X(1) - [g'_X(1)]^2 = \lambda^2 g_X(1) + \lambda - \lambda^2 = \lambda$.   ◁

Note that if $\mathbb{E}\,X^{n-1}$ is finite and if $g_X^{(n-1)}$ is differentiable from the left at 1, then the factorial moments of X are given by

$$\mathbb{E}\,[X(X-1)\ldots(X-n+1)] = \lim_{t \to 1^-} g_X^{(n)}(t).$$

1.2.2 Characteristic Functions

The Fourier transform is used in probability theory as follows.

Definition 1.35 Let X be a random variable. The characteristic function of X is the Fourier transform φ_X of its distribution, that is

$$\varphi_X(t) = \mathbb{E}\,(e^{itX}), \quad t \in \mathbb{R},$$

where $i^2 = -1$.

Main properties of characteristic functions derive from properties of Fourier transforms of measures.

Properties of Characteristic Functions

1. $\varphi_X(0) = 1$ and $|\varphi_X(t)| \le 1$ for all $t \in \mathbb{R}$.
2. The characteristic function characterizes the distribution: two variables with the same characteristic function have the same distribution.
3. If X has a density f, then φ_X is the Fourier transform $\widehat{f}$ of f_X,

$$\phi_X(t) = \widehat{f}(t) = \int_{\mathbb{R}} f_X(x) e^{itx} dx,$$

because the distribution $\mathbb{P}_X$ has density f with respect to Lebesgue measure on $\mathbb{R}$.
4. $\varphi_{-X}(t) = \varphi_X(-t) = \overline{\varphi_X(t)}$ so φ_X takes values in $\mathbb{R}$ if and only if X is symmetric, and then φ_X is an even function.
5. $\varphi_{aX+b}(t) = e^{itb}\varphi_X(at)$ for all $a \in \mathbb{R}^*$ and $b \in \mathbb{R}$.

▷ *Example 1.36 (Characteristic Function of Binomial Distributions)* Let $X \sim \mathcal{B}(n, p)$, a binomial distribution with parameters $n \in \mathbb{N}^*$ and $p \in [0, 1]$, that takes the integer values $k \in [\![0, n]\!]$ with probability $\mathbb{P}(X = k) = \binom{n}{k} p^k (1 - p)^{n-k}$. The binomial distribution is in particular the distribution of a repetition of independent Bernoulli experiments. We compute

$$\varphi_X(t) = \sum_{k=0}^{n} e^{itk} \binom{n}{k} p^k (1 - p)^{n-k} = [pe^{it} + (1 - p)]^n,$$

a well-defined quantity for all $t \in \mathbb{R}$.

For $n = 1$, we get the characteristic function of a Bernoulli distribution with parameter p, that is $\varphi_X(t) = pe^{it} + (1 - p)$.   ◁

The following result derives from Fourier's inversion formula.

Proposition 1.37 *If φ_X is integrable, then X has a density with respect to the Lebesgue measure, say f_X, given by*

$$f_X(x) = \frac{1}{2\pi} \int_{\mathbb{R}} e^{-itx} \varphi_X(t) \, dt, \quad \lambda - a.e..$$

▷ *Example 1.38 (Characteristic Function of Gaussian Distributions)* Let X have the standard Gaussian distribution $\mathcal{N}(0, 1)$. Then, by definition,

$$\varphi_X(t) = \frac{1}{\sqrt{2\pi}} \int_{\mathbb{R}} e^{itx} e^{-x^2/2} \, dx.$$

Differentiating with respect to t yields

$$\varphi_X'(t) = \frac{1}{\sqrt{2\pi}} \int_{\mathbb{R}} ixe^{itx}e^{-x^2/2}\,dx = -\frac{1}{\sqrt{2\pi}}t \int_{\mathbb{R}} e^{itx}e^{-x^2/2}\,dx = -t\varphi_X(t).$$

Therefore, φ_X is solution of the differentiable equation $\varphi' = -t\varphi$ with $\varphi(0) = 1$, and hence $\varphi_X(t) = e^{-t^2/2}$.

If $Y \sim \mathcal{N}(m, \sigma^2)$, then

$$\varphi_Y(t) = \frac{1}{\sqrt{2\pi}\sigma} \int_{\mathbb{R}} e^{ity}e^{-(y-m)^2/2\sigma^2}\,dx = \frac{1}{\sqrt{2\pi}}e^{itm} \int_{\mathbb{R}} e^{it\sigma x}e^{-x^2/2}\,dx$$

$$= e^{itm}\varphi_X(\sigma t) = e^{itm}e^{-\sigma^2 t^2/2},$$

for all $t \in \mathbb{R}$.   ◁

The moments of an integer valued random variable can be deduced from its generating function. Similarly, under conditions, the moments of a general random variable can be deduced from its characteristic function.

Theorem 1.39 *Let X be a random variable. If $\mathbb{E}\left(|X|^n\right) < +\infty$, then $\varphi_X \in C^n(\mathbb{R})$, and $\varphi_X^{(m)}(t) = i^m \mathbb{E}\left(X^m e^{itX}\right)$, for $1 \leq m \leq n$. In particular, $\mathbb{E}\left(X^m\right) = i^m \varphi_X^{(m)}(0)$.*

Proof Proof by induction. The equality holds true for $n = 0$. Assume that $\mathbb{E}\left(|X|^{n+1}\right) < +\infty$. By induction, $\varphi_X^{(n)}(t) = \int_{\mathbb{R}} x^n e^{itx}\,d\mathbb{P}_X(x)$.

If $h \neq 0$, then

$$\frac{1}{h}\left[\varphi_X^{(n)}(t+h) - \varphi_X^{(n)}(t)\right] = \int_{\mathbb{R}} \frac{x^n}{h}[e^{ix(t+h)} - e^{itx}]\,d\mathbb{P}_X(x).$$

Let us set $g_h(x) = x^n[e^{itx(t+h)} - e^{itx}]/h$ and $G(x) = x^{n+1}t$. We have

$$|g_h(x)| \leq \frac{1}{|h|}|x^n||e^{ixh} - 1|,$$

but $|e^{ixh} - 1| \leq |xh|$, so that $|g_h| \leq |G|$. Hence, the dominated convergence theorem applies for h tending to zero and yields

$$\varphi_X^{(n+1)}(t) = \int_{\mathbb{R}} tx^n e^{itx}\,d\mathbb{P}_X(x) = \mathbb{E}\left(X^n e^{itX}\right),$$

and the result is proven. Finally, for $t = 0$, we get $\mathbb{E}\left(X^m\right) = i^m \varphi_X^{(m)}(0)$.   □

Moreover, Taylor-Young formula yields

$$\varphi_X(t) = \sum_{k=0}^{n} \frac{i^k t^k}{k!} \mathbb{E}(X^k) + o(t^n).$$

If φ_X is indefinitely differentiable at 0, then $\mathbb{E}(|X|^n) < +\infty$, for all integers n. On the contrary, φ_X may be n-times continuously differentiable on $\mathbb{R}$ while $\mathbb{E}(X^m)$ is not finite for all $m \in [\![1, n]\!]$.

▷ *Example 1.40 (Moments of Gaussian Distributions)* If $X \sim \mathcal{N}(0, \sigma^2)$, its characteristic function is $\varphi_X(t) = e^{-\sigma^2 t^2/2}$.
We compute by induction $\varphi_X^{(2n+1)}(0) = 0$ and

$$\varphi_X^{(2n)}(0) = (-1)^n \sigma^{2n} (2n - 1)(2n - 3)\ldots 3.1, \quad n \geq 0.$$

Theorem 1.39 yields that $\mathbb{E}(X^{2n+1}) = 0$ and $\mathbb{E}(X^{2n}) = \sigma^{2n}(2n)!/n! \, 2^n$. ◁

1.2.3 Laplace Transforms

Laplace transform is classically defined for nonnegative random variables.

Definition 1.41 Let X be a nonnegative random variable. The Laplace transform ψ_X of X is defined by

$$\psi_X(t) = \mathbb{E}(e^{-tX}), \quad t \in \mathbb{R}_+.$$

The definition carries over to possibly negative random variables, but the support of ψ_X would then be an interval of $\mathbb{R}$ possibly reduced to $\{0\}$.
The Laplace transform of a random variable X is the Laplace transform ψ_P of its distribution $\mathbb{P}_X$. The transform $\log \psi_{\mathbb{P}_X}$ is sometimes referred to as the cumulant generating function.
The Laplace transform is related to the characteristic function through $\psi_X(t) = \varphi_X(-it)$, and, when X is integer valued, to the generating function through $\psi_X(t) = g_X(e^{-t})$. Therefore, Laplace transform also characterizes distributions.

Properties of Laplace Transforms

1. $\psi_X(0) = 1$.
2. If X has density f_X, then φ_X is the Laplace transform of f_X,

$$\phi_X(t) = \int_{\mathbb{R}} f_X(x) e^{-tx} dx.$$

3. $\psi_{aX}(t) = \psi_X(at)$ for all $a \in \mathbb{R}_+^*$ and $\psi_{X+b}(t) = e^{-bt} \psi_X(t)$ for all $b \in \mathbb{R}$.

No simple inversion formula exists, but Laplace transforms of the classical distributions are tabulated.

▷ *Example 1.42 (Laplace Transforms of Classical Distributions)* A random variable X has a gamma distribution with parameters $a \in \mathbb{R}_+^*$ and $b \in \mathbb{R}_+^*$ if its density is

$$\frac{b^a}{\Gamma(a)}\, e^{-bx} x^{a-1} \mathbb{1}_{\mathbb{R}_+}(x),$$

where

$$\Gamma(a) = \int_0^{+\infty} e^{-x} x^{a-1}\, dx.$$

Laplace transforms of Poisson distribution $\mathcal{P}(\lambda)$ and gamma distribution $\gamma(a, \lambda)$ are linked through

$$\log \psi_\gamma(t) = a(\log \psi_\mathcal{P})^{-1}(t), \quad t \in \mathbb{R}.$$

Indeed, on the one hand, if $X \sim \gamma(a, \lambda)$,

$$\psi_X(t) = \frac{\lambda^a}{\Gamma(a)} \int_0^{+\infty} e^{-(\lambda+t)x} x^{a-1}\, dx.$$

For $t < b$, using the change of variable $u = (\lambda + t)x$, we get

$$\psi_X(t) = \frac{\lambda^a}{\Gamma(a)} \int_0^{+\infty} e^{-u} \frac{u^{a-1}}{(\lambda+t)^a} du = \left(\frac{\lambda}{\lambda+t}\right)^a.$$

On the other hand, if $X \sim \mathcal{P}(\lambda)$, then

$$\psi_X(t) = \sum_{n \geq 0} e^{-\lambda} \frac{\lambda^n e^{-nt}}{n!} = e^{-\lambda} \sum_{n \geq 0} \frac{(\lambda e^{-t})^n}{n!} = e^{\lambda(e^{-t}-1)},$$

which induces $(\log \psi_\mathcal{P})^{-1}(t) = \log[\lambda/(\lambda+t)]$.

Binomial and negative binomial distributions are similarly linked through their Laplace transforms.   ◁

Moments can be deduced from Laplace transforms. The following result can be proven by induction from the definition.

Proposition 1.43 *If* $\mathbb{E}(X^n) < +\infty$ *for* $n \in \mathbb{N}^*$, *then* ψ_X *is n-times differentiable at 0, with* $\mathbb{E}(X^n) = (-1)^n \psi_X^{(n)}(0)$ *and* $\mathbb{E}(X^n e^{-tX}) = (-1)^n \psi_X^{(n)}(1)$.

An indefinitely differentiable function $\psi : \mathbb{R}_+ \longrightarrow \mathbb{R}$ is said to be completely monotonous if $(-1)^n \psi^{(n)}(t)$ is nonnegative for all n. We state the next result without proof.

Proposition 1.44 *A function* $\psi : \mathbb{R}_+ \longrightarrow \mathbb{R}$ *is the Laplace transform of a nonnegative random variable if and only if is completely monotonous.*

▷ *Example 1.45 (Laplace Transforms of Exponential Distributions)* A distribution $\gamma(1, \lambda)$ is called an exponential distribution—denoted by $\mathcal{E}(\lambda)$, with density $\lambda e^{-\lambda x} \mathbb{1}_{\mathbb{R}_+}(x)$. Exponential distributions model accurately lifetimes of radio-active substances or of electronic components, thanks to their absence of memory property.

The completely monotonous function $t \longrightarrow \lambda/(\lambda + t)$ is the Laplace transform of the exponential distribution with parameter λ.   ◁

1.2.4 Moment Generating Functions and Cramér Transforms

Together with the characteristic function and the Laplace transform, the so-called moment generating function also yields moments of random variables.

Definition 1.46 The moment generating function M_X of a random variable X is defined by

$$M_X(t) = \mathbb{E}(e^{tX}),$$

for all real numbers t such that this quantity is finite.

The moment generating function is defined on an interval I_X of $\mathbb{R}$, possibly reduced to $\{0\}$. This is a convex function, indefinitely differentiable when $I_X \neq \{0\}$. Its name comes from the following result proven by induction using the definition.

Proposition 1.47 *If* M_X *is defined on* $[-t_0, t_0]$ *with* $t_0 > 0$, *then* X *has finite moments of all orders, with* $\mathbb{E}(X^n) = M_X^{(n)}(0)$. *Moreover,* $M_X^{(n)}(1) = \mathbb{E}(X^n e^X)$, *and*

$$M_X(t) = \sum_{n \geq 0} \frac{t^n}{n!} \mathbb{E}(X^n).$$

▷ *Example 1.48 (Moment Generating Functions of Binomial Distributions)* Let $X \sim \mathcal{B}(n, p)$. We have

$$M_X(t) = \sum_{k=0}^{n} e^{tk} \binom{n}{k} p^k (1 - p)^{n-k} = (1 - p + pe^t)^n,$$

and hence $M'(t) = npe^t (1 - p + pe^t)^{n-1}$, so that $\mathbb{E}\,X = np$.   ◁

The Cramér transform relies on the moment generating function.

Definition 1.49 Let X be a non constant random variable such that M_X is defined on $I_X \neq \{0\}$. The Cramér transform $h_X : I_X \longrightarrow \mathbb{R}$ of X is defined by

$$h_X(t) = \sup_{u \in I_X} [ut - \log M_X(u)].$$

Properties of Cramér Transform

1. h_X is convex, as the supremum of affine functions.
2. h_X is nonnegative, because $h_X(t) \geq ut - \log M_X(u)$ for all u, in particular for $u = 0$.
3. The minimum value of h_X is 0, obtained at $t = \mathbb{E}X$. Indeed, by Jensen's inequality, $M_X(u) \geq e^{u\mathbb{E}X}$, so $u\mathbb{E}X - \log M_X(u) \leq 0$ for all u; hence $h_X(\mathbb{E}X) = 0$ and $h_X(t) \geq h_X(\mathbb{E}X)$.

▷ *Example 1.50 (Cramér Transform of the Standard Normal Distribution)* Let $X \sim \mathcal{N}(0, 1)$. We have

$$M_X(t) = \frac{1}{\sqrt{2\pi}} \int_{\mathbb{R}} e^{tx} e^{-x^2/2} \, dx = e^{t^2/2},$$

and hence $h_X(t) = \sup_{u \in \mathbb{R}} (ut - u^2/2) = t^2/2$, for all $t \in \mathbb{R}$. See Fig. 1.1 for $t = 3$.
◁

The following inequalities bound the distance between a random variable and its expected value.

Theorem 1.51 (Chernoff) *If $I_X = \mathbb{R}$, for all $(a, b) \in \mathbb{R}^2$, then*

$$\mathbb{P}(X \geq a) \leq \exp[-h_X(a)] \quad and \quad \mathbb{P}(X \leq b) \leq \exp[-h_X(b)].$$

Clearly, these inequalities are mainly interesting for $a \geq \mathbb{E}X$ and $b \leq \mathbb{E}X$.

Proof We have $(X > a) = (e^{tX} > e^{ta})$ for all $t \in \mathbb{R}_+^*$, and hence by Markov's inequality,

$$\mathbb{P}(X \geq a) = \mathbb{P}(e^{tX} \geq e^{ta}) \leq \frac{M_X(t)}{e^{ta}} = e^{-at + \log M_X(t)}, \quad t > 0,$$

from which the first inequality derives. The second can be proven similarly.   □

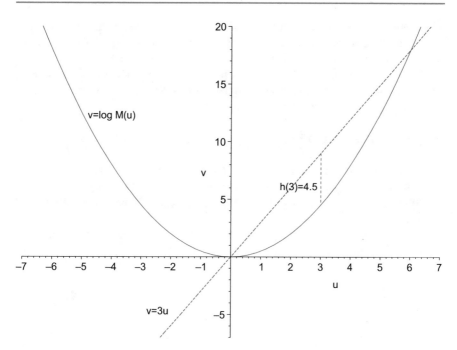

Fig. 1.1 Cramér transform of distribution $\mathcal{N}(0, 1)$ at $t = 3$

1.2.5 From Entropy to Entropy Rate

Entropy measures the quantity of information contained in a random system, that is to say the uncertainty of a random phenomenon.

▷ *Example 1.52 (Information Theory)* A source of information produces randomly sequences of symbols drawn from a given set. The source is said to be without memory if each occurring symbol does not depend on the preceding one. A source is modeled by the set of all possible sequences and is identified to a probability space. For a discrete source, the sample space is called the alphabet, outcomes are letters, and events are words. ◁

Definition 1.53 Let $\Omega = \{\omega_i : i \in I\}$ be a discrete sample space and let P be a probability on $(\Omega, \mathcal{P}(\Omega))$. Set $p_i = P(\{\omega_i\})$. The Shannon entropy (or quantity of information) of P is defined by

$$\mathcal{S}(P) = -\sum_{i \in I} p_i \log p_i,$$

with the convention $0 \log 0 = 0$ and for any base of logarithm.

The quantity of information linked to $A \in \mathcal{P}(\Omega)$ is $-\log P(A)$; especially, P associates to each outcome ω_i the quantity of information $-\log p_i$ and $\mathcal{S}(P)$ is the mean of these quantities, weighted by their probability of occurrence.

The same procedure operates on a partition $\mathcal{A} = (A_i)_{i=1}^n$ of any sample space Ω; then, the system $(\mathcal{A}, P)$ associated with the entropy $\mathcal{S}(\mathcal{A}, P) = \mathcal{S}(p_1, \ldots, p_n) = -\sum_{i=1}^n p_i \log p_i$ is considered, where $p_i = P(A_i)$.

Properties of Shannon Entropy

1. $\mathcal{S}(P) \geq 0$, with equality if $p_{i_0} = 1$ for some i_0, since $p_i \in [0, 1]$.
2. **(Continuous)** $\mathcal{S}(p_1, \ldots, p_n)$ is continuous with respect to each p_i.
3. **(Symmetric)** $\mathcal{S}(p_1, \ldots, p_n) = \mathcal{S}(p_{\sigma(1)}, \ldots, p_{\sigma(n)})$ for any permutation σ of $\{1, \ldots, n\}$.
4. **(Maximum)** For a finite space, the entropy is maximum for the uniform probability. Indeed, $\mathcal{S}(1/n, \ldots, 1/n) = \log n$ and

$$S(p_1, \ldots, p_n) - \log n = -\sum_{i=1}^n p_i \log p_i + \log \frac{1}{n} = \sum_{i=1}^n p_i \log \frac{1}{np_i}.$$

Since $\log x \leq x - 1$ for any positive real number x, we get

$$S(p_1, \ldots, p_n) - \log n \leq \sum_{i=1}^n p_i \left(\frac{1}{np_i} - 1 \right) = 0.$$

5. **(Increasing)** $\mathcal{S}(1/n, \ldots, 1/n)$ increases with n; indeed the uncertainty of a system increases with the number of its elements.
6. **(Additive)** For $p_n = \sum_{k=1}^m q_k$,

$$S(p_1, \ldots, p_{n-1}, q_1, \ldots, q_m) = S(p_1, \ldots, p_n) + p_n S\left(\frac{q_1}{p_n}, \ldots, \frac{q_m}{p_n} \right).$$

Indeed,

$$S(p_1, \ldots, p_{n-1}, q_1, \ldots, q_m) - S(p_1, \ldots, p_n) =$$

$$= p_n \log p_n - \sum_{k=1}^m q_k \log q_k = p_n \sum_{k=1}^m \frac{q_k}{p_n} \log p_n - \sum_{k=1}^m q_k \log q_k$$

$$= -p_n \sum_{k=1}^m \frac{q_k}{p_n} \log \frac{q_k}{p_n} = p_n S\left(\frac{q_1}{p_n}, \ldots, \frac{q_m}{p_n} \right).$$

A functional satisfying 1., 3. and 4. or 1., 2. and 4. can be proven to be the Shannon entropy up to a multiplicative constant (depending on the chosen base of logarithm). These conditions are linked to the physical interpretation of the concept of entropy.

▷ *Example 1.54 (Binary Systems)* Suppose a system contains two elements.

If $P(A_1) = 1 - P(A_2) = 1/200$, then $S(\mathcal{A}, P) \cong 0.0215$ (for the natural logarithm). Guessing which of the two events occurs is easy.

If $P(A_1) = 1 - P(A_2) = 1/20$, then $S(\mathcal{A}, P) = 0.101$, prediction is less easy.

If $P(A_1) = 1 - P(A_2) = 1/2$, then $S(\mathcal{A}, P) = 0.693$, uncertainty is maximum and prediction impossible. The uncertainty is maximum for a uniform distribution of the events.   ◁

The uniform probability has maximum entropy among all probabilities defined on a finite set. Further, the method of maximum entropy consists of choosing among all probabilities satisfying a finite number of constraints the solution that maximizes the entropy—meaning the one modelling the most uncertain system.

Further, the entropy of a discrete random variable X taking values $\{x_i : i \in I\}$ is the (Shannon) entropy of X is that of its distribution, that is

$$S(X) = - \sum_{i \in I} \mathbb{P}(X = x_i) \log \mathbb{P}(X = x_i).$$

▷ *Example 1.55 (Entropy of a Geometric Distribution)* A negative binomial distribution $\mathcal{B}_-(1, p)$ is called a geometric distribution $\mathcal{G}(p)$, modelling for instance the first success in an infinite coin tossing game.

The entropy of a random variable $X \sim \mathcal{G}(p)$, such that $\mathbb{P}(X = k) = p(1-p)^{k-1}$ for $k \geq 1$, is

$$- \sum_{k \geq 1} pq^{k-1} \log(pq^{k-1}) = -p \log p \sum_{k \geq 1} q^{k-1} - p \log q \sum_{k \geq 1} (k-1)q^{k-1}$$

$$= - \log p - \frac{q}{p} \log q,$$

where $q = 1 - p$.   ◁

Many of the properties of the entropy of random variables derive from the properties of entropy of probabilities. In particular, the entropy of a random variable X taking N values is maximum, equal to $\log N$, for a uniform distribution.

The notion of Shannon entropy extends to random variables with densities.

Definition 1.56 Let X be a random variable with density f_X that is positive on $I \subset \mathbb{R}$. The entropy of X is the entropy of its distribution, or

$$S(X) = \mathbb{E}[-\log f_X(X)] = - \int_I f_X(u) \log f_X(u) du.$$

▷ *Example 1.57 (Entropy of Exponential Distributions)* The entropy of a random variable $X \sim \mathcal{E}(\lambda)$ is

$$\int_{\mathbb{R}_+} \lambda e^{-\lambda x} \log(\lambda e^{-\lambda x})\, dx = \log \lambda \int_{\mathbb{R}_+} \lambda e^{-\lambda x}\, dx - \lambda \int_{\mathbb{R}_+} x \lambda e^{-\lambda x}\, dx = \log \lambda + 1,$$

because $\displaystyle\int_{\mathbb{R}_+} f_X(x)\, dx = 1$ and $\mathbb{E}\, X = 1/\lambda$.   ◁

For continuous random variables, the entropy $\mathcal{S}(X)$ can be negative or infinite. Thus, the notion of maximum can be considered only under constraints; maximum entropy methods can be developed in this context.

▷ *Example 1.58 (Maximum Entropy)* The maximum of $\mathcal{S}(X)$ for $a \leq X \leq b$, where $a < b$ are fixed real numbers, is obtained for a variable with uniform distribution $\mathcal{U}(a, b)$; its value is $\log(b - a)$.

The maximum of $\mathcal{S}(X)$ for $X \geq 0$ and $\mathbb{E}\, X = a$, where a is a fixed positive real number, is obtained for a variable with exponential distribution $\mathcal{E}(1/a)$; its value is $\log(ea)$.

The maximum of $\mathcal{S}(X)$ for $\mathbb{E}\, X = 0$ and $\mathbb{E}\,(X^2) = \sigma^2$, where σ^2 is a fixed positive real number is obtained for a variable with normal distribution $\mathcal{N}(0, \sigma^2)$; its value is $\log(\sqrt{2\pi e}\,\sigma)$.

The proof of these results is based on the Lagrange multipliers method. Their extension to random vectors is presented in Exercise 1.6.   ◁

The entropy of a random vector $(X_1, \ldots, X_n)$ is the entropy of its distribution, that is

$$\mathcal{S}(X_1, \ldots, X_n) = -\sum \mathbb{P}(X_0 = x_0, \ldots, X_n = x_n) \log \mathbb{P}(X_0 = x_0, \ldots, X_n = x_n),$$

if X_n takes values in a discrete set $E \subset \mathbb{R}$ for all n, where the sum is taken on all $(x_1, \ldots, x_n) \in E^n$, and

$$\mathcal{S}(X_1, \ldots, X_n) = -\int_I f(x_0, \ldots, x_n) \log f(x_0, \ldots, x_n) dx_0 \ldots dx_n,$$

if $(X_1, \ldots, X_n)$ has a density f, positive on $I \subset \mathbb{R}^n$.

Definition 1.59 Let $X = (X_n)$ be a random sequence. If

$$\frac{1}{n}\mathcal{S}(X_1, \ldots, X_n) \longrightarrow \mathbb{S}(X), \quad n \to +\infty,$$

the limit $\mathbb{S}(X)$ is called the (Shannon) entropy rate of the sequence.

Note that if $\mathbb{S}(X)$ is finite, then $\mathbb{S}(X) = \inf \mathbb{S}_n(X)/n$.

▷ *Example 1.60* Among the random sequences with the same sequence of covariance matrices, the maximum entropy rate is obtained for a Gaussian sequence; see Exercise 1.6.   ◁

▷ *Example 1.61* The entropy rate of an i.i.d. random sequence with distribution P is equal to the entropy of P. Indeed, one can show by induction that the entropy of the sequence up to time n is n times its entropy at time 1.   ◁

1.3   Sums and Random Sums

Sums of a random number of random variables are called random sums. Their properties are based on the properties of the sums of a fixed number of random variables.

1.3.1   Sums of Independent Variables

The determination of the distributions of sums of random variables or vectors is based on the notion of convolution.

Definition 1.62 The convolution of two elements is defined as follows.

1. Let f and g be two nonnegative or integrable Borel functions defined on $\mathbb{R}$. The convolution of f and g is the function defined on $\mathbb{R}$ by

$$f * g(x) = \int_{\mathbb{R}} f(x - y)g(y)dy = \int_{\mathbb{R}} f(y)g(x - y)dy, \quad x \in \mathbb{R}.$$

2. Let μ and v be two measures on $(\mathbb{R}, \mathcal{B}(\mathbb{R}))$, with $d \geq 1$. The convolution of μ and v is the measure $\mu * v$ on $\mathbb{R}$ defined by

$$\mu * v(B) = \int_{\mathbb{R}^2} \mathbb{1}_B(x + y)d\mu(x)d\mu(y), \quad B \in \mathcal{B}(\mathbb{R}).$$

Point 2. applies to bounded Borel functions $h : \mathbb{R} \longrightarrow \mathbb{R}$ under the form

$$\int_{\mathbb{R}} h(z)d\mu * v(z) = \int_{\mathbb{R}^2} h(x + y)d\mu(x)dv(y). \tag{1.4}$$

Proposition 1.63 *If μ and v are absolutely continuous with respect to the Lebesgue measure λ on $\mathbb{R}$, with respective densities f and g, then $\mu * v$ has the density $f * g$ with respect to λ.*

Proof Indeed,

$$\int_{\mathbb{R}} h(z)d\mu * \nu(z) \overset{(1)}{=} \iint_{\mathbb{R} \times \mathbb{R}} h(x+y)f(x)\,dxg(y)dy$$

$$= \int_{\mathbb{R}} h(z)\Big[\int_{\mathbb{R}} f(z-y)g(y)dy\Big]dz.$$

(1) according to (1.4).   □

Properties of Convolution

1. Convolution is commutative, associative and distributive with respect to addition.
2. If $f \in L^1$ and $g \in L^p$, with $p \geq 1$ then $f * g$ is λ-a.e. finite, belongs to L^p, and $\|f * g\| \leq \|f\|_1 \|g\|_p$.
3. The Fourier (Laplace) transform of the convolution product of two elements is the usual product of their Fourier (Laplace) transforms.

The convolution of two discrete measures takes the form of a sum. Let μ and ν be two measures defined on a countable space $\Omega = \{\omega_n : n \in \mathbb{N}\}$ with respective values $\mu(\omega_n) = \mu_n$ and $\nu(\omega_n) = \nu_n$. Their convolution is given by

$$(\mu * \nu)_n = \sum_{k=0}^n \mu_k \nu_{n-k}, \quad n \in \mathbb{N}. \tag{1.5}$$

The generating function of the convolution of two elements is the product of their generating functions.

An example of indefinitely divisible distribution will be given in Exercise 1.3.

Definition 1.64 A probability measure—or distribution—P on $(\mathbb{R}, \mathcal{B}(\mathbb{R}))$ is said to be indefinitely divisible if for all $n \in \mathbb{N}$, a probability measure Q_n exists on $(\mathbb{R}, \mathcal{B}(\mathbb{R}))$ such that P is the order n convolution of Q_n, in other words $P = Q_n^{*n} = Q_n * \cdots * Q_n$.

The distribution of a sum of independent random variables is obtained as the convolution of their distributions.

Theorem 1.65 *Let $X_1, \ldots, X_n$ be n independent random variables. We have*

1. $\mathbb{P}_{X_1 + \cdots + X_n} = \mathbb{P}_{X_1} * \cdots * \mathbb{P}_{X_n}$.
2. *If, moreover, all $X_1, \ldots, X_n$ have densities $f_1, \ldots, f_n$, then their sum $\sum_{i=1}^n X_i$ has a density too, say f, with $f = f_1 * \cdots * f_n$.*

Proof For two variables.

1. The result is an immediate consequence of the definition of convolution.

2. Let $h \colon \mathbb{R} \longrightarrow \mathbb{R}$ be any bounded Borel function. We have

$$
\begin{aligned}
\mathbb{E}\left[h(X_1 + X_2)\right] &= \iint_{\mathbb{R}^2} h(x_1 + x_2) d\mathbb{P}_{(X_1, X_2)}(x_1, x_2) \\
&= \iint_{\mathbb{R}^2} h(x_1 + x_2) d\mathbb{P}_{X_1}(x_1) d\mathbb{P}_{X_2}(x_2) \\
&\overset{(1)}{=} \iint_{\mathbb{R}^2} h(x) d(\mathbb{P}_{X_1} * \mathbb{P}_{X_2})(x),
\end{aligned}
$$

(1) by (1.4). Applying Theorem 1.63 yields the conclusion. □

▷ *Example 1.66 (Sums of Gamma Distributed Variables)* For $n \in \mathbb{N}^*$, a distribution $\gamma(n, \lambda)$ is called an Erlang distribution $\mathcal{E}(n, \lambda)$. The distribution $\mathcal{E}(1, \lambda)$ is just an exponential distribution $\mathcal{E}(\lambda)$.

Let $X_1, \ldots, X_n$ be i.i.d. random variables with distribution $\mathcal{E}(\lambda)$, and set $S_k = X_1 + \cdots + X_k$, for $1 \le k \le n$. The density of S_k can be computed by iterated convolution. An alternative method comes from noting that the density of $(X_1, \ldots, X_n)$ is $\lambda^n e^{-\lambda(x_1 + \cdots + x_n)} \mathbb{1}_{\mathbb{R}_+^n}(x_1, \ldots, x_n)$. Thus, the density of $(S_1, \ldots, S_n)$ is $\lambda^n e^{-\lambda s_n} \mathbb{1}_{(0 < s_1 < \cdots < s_n)}$, easily obtained through the linear change of variables $s_1 = x_1, s_2 = x_1 + x_2, \ldots, s_n = x_1 + \cdots + x_n$.

The distribution of S_n follows by taking the marginal distribution, that is

$$
f_n(s) = \frac{(\lambda s)^{n-1}}{(n-1)!} \lambda e^{-\lambda t} \mathbb{1}_{\mathbb{R}_+}(s),
$$

sayinf that $S_n \sim \mathcal{E}(n, \lambda)$.

Similarly, if $X \sim \gamma(a_1, b)$ and $Y \sim \gamma(a_2, b)$ are independent, then $X + Y \sim \gamma(a_1 + a_2, b)$.

The distribution $\mathcal{E}(k, 1/2)$ is called a chi-squared distribution $\chi^2(2k)$, the distribution of the square of random variable with a standard normal distribution. If $X \sim \chi^2(m)$ and $Y \sim \chi^2(n)$ then $X + Y \sim \chi^2(m + n)$. If X_1 and X_2 are two i.i.d. standard normal variables, then X_1^2 and X_2^2 are i.i.d. with distribution $\chi^2(1)$, so $X_1^2 + X_2^2 \sim \chi^2(2)$. ◁

The distribution of the sum of discrete independent variables is also given by the convolution of their distributions. For example, for two random variables taking integer values, (1.5) becomes

$$
\mathbb{P}(X + Y = n) = \sum_{k=0}^{n} \mathbb{P}(X = k, Y = n - k) = \sum_{k=0}^{n} \mathbb{P}(X = k)\mathbb{P}(Y = n - k).
$$

The variance of the sum of any variables can be written in terms of variances and covariances of the variables. The covariance of two random variables X_1 and X_2 is

$$\mathbb{C}\mathrm{ov}\,(X_1, X_2) = \mathbb{E}\,[(X_1 - \mathbb{E}\,X_1)(X_2 - \mathbb{E}\,X_2)],$$

and the variance of X is simply $\mathbb{V}\mathrm{ar}\,X = \mathbb{C}\mathrm{ov}\,(X, X)$. If $\mathbb{C}\mathrm{ov}\,(X_1, X_2) = 0$, then X_1 and X_1 are said to be uncorrelated.

Proposition 1.67 *Let* $X_1, \ldots, X_d$ *be random variables with finite variances. We have*

$$\mathbb{V}\mathrm{ar}\,\sum_{i=1}^{d} X_i = \sum_{i=1}^{d} \mathbb{V}\mathrm{ar}\,X_i + \sum_{i \neq j} \mathbb{C}\mathrm{ov}\,(X_i, X_j).$$

If moreover $X_1, \ldots, X_d$ *are uncorrelated, then*

$$\mathbb{V}\mathrm{ar}\,(X_1 + \cdots + X_d) = \mathbb{V}\mathrm{ar}\,X_1 + \cdots + \mathbb{V}\mathrm{ar}\,X_d.$$

In particular, for two variables,

$$\mathbb{V}\mathrm{ar}\,(X_1 + X_2) = \mathbb{V}\mathrm{ar}\,X_1 + \mathbb{V}\mathrm{ar}\,X_2 + 2\mathbb{C}\mathrm{ov}\,(X_1, X_2).$$

Proof We prove the proposition for two variables. We develop

$$\mathbb{V}\mathrm{ar}\,(X_1 + X_2) = \mathbb{E}\,([X_1 + X_2 - \mathbb{E}\,(X_1 + X_2)]^2) =$$
$$= \mathbb{E}\,[(X_1 - \mathbb{E}\,X_1)^2 + \mathbb{E}\,(X_2 - \mathbb{E}\,X_2)^2 + 2\mathbb{E}\,(X_1 - \mathbb{E}\,X_1)(X_2 - \mathbb{E}\,X_2)].$$

The case of uncorrelated variables follows immediately.   □

The generating function, the Laplace transform and the characteristic function transform densities of sums into products of transforms, and characterize distributions. Thanks to both these properties, the distribution of sums of random variables can be determined by using ordinary product of functions instead of convolution.

Theorem 1.68 *Let* $X_1, \ldots, X_n$ *be independent random variables.*

1. *The characteristic function of their sum is the product of their characteristic functions.*
2. *If* $X_1, \ldots, X_n$ *are nonnegative, the Laplace transform of their sum is the product of their Laplace transforms.*
3. *If* $X_1, \ldots, X_n$ *take integer values, the generating function of their sum is the product of their generating functions.*

In particular, for i.i.d. variables,

$$g_{\sum_{i=1}^n X_i}(t) = [g_{X_1}(t)]^n \quad \text{and} \quad \varphi_{\sum_{i=1}^n X_i}(t) = [\varphi_{X_1}(t)]^n, \quad t \in \mathbb{R}.$$

Proof

1. For two random variables with densities f and g,

$$\widehat{f * g}(t) = \int_{\mathbb{R}} e^{itx} f * g(x)\, dx = \int_{\mathbb{R}} e^{itx} \int_{\mathbb{R}} f(x - y) g(y) dy\, dx$$

$$\overset{(1)}{=} \int_{\mathbb{R}} g(y) e^{ity} dy \int_{\mathbb{R}} f(x - y) e^{it(x-y)}\, dx$$

$$= \widehat{g}(t) \int_{\mathbb{R}} f(u) e^{itu} du = \widehat{g}(t)\widehat{f}(t).$$

(1) by Fubini's theorem.
2. can be proven in the same way.
3. For two variables,

$$g_{X_1+X_2}(t) = \mathbb{E}\,(t^{X_1+X_2}) = \mathbb{E}\,(t^{X_1})\mathbb{E}\,(t^{X_2}) = g_{X_1}(t) g_{X_2}(t),$$

for all $t \in \mathbb{R}$ for which these quantities are finite.   □

▷ *Example 1.69 (Distributions of Sums of Discrete Distributions)* Let $X \sim \mathcal{P}(\lambda)$ and $Y \sim \mathcal{P}(\mu)$ be two independent variables. Let us show that $X + Y \sim \mathcal{P}(\lambda + \mu)$ by using two different methods:
First, by convolution,

$$\mathbb{P}(X + Y = n) = \sum_{k=0}^{n} \mathbb{P}(X = k)\mathbb{P}(Y = n - k) = \sum_{k=0}^{n} e^{-\lambda}\frac{\lambda^k}{k!} e^{-\mu}\frac{\mu^{n-k}}{(n-k)!}$$

$$= \frac{e^{-(\lambda+\mu)}}{n!} \sum_{k=0}^{n} \binom{n}{k} \lambda^k \mu^{n-k} = e^{-(\lambda+\mu)}\frac{(\lambda + \mu)^n}{n!}.$$

Second, the generating function of $X + Y$ characterizes the distribution. Indeed, using Example 1.34, we compute

$$g_{X+Y}(t) = \mathbb{E}\,(t^X t^Y) = \mathbb{E}\,(t^X)\mathbb{E}\,(t^Y) = e^{\lambda(t-1)} e^{\mu(t-1)} = e^{(\lambda+\mu)(t-1)},$$

that is the generating function of a distribution $\mathcal{P}(\lambda + \mu)$.

Let $X \sim \mathcal{B}(n, p)$ and $Y \sim \mathcal{B}(m, p)$ be two independent binomial variables. Using Example 1.36, we get

$$g_{X+Y}(t) = [pt + (1 - p)]^n [pt + (1 - p)]^m = [pt + (1 - p)]^{n+m}, \quad t \in \mathbb{R},$$

and hence $X + Y \sim \mathcal{B}(m + n, p)$. Similarly, if $X \sim \mathcal{B}_-(r_1, p)$ and $Y \sim \mathcal{B}_-(r_2, p)$ are independent, then $X + Y \sim \mathcal{B}_-(r_1 + r_2, p)$.   ◁

▷ *Example 1.70 (Distributions of Sums of Normal Distributions)* Let $X \sim \mathcal{N}(m_1, \sigma_1^2)$ and $Y \sim \mathcal{N}(m_2, \sigma_2^2)$ be two independent variables. Using Example 1.38, the characteristic function of $X + Y$ is

$$\varphi_{X+Y}(t) = e^{tm_1} e^{-\sigma_1^2 t^2/2} e^{tm_2} e^{-\sigma_2^2 t^2/2} = e^{t(m_1+m_2)} e^{-(\sigma_1^2 + \sigma_2^2)t^2/2}, \quad t \in \mathbb{R},$$

and hence $X + Y \sim \mathcal{N}(m_1 + m_2, \sigma_1^2 + \sigma_2^2)$.

This does not carry over to dependent variables, as the following example shows. Let $X_1 \sim \mathcal{N}(0, 1)$. Let $\varepsilon \sim \mathcal{B}(1/2)$ take values in $\{-1, 1\}$ and be independent of X_1. Set $X_2 = \varepsilon X_1$. We have

$$\mathbb{P}(X_2 \leq x) = \mathbb{P}(X_1 \leq x, \varepsilon = 1) + \mathbb{P}(-X_1 \leq x, \varepsilon = -1)$$

$$= \frac{1}{2}\mathbb{P}(X_1 \leq x) + \frac{1}{2}\mathbb{P}(-X_1 \leq x).$$

Since the standard Gaussian distribution is symmetric, $-X_1 \sim \mathcal{N}(0, 1)$, and hence $X_2 \sim \mathcal{N}(0, 1)$ too. Moreover, since ε and X_1 are independent, we compute $\mathbb{Cov}(X_1, X_2) = \mathbb{E}(\varepsilon X_1^2) = \mathbb{E}(\varepsilon)\mathbb{E}(X_1^2) = 0$, so X_1 and X_2 are uncorrelated and Gaussian. Still, for instance $\mathbb{P}(X_1 + X_2 = 0) = \mathbb{P}(\varepsilon = -1) = 1/2$, showing that the distribution of $X_1 + X_2$ is not Gaussian.   ◁

Further, note that a normal random variable can be shown to decompose into a sum of i.i.d. variables only if these variables also have normal distributions.

▷ *Example 1.71 (The Stable Family of Cauchy Distributions)* A random variable X has a Cauchy distribution $\mathcal{C}(\alpha)$ with parameter $\alpha > 0$ if it has the density $f(x) = \frac{1}{\pi} \frac{\alpha^2}{\alpha^2 + x^2}$ on $\mathbb{R}$, with median zero and infinite expected value.

Using characteristic functions, one can easily show that if X and Y are two i.i.d. variables with Cauchy distribution $\mathcal{C}(\alpha)$, then $aX + bY$ has also a Cauchy distribution $\mathcal{C}((a + b)\alpha)$; this is an example of a stable distribution.   ◁

1.3.2 Random Sums

Randomly indexed random variables often appear in applications.

Definition 1.72 Let (S_n) be a random sequence defined on a probability space $(\Omega, \mathcal{F}, \mathbb{P})$. Let N be an almost surely finite random variable defined on the same space and taking integer values. The randomly indexed variable defined by $\omega \longrightarrow S_{N(\omega)}(\omega)$ is denoted by S_N.

Randomly indexed sums are especially considered, that is $S_N = \sum_{i=1}^{N} X_i$, where $S_n = X_1 + \cdots + X_n$ for a random sequence (X_n) and an integer valued random variable N.

The above results extend to random sums as follows.

Proposition 1.73 *Let (X_n) be an i.i.d. random sequence; let P denote its distribution, with distribution function F, mean M and variance s^2. Let N be an almost surely finite random variable independent of (X_n) and taking positive integer values; let m denote its expected value and σ^2 its variance. Then:*

1. *S_N has expected value mM and variance $s^2m + M^2(\sigma^2 + m^2)$.*
2. *The distribution function of S_N is*

$$F_{S_N}(x) = \sum_{n \geq 1} F^{*(n)}(x)\mathbb{P}(N = n).$$

3. *The Laplace transform (characteristic function) of S_N is the composition of the generating function of N and of the Laplace (Fourier) transform of P.*
4. *If X_n takes integer values, the generating function of S_N is the composition of the generating functions of N and P.*

Proof

1. We compute

$$\mathbb{E}\, S_N = \mathbb{E}\left[\sum_{n \geq 1} \mathbb{1}_{(N=n)} S_n\right] = \sum_{n \geq 1} \mathbb{E}\,[\mathbb{1}_{(N=n)} S_n].$$

Moreover, N and S_n are independent, so

$$\mathbb{E}\, S_N = \sum_{n \geq 1} \mathbb{P}(N = n)\mathbb{E}\, S_n = \sum_{n \geq 1} \mathbb{P}(N = n)n\mathbb{E}\, X_1 = (\mathbb{E}\, N)(\mathbb{E}\, X_1) = mM.$$

Similarly, $\mathbb{E}\,(S_N^2) = \sum_{n \geq 1} \mathbb{P}(N = n)\mathbb{E}\,(S_n^2)$. Since $\mathbb{E}\,(S_n^2) = n\mathrm{Var}\, X_1 + n(\mathbb{E}\, X_1)^2$, we get

$$\mathbb{E}\,(S_N^2) = s^2\left[\sum_{n \geq 1} n\mathbb{P}(N = n)\right] + M^2\left[\sum_{n \geq 1} n^2\mathbb{P}(N = n)\right]$$

$$= s^2\mathbb{E}\, N + M^2\mathbb{E}\,(N^2).$$

Since $\mathbb{E}\,(N^2) = \sigma^2 + m^2$, we get $\mathbb{V}\mathrm{ar}\,S_N = s^2m + M^2(\sigma^2 + m^2)$.

2. We compute

$$\mathbb{P}(S_N \leq s) = \sum_{n \geq 1} \mathbb{P}(S_n \leq s, N = n) = \sum_{n \geq 1} \mathbb{P}(S_n \leq s)\mathbb{P}(N = n),$$

from which the result follows by Point 2. of Theorem 1.65.

3. According to Theorem 1.68, $\psi_{S_n}(t) = \psi_P(t)^n$, so

$$\psi_{S_N}(t) = \sum_{n \geq 1} \mathbb{E}\,[e^{-tS_n}\mathbb{1}_{(N=n)}] = \sum_{n \geq 1} \mathbb{E}\,(e^{-tS_n})\mathbb{P}(N = n)$$

$$= \sum_{n \geq 1} \psi_P(t)^n\mathbb{P}(N = n) = g_N(\psi_P(t)).$$

Similarly, $\phi_{S_N}(t) = g_N(\widehat{P}(t))$.

4. We have

$$\mathbb{P}(S_N = k) = \sum_{n \geq 1} \mathbb{P}(N = n, S_n = k) = \sum_{n \geq 1} \mathbb{P}(N = n)\mathbb{P}(S_n = k).$$

According to Theorem 1.68, $g_{S_n}(s) = g_P(s)^n$, so

$$g_{S_N}(s) = \sum_{k \geq 1} \mathbb{P}(S_N = k)s^k = \sum_{n \geq 1} \mathbb{P}(N = n) \sum_{k \geq 1} \mathbb{P}(S_n = k)s^k$$

$$= \sum_{n \geq 1} \mathbb{P}(N = n)g_{S_n}(s) = \sum_{n \geq 1} \mathbb{P}(N = n)g_P(s)^n,$$

or $g_{S_N}(s) = g_N(g_P(s))$.   □

Definition 1.74 Let (X_n) be an i.i.d. random sequence with distribution P. Let $N \sim \mathcal{P}(\lambda)$ be independent of (X_n). The compound Poisson distribution is defined as the distribution of the random sum $S_N = \sum_{n=1}^{N} X_n$; we will write $S_N \sim \mathcal{CP}(\lambda; P)$.

▷ *Example 1.75* Suppose that the number of accidents per year in a certain type of factory has a Poisson distribution with parameter λ and that the probability of one or more casualties per accident is p. Let us determine the distribution of the number of casualties per year.

We have $g_N(s) = e^{\lambda(s-1)}$ and $X_n \sim \mathcal{B}(p)$, so $g_{X_n}(s) = ps + 1 - p$. According to Proposition 1.73, $g_{S_N}(s) = e^{\lambda p(s-1)}$, so $S_N \sim \mathcal{P}(p\lambda)$. Thus we have shown that the compound Poisson distribution $\mathcal{CP}(\lambda; \mathcal{B}(p))$ reduces to the ordinary Poisson distribution $\mathcal{P}(p\lambda)$.   ◁

1.3.3   Random Walks

Random walks are random sequences defined as sums, which are not i.i.d. but have i.i.d. increments.

Definition 1.76 Let (X_n) be an i.i.d. random sequence taking values in $\mathbb{Z}$. Let S_0 be a random variable also taking values in $\mathbb{Z}$, and independent of (X_n). The sequence (S_n) of random variables

$$S_n = S_0 + X_1 + \cdots + X_n, \quad n \geq 1,$$

is called a random walk.

When $X_n \sim \mathcal{B}(p)$ for all n, the random walk is called a Bernoulli random walk. When X_n takes only the values -1 and 1, the random walk is said to be simple, and symmetric if $\mathbb{P}(X_n = -1) = \mathbb{P}(X_n = 1)$.

A trajectory of a simple random walk is shown in Fig. 1.2.

An equivalent definition of random walks will be given in Chap. 3 in terms of Markov chains. An example of a random walk is given in Exercise 1.8. The next example will allow us to give a simple construction of the Brownian motion in Chap. 4.

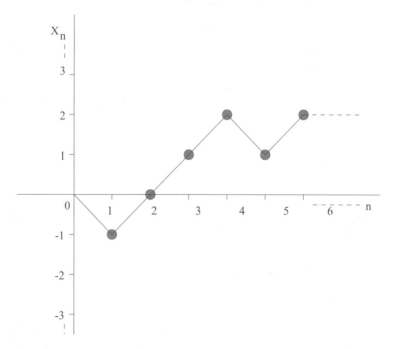

Fig. 1.2 A simple random walk trajectory

▷ *Example 1.77* A coin is tossed every T seconds, or, equivalently, an i.i.d. random sequence (ζ_n) with distribution $\mathbb{P}(\zeta_n = 1) = p = 1 - \mathbb{P}(\zeta_n = -1)$ is considered. Then

$$S_n = s \sum_{j=1}^{n} \zeta_{jT},$$

where $s \in \mathbb{R}_+^*$ and $T \in \mathbb{N}^*$ are fixed, is a random walk taking the values $(2k - n)s$ for $k \in \{0, \ldots, n\}$ with $\mathbb{P}[S_n = (2k - n)s] = \binom{n}{k} p^k (1 - p)^{n-k}$.

For instance, taking $p = 1/2$, we have $\mathbb{E}\, S_n = 0$ and $\mathbb{V}\mathrm{ar}\, S_n = ns^2$.

A particular case (for $s = T = 1$) is studied in details in Exercise 1.9.   ◁

1.4 Convergence of Random Sequences

The basis of stochastic topology is presented here, that is the main types of convergence of random sequences: almost sure, mean, quadratic mean, probability, and distribution. We also prove the weak and strong large numbers laws and the central limit theorem under weak hypothesis.

1.4.1 Different Types of Convergence

We define in this section the four main types of convergence of random sequences, together with some practical criteria of convergence. All the random variables are supposed to be defined on a probability space $(\Omega, \mathcal{F}, \mathbb{P})$.

Definition 1.78 A random sequence (X_n) is said to converge to a a random variable X when n tends to infinity:

- almost surely or with probability 1, denoted by $X_n \overset{\text{a.s.}}{\longrightarrow} X$, if $\mathbb{P}[\{\omega \in \Omega : X_n(\omega) \longrightarrow X(\omega)\}] = 1$.
- in probability, denoted by $X_n \overset{\mathbb{P}}{\longrightarrow} X$, if for all $\varepsilon > 0$, $\mathbb{P}(|X_n - X| > \varepsilon)$ converges to zero.
- in distribution, denoted by $X_n \overset{\mathcal{D}}{\longrightarrow} X$, if $\mathbb{E}[h(X_n)]$ converges to $\mathbb{E}[h(X)]$ for all bounded continuous functions $h : \mathbb{R} \longrightarrow \mathbb{R}$.
- in L^p-norm with $0 < p \leq +\infty$, denoted by $X_n \overset{L^p}{\longrightarrow} X$, if $X_n \in L^p$ for all $n \in \mathbb{N}$, $X \in L^p$ and $\|X_n - X\|_p$ converges to zero.

Note that all above notions of convergence is dependent on the considered probability $\mathbb{P}$.

The convergence in L^p-norm is the topological convergence of the space $L^p(\Omega, \mathcal{F}, \mathbb{P})$ equipped with the norm $\|X\|_p = (\int |X|^p d\mathbb{P})^{1/p}$. Convergence in

L^1-norm is referred to as convergence in mean, and convergence in L^2-norm as convergence in quadratic mean.

The limit random variable X is almost surely unique for the almost sure, probability and L^p convergences. Only distributions are involved for the convergence in distribution: a sequence can indifferently be said to converge to a distribution or to a random variable, but this random variable is not unique, as shown by the following example. The variables of the sequence and their limit in distribution may even be defined on different probability spaces.

▷ *Example 1.79* Let X be a random variable taking values in 1, 1}, with $\mathbb{P}(X = -1) = \mathbb{P}(X = 1) = 1/2$. Set $X_n = X$ for all n. The random sequence (X_n) converges in distribution to X but also to any Y such that $\mathbb{P}(Y = 0) = \mathbb{P}(Y = 1) = 1/2$, and more generally to any random variable with distribution $\mathcal{B}(1/2)$.   ◁

The definitions of the different type of convergence for a sequence of d-dimensional vectors are similar, by replacing the difference $X_n - X$ by a norm of this quantity, for example the Euclidean norm

$$\|X_n - X\| = \sqrt{\sum_{i=1}^{d} [X_n(i) - X(i)]^2}.$$

Thus, the convergence of a sequence of random vectors is equivalent to the convergence of the sequences of its coordinates for all types of convergence but convergence in distribution. Indeed, convergence in distribution of a sequence of vectors induces convergence of the sequences of its coordinates but convergence of all linear combinations is necessary for the converse to hold.

The links between the different convergences are shown by the following diagram.

$$\begin{array}{c} \text{a.s.} \\ \Downarrow \\ L^\infty \Rightarrow L^{p'} \Rightarrow L^p \Rightarrow L^1 \Rightarrow \mathbb{P} , \quad p' \geq p \geq 1. \\ \Downarrow \\ \mathcal{L} \end{array}$$

Some practical criteria simplify proofs of convergence.

First, (X_n) converges almost surely to X if the event $(X_n \longrightarrow X)$ has probability 1. Hence (X_n) converges almost surely if and only if $\mathbb{P}(\underline{\lim} X_n = \overline{\lim} X_n) = 1$. Thus, thanks to Borel-Cantelli lemma, the following result says that an i.i.d. sequence (X_n) will converge almost surely to X if $\sum_{n \geq 0} \mathbb{P}(|X_n - X| > \varepsilon) < +\infty$ for all $\varepsilon > 0$.

Proposition 1.80 *A random sequence* (X_n) *converges almost surely to a variable* X *if and only if* $\mathbb{P}(\overline{\lim}\,|X_n - X| > \varepsilon) = 0$ *for all* $\varepsilon > 0$.

Proof We have

$$(X_n \longrightarrow X) =$$

$$= \{\omega \in \Omega \ : \ \forall N, \exists n \text{ such that if } k \geq n \text{ then } |X_k(\omega) - X(\omega)| < 1/N\}$$

$$= \bigcap_{N \geq 1} \bigcup_{n \geq 0} \bigcap_{k \geq n} (|X_k - X| < 1/N) = \bigcap_{N \geq 1} \underline{\lim}_n (|X_n - X| < 1/N).$$

The sequence of events $\underline{\lim}_n(|X_n - X| < 1/N)$ is decreasing. Their intersection has probability one if and only if all events have probability one, that is if $\mathbb{P}(\underline{\lim}(|X_n - X| < \varepsilon) = 1$, or, by taking complements, if $\mathbb{P}(\overline{\lim}\,|X_n - X| > \varepsilon) = 0$, for all $\varepsilon > 0$.
□

Any Cauchy sequence for the convergence in probability is a convergent sequence because $\mathbb{R}$ is a complete topological space. In other words, if

$$\forall \varepsilon > 0, \forall a > 0, \ \exists N(a, \varepsilon), \quad n, m > N(a, \varepsilon) \Rightarrow \mathbb{P}(|X_n - X_m| > \varepsilon) < a,$$

then a random variable X exists such that $\mathbb{P}(|X_n - X| > \varepsilon)$ converges to zero. Similarly, for a.s. and L^p convergences, Cauchy sequences are convergent.

Convergence in distribution of sequences of random variables is a property of strong convergence of their distributions. Since the generating function, characteristic function and Laplace transform all characterize distributions, convergence criteria follow.

Theorem 1.81 *Let* (X_n) *be a sequence of random variables.*

If the sequence of characteristic functions of X_n *converges to the characteristic function of a random variable* X, *then* (X_n) *converges in distribution to* X.

If the sequence is nonnegative valued, the criterion also holds true for Laplace transforms. If the random sequence takes integer values, the criterion holds true for generating functions.

Moreover, the characteristic function gives a criterion of convergence in distribution that requires no previous knowledge of the limit.

Theorem 1.82 (Lévy's Continuity) *Let* (X_n) *be a sequence of random variables. If the sequence of characteristic functions of* (X_n) *converges to a function* g *continuous at* 0, *then* g *is the characteristic function of a random variable, say* X, *and* (X_n) *converges in distribution to* X.

A similar criterion can be stated for nonnegative variables with the Laplace transform.

1.4.2 Limit Theorems

Numerous criteria of convergence of weighted sums of random variables exist. Here are presented some which are both interesting by themselves, and necessary to the proof of the strong law of large numbers. A large numbers law is a theorem giving conditions under which $\sum_{i=1}^{n}(X_i/n) - \sum_{i=1}^{n} \mathbb{E}(X_i/n)$ converges to zero. It is weak if convergence holds in probability and strong if convergence is almost sure. All following convergence results mainly concern sums of i.i.d. random variables.

Theorem 1.83 (Kolmogorov Inequality) *Let* $X_1, \ldots, X_n$ *be independent centered random variables with finite variances* σ_k^2. *Then, for all* $\varepsilon > 0$,

$$\mathbb{P}\Big[\max(|X_1|, \ldots, |X_1 + \cdots + X_n|) > \varepsilon\Big] \leq \sum_{k=1}^{n} \frac{\sigma_k^2}{\varepsilon^2}.$$

Proof Let us define the events $E = (\max(|S_1|, \ldots, |S_n|) > \varepsilon)$ and $E_k = (|S_k| > \varepsilon, |S_i| < \varepsilon, i = 1, \ldots, k-1)$ for $2 \leq k < n$, where $S_n = X_1 + \cdots + X_n$. We compute

$$\mathbb{E}(\mathbb{1}_{E_k} S_n^2) = \mathbb{E}(\mathbb{1}_{E_k} S_k^2) + \mathbb{E}[\mathbb{1}_{E_k}(S_n - S_k)^2] + 2\mathbb{E}[\mathbb{1}_{E_k} S_k(S_n - S_k)]$$

$$\overset{(1)}{=} \mathbb{E}(\mathbb{1}_{E_k} S_k^2) + \mathbb{P}(E_k)\mathbb{E}[(S_n - S_k)^2].$$

(1) since $\mathbb{1}_{E_k}$ and $\mathbb{1}_{E_k} S_k$ do not depend on $S_n - S_k = X_{k+1} + \cdots + X_n$.
  Therefore, $\mathbb{E}(\mathbb{1}_{E_k} S_k^2) \leq \mathbb{E}(\mathbb{1}_{E_k} S_n^2)$, for all k. Moreover, $|S_k| > \varepsilon$ on E_k, so

$$\mathbb{P}(E_k) \leq \frac{1}{\varepsilon^2}\mathbb{E}(\mathbb{1}_{E_k} S_k^2) \leq \frac{1}{\varepsilon^2}\mathbb{E}(\mathbb{1}_{E_k} S_n^2).$$

Finally, since $E = (\max(|S_1|, \ldots, |S_n|) > \varepsilon)$ is the disjoint union of the E_k, we have

$$\mathbb{P}(E) = \sum_{k \geq 0}\mathbb{P}(E_k) \leq \frac{1}{\varepsilon^2}\sum_{k \geq 0}\mathbb{E}(\mathbb{1}_{E_k} S_n^2) = \frac{1}{\varepsilon^2}\mathbb{E}(\mathbb{1}_E S_n^2) \leq \frac{1}{\varepsilon^2}\mathbb{E}(S_n^2).$$

The conclusion follows because $\mathbb{E}(S_n^2) = \mathbb{V}\text{ar}\, S_n = \sum_{k=1}^{n}\sigma_k^2$. □

Proposition 1.84 *If* (X_n) *is an i.i.d. sequence such that* $\sum_{n \geq 1}\mathbb{V}\text{ar}\, X_n$ *is finite, then* $\sum_{n \geq 1} X_n$ *converges almost surely.*

Proof Let us show that $(X_1 + \cdots + X_n)$ is an almost surely Cauchy sequence, i.e., that for any $\varepsilon > 0$,

$$\mathbb{P}\Big[\bigcup_{m \geq 1}(|X_{n+1} + \cdots + X_{n+m}| > \varepsilon)\Big] \longrightarrow 0, \quad n \to +\infty.$$

We compute

$$\mathbb{P}\left[\bigcup_{m=1}^{N}(|X_{n+1}+\cdots+X_{n+m}|>\varepsilon)\right]=$$

$$=\mathbb{P}\left[\max_{1\le m\le N}(|X_{n+1}+\cdots+X_{n+m}|)>\varepsilon\right]$$

$$\overset{(1)}{\le}\frac{1}{\varepsilon^2}\mathbb{V}\mathrm{ar}\,(X_{n+1}+\cdots+X_{n+N}).$$

(1) by Kolmogorov inequality.

Finally, $\mathbb{V}\mathrm{ar}\,(X_{n+1}+\cdots+X_{n+N})=\sum_{m\ge 1}\mathbb{V}\mathrm{ar}\,X_{n+m}$, converges to 0 when n tends to infinity as the rest of a convergent series.   □

The following admitted lemma is necessary to the proof of Kolmogorov criterion below.

Lemma 1.85 (Stochastic Kronecker Lemma) *Let (X_n) be a random sequence such that $\sum_{n\ge 1}X_n$ is almost surely finite. If (u_n) is an increasing sequence of real numbers converging to infinity, then $\frac{1}{u_n}\sum_{i=1}^{n}u_i X_i$ converges a.s. to zero.*

Proposition 1.86 (Kolmogorov Criterion) *Let (X_n) be an independent random sequence such that*

$$\sum_{n\ge 1}\mathbb{V}\mathrm{ar}\,\left(\frac{X_n}{n}\right)<+\infty. \tag{1.6}$$

Then

$$\frac{1}{n}\sum_{i=1}^{n}(X_i-\mathbb{E}\,X_i)\overset{\text{a.s.}}{\longrightarrow}0.$$

Proof Set $Y_n=X_n/n$. Since $\sum_{n\ge 1}\mathbb{V}\mathrm{ar}\,(Y_n-\mathbb{E}\,Y_n)<+\infty$, by applying Proposition 1.84, the series $\sum_{n\ge 1}(Y_n-\mathbb{E}\,Y_n)$ converges almost surely to Y. Necessarily $\mathbb{E}\,Y=0$, and hence Y is almost surely finite.

Applying the stochastic Kronecker lemma to $(Y_n-\mathbb{E}\,Y_n)$ with $u_n=n$ yields the conclusion.   □

If (1.6) is not satisfied, then the result does not hold any more. Similarly, independence of the variables X_n with equality of means is not sufficient, as shown by the following example.

▷ *Example 1.87* Let (X_n) be an independent sequence such that $\mathbb{P}(X_n=-n)=1-1/n^2=1-\mathbb{P}(X_n=n-n^3)$.

Then $\mathbb{E}\,X_n = 0$ for all $n \geq 1$, but (X_n/n) converges almost surely to -1, and hence $\sum_{i=1}^{n} X_i/n$ can converge almost surely to no finite random variable. $\lhd$

On the contrary, Kolmogorov criterion induces the almost surely convergence of weighted sums of i.i.d. variables.

Theorem 1.88 (Strong Law of Large Numbers) *If (X_n) is an integrable i.i.d. sequence, then*

$$\frac{1}{n}(X_1 + \cdots + X_n) \xrightarrow{\text{a.s.}} \mathbb{E}\,X_1.$$

If moreover the sequence is square integrable, convergence also holds in quadratic mean.

Proof We can suppose with no loss of generality that the variables are centered. Set $X_n^* = X_n \mathbb{1}_{(|X_n| \leq n)}$. Since (X_n) is i.i.d.,

$$\sum_{n \geq 1} \mathbb{P}(X_n^* \neq X_n) = \sum_{n \geq 1} \mathbb{P}(|X_n| > n) = \sum_{n \geq 1} \mathbb{P}(|X_1| > n).$$

Since $\sum_{n \geq 1} \mathbb{P}(|X_1| > n) \leq \mathbb{E}\,|X_1|$, by Borel-Cantelli lemma,

$$\mathbb{P}(X_n^* \neq X_n \text{ infinitely often}) = 0.$$

For proving the searched convergence, it is enough to prove that the sequence $(\sum_{i=1}^{n} X_i^*/n)$ converges almost surely to 0. We have

$$\sum_{n \geq 1} \mathbb{V}\text{ar}\left(\frac{X_n^*}{n}\right) \leq \sum_{n \geq 1} \frac{1}{n^2} \mathbb{E}\,(X_n^{*2}),$$

with

$$\sum_{n \geq 1} \frac{1}{n^2} \mathbb{E}\,(X_n^{*2}) = \sum_{n \geq 1} \frac{1}{n^2} \sum_{m=1}^{n} \mathbb{E}\,[(X_1)^2 \mathbb{1}_{(m-1<|X_1|\leq m)}]$$

$$= \sum_{m \geq 1} \mathbb{E}\,[(X_1)^2 \mathbb{1}_{(m-1<|X_1|\leq m)}] \sum_{n \geq m} \frac{1}{n^2}.$$

Comparing $\sum_{n \geq 1} (1/n^2)$ to $\int_1^{+\infty} x^{-2} dx$, we get that a nonnegative constant K exists such that

$$\sum_{n \geq 1} \mathbb{V}\text{ar}\left(\frac{X_n^*}{n}\right) \leq \sum_{m \geq 1} \mathbb{E}\,\left[|X_1||X_1|\mathbb{1}_{(m-1<|X_1|\leq m)}\right] \frac{K}{m}$$

$$\leq K \sum_{m \geq 1} \frac{1}{m} \mathbb{E}\,\left[m|X_1|\mathbb{1}_{(m-1<|X_1|\leq m)}\right] = K\mathbb{E}\,|X_1| < +\infty.$$

Thus, by Kolmogorov criterion, $\sum_{i=1}^{n}(X_i^* - \mathbb{E}\,X_i^*)/n$ converges almost surely to 0. But $\mathbb{E}\,X_n^* = \mathbb{E}[X_1 \mathbb{1}_{(|X_1| \le n)}]$ converges to 0 by the dominated convergence theorem, and hence, by Cesaro Lemma, the sequence $\sum_{i=1}^{n}\mathbb{E}\,X_i^*/n$ also converges almost surely to 0, and the conclusion follows.

For square integrable variables, it remains to prove that $\overline{X}_n$ converges to 0 in L^2, where $\overline{X}_n = \sum_{i=1}^{n} X_i/n$. We compute

$$\mathbb{E}(|\overline{X}_n|^2) = \frac{1}{n^2}\sum_{i=1}^{n}\mathbb{E}(X_i^2) + \sum_{i \ne j}\mathbb{E}(X_i X_j) = \frac{1}{n^2}\sum_{i=1}^{n}\mathbb{E}(X_1^2) = \frac{1}{n}\mathbb{E}(X_1^2).$$

Therefore, $\mathbb{E}(|\overline{X}_n|^2)$ converges to 0. $\qquad\square$

Moreover, if X_1 is not integrable, then the sequence $(\sum_{i=1}^{n} X_i/n)$ is divergent, thanks of the next proposition.

Proposition 1.89 *If (X_n) is an i.i.d. random sequence such that the random sequence $(\sum_{i=1}^{n} X_i/n)$ converges almost surely, then (X_n) is integrable.*

Proof We have—see Exercise 1.1,

$$\mathbb{E}(|X_1|) \le \sum_{n \ge 0}\mathbb{P}(|X_1| > n) = \sum_{n \ge 0}\mathbb{P}(|X_n| > n).$$

If $(\sum_{i=1}^{n} X_i/n)$ converges almost surely, then (X_n/n) converges to 0 almost surely. Since the variables are independent, Borel-Cantelli induces that necessarily $\sum_{n \ge 1}\mathbb{P}(|X_n| > n) < +\infty$. $\qquad\square$

The weak law of large numbers for i.i.d. sequences appears as a corollary of the strong one, because almost sure convergence implies convergence in probability.

Corollary 1.90 (Weak Law of Large Numbers) *If (X_n) is an integrable i.i.d. random sequence, then*

$$\frac{1}{n}(X_1 + \cdots + X_n) \xrightarrow{\mathbb{P}} \mathbb{E}\,X_1.$$

The following result is more general, the random sequence is not required to be i.i.d., but only to have two finite moments.

Theorem 1.91 *Let (X_n) be a random sequence such that $\mathbb{E}(X_n) = m_n$ and $\mathbb{V}\mathrm{ar}(X_n) = \sigma_n^2$, where $(m_n, \sigma_n^2) \in \mathbb{R} \times \mathbb{R}_+^*$ for all n. If*

$$\frac{1}{n^2}\mathbb{V}\mathrm{ar}\left(\sum_{k=1}^{n} X_k\right) \longrightarrow 0,$$

then

$$\frac{1}{n}\sum_{k=1}^{n}(X_k - m_k) \xrightarrow{\mathbb{P}} 0.$$

Proof Let us set $Y_n = \sum_{k=1}^{n}(X_k - m_k)/n$ for $n \geq 1$. These random variables are centered, with variance

$$\mathbb{E}(Y_n^2) = \mathbb{E}\left[\frac{1}{n}\sum_{k=1}^{n}(X_k - m_k)\right]^2 = \frac{1}{n^2}\mathrm{Var}\left(\sum_{k=1}^{n}X_k\right).$$

Therefore, thanks to Chebyshev's inequality,

$$\mathbb{P}(|Y_n| > \varepsilon) \leq \frac{1}{\varepsilon^2 n^2}\mathrm{Var}\left(\sum_{k=1}^{n}X_k\right),$$

from it follows that $\mathbb{P}(|Y_n| > \varepsilon)$ converges to 0. □

▷ *Example 1.92 (Frequencies and Probabilities)* Let (X_n) be an i.i.d. random sequence with distribution P. The random variables $\mathbb{1}_{(X_n \in B)}$ are i.i.d. too for all Borel sets B. The mean of their common distribution is $P(B)$. The law of large numbers yields

$$P(B) = \lim_{n \to +\infty}\frac{1}{n}\sum_{i=1}^{n}\mathbb{1}_{(X_i \in B)} = \lim_{n \to +\infty}\frac{\text{number of } i \leq n \text{ such that } X_i \in B}{n}.$$

This result bring a mathematical justification to the estimation of probabilities by frequencies. ◁

The law of large numbers extends to i.i.d. random sums as follows.

Theorem 1.93 *Let (X_n) be an integrable i.i.d. random sequence. Let (N_n) be a random sequence taking values in $\mathbb{N}^*$, almost surely finite for all n and converging to infinity when n tends to infinity. If (X_n) and (N_n) are independent, then*

$$\frac{1}{N_n}(X_1 + \cdots + X_{N_n}) \xrightarrow{a.s.} \mathbb{E}X_1, \quad n \to +\infty.$$

Proof Set $S_n = X_1 + \cdots + X_n$. The strong law of large numbers induces that S_n/n converges almost surely to $\mathbb{E}X_1$, or

$$\forall \varepsilon > 0, \exists n_\varepsilon \in \mathbb{N}^*, n > n_\varepsilon \implies \left|\frac{S_n}{n} - \mathbb{E}X_1\right| < \varepsilon, \quad a.s..$$

Since N_n converges almost surely to infinity, some $n_\varepsilon > 0$ exists such that if $n > n_\varepsilon$ then $N_n > n_\varepsilon$ almost surely; the conclusion follows.   □

Central limit theorems are of paramount importance as well in probability theory as in statistics. The simplest one is a weak convergence result.

Theorem 1.94 (Central Limit) *Let (X_n) be an i.i.d. square integrable random sequence. Then*

$$\sqrt{n}\left(\frac{X_1 + \cdots + X_n}{n} - \mathbb{E}\, X_1\right) \xrightarrow{\mathcal{D}} \mathcal{N}(0, \mathbb{V}\mathrm{ar}\, X_1).$$

Proof Set $\overline{X}_n = \sum_{i=1}^{n} X_i / n$. The characteristic function of $\sqrt{n}\,\overline{X}_n$ is

$$\varphi_{\sqrt{n}\overline{X}_n}(t) = \mathbb{E}\left[\exp\left(\frac{it}{\sqrt{n}} \sum_{i=1}^{n} X_i\right)\right] = \varphi_{X_1}\left(\frac{t}{\sqrt{n}}\right)^n.$$

We can suppose all variables centered. Set $\mathbb{V}\mathrm{ar}\, X_1 = \sigma^2$. We have $\varphi'_{X_1}(0) = 0$ and $\varphi''_{X_1}(0) = -\sigma^2$, so, using Taylor's formula, $\varphi_{X_1}(t/\sqrt{n}) = 1 - \sigma^2 t^2 / 2n + O(1/n)$. Therefore,

$$\log \varphi_{\sqrt{n}\overline{X}_n}(t) = n \log \varphi_{X_1}(t/\sqrt{n}) = n \log[1 - \sigma^2 t^2 / 2n + O(1/n)],$$

and $\varphi_{\sqrt{n}\overline{X}_n}(t)$ converges to $e^{-\sigma^2 t^2 / 2}$, for all real numbers t.

The conclusion follows because the latter is the characteristic function of the distribution $\mathcal{N}(0, \sigma^2)$.   □

The weak central limit theorem extends to random sums as follows. A more general statement for random processes—called Anscombe theorem, will be proven in Chap. 4.

Theorem 1.95 *Let (X_n) be an integrable i.i.d. random sequence. Let (N_n) be a random sequence taking values in $\mathbb{N}^*$, almost surely finite for all n, and converging to infinity when n tends to infinity.*

If (X_n) and (N_n) are independent, then

$$\frac{1}{\sqrt{N_n}}(X_1 + \cdots + X_{N_n}) \xrightarrow{\mathcal{D}} \mathcal{N}(0, \mathbb{V}\mathrm{ar}\, X_1), \quad n \to +\infty.$$

The following admitted result is a strong convergence result. For simplifying notation, we state it only for standard variables.

Theorem 1.96 (Almost Sure Central Limit Theorem) *If (X_n) is an i.i.d. random sequence such that $\mathbb{E}\, X_1 = 0$ and $\mathbb{V}\mathrm{ar}\, X_1 = 1$, then*

$$\frac{1}{\log n} \sum_{k=1}^{n} \frac{1}{k} \mathbb{1}_{]-\infty, x]} \left(\frac{X_1 + \cdots + X_k}{\sqrt{k}} \right) \xrightarrow{\text{a.s.}} \Phi(x), \quad x \in \mathbb{R},$$

where Φ is the distribution function of the standard Gaussian distribution $\mathcal{N}(0, 1)$.

Setting $S_n = X_1 + \cdots + X_n$, the weak central limit theorem says that for any continuous bounded function $h : \mathbb{R} \to \mathbb{R}$, when n tends to infinity,

$$\mathbb{E}\left[h(n^{-1/2} S_n) \right] \longrightarrow \frac{1}{2\pi} \int_{\mathbb{R}} h(x) e^{-x^2/2} dx$$

while the strong one says that

$$\frac{1}{\log n} \sum_{k=1}^{n} \frac{1}{k} h(k^{-1/2} S_k) \xrightarrow{\text{a.s.}} \frac{1}{2\pi} \int_{\mathbb{R}} h(x) e^{-x^2/2} dx.$$

The following result gives extremes for the speed of convergence in the limit central theorem, by bounding it. Note that it is a divergence result.

Theorem 1.97 (Law of Iterated Logarithm) *If (X_n) is an i.i.d. square integrable random sequence, then*

$$\overline{\lim} \frac{\sqrt{n}(\overline{X}_n - \mathbb{E}\, X_1)}{\sqrt{\mathbb{V}\mathrm{ar}\, X_1} \sqrt{2 \log \log n}} = 1 \quad and \quad \underline{\lim} \frac{\sqrt{n}(\overline{X}_n - \mathbb{E}\, X_1)}{\sqrt{\mathbb{V}\mathrm{ar}\, X_1} \sqrt{2 \log \log n}} = -1,$$

where $\overline{X}_n = \sum_{i=1}^{n} X_i / n$.

Let (X_n) be an i.i.d. square integrable random sequence and set $S_n = \sum_{i=1}^{n} X_i$. According to the law of large numbers, $S_n \approx n\mathbb{E}\, X_1$, and, according to the central limit theorem, $S_n - n\mathbb{E}\, X_1 = o(n^{1/2})$ when n tends to infinity. Of course, the order of the deviation between S_n and $n\mathbb{E}\, X_1$ can be more than $n^{1/2}$ for some n. The theory of large deviations investigates the asymptotic behavior of the quantity $\mathbb{P}(|S_n - n\mathbb{E}\, X_1| > n^\alpha)$ for $\alpha > 1/2$. The next result, for $\alpha = 1$, is the simplest of the so-called large deviations theorems.

Theorem 1.98 (Chernoff) *Let (X_n) be an i.i.d. square integrable random sequence and let h denote its Cramér transform. Let a be a nonnegative real* *number. If the moment generating function M of X_n is finite on an interval $I \neq \{0\}$, then*

$$\mathbb{P}\left(\frac{X_1 + \cdots + X_n}{n} - m > a\right) \leq e^{-nh(m+a)}, \quad n \in \mathbb{N}^*, \tag{1.7}$$

$$\mathbb{P}\left(\frac{X_1 + \cdots + X_n}{n} - m < a\right) \leq e^{-nh(m-a)}, \quad n \in \mathbb{N}^*. \tag{1.8}$$

Proof We prove the theorem for centered variables.

Set $S_n = X_1 + \cdots + X_n$. According to Chernoff's inequality, $\mathbb{P}(S_n \leq an) \leq e^{-h_{S_n}(na)}$. Moreover, since $M_{S_n}(u) = \mathbb{E}(e^{uS_n}) = \mathbb{E}(\prod_{i=1}^{n} e^{uX_i}) = M(u)^n$, by definition

$$h_{S_n}(na) = \sup_{u \in I}[nau - \log M_{S_n}(u)] = \sup_{u \in I}[nau - n \log M(u)] = nh(a).$$

The inequality (1.7) follows. The inequality (1.8) can be proven in the same way.

□

▷ *Example 1.99 (A Large Deviation Inequality for the Normal Distribution)* If (X_n) is a standard Gaussian sequence, Example 1.50 says that $h(t) = t^2/2$. Hence

$$\mathbb{P}\left(\frac{X_1 + \cdots + X_n}{n} > a\right) \leq e^{-na^2/2}, \quad n \in \mathbb{N}^*,$$

with a similar inequality for $\mathbb{P}[(X_1 + \cdots + X_n)/n < a]$.   ◁

1.5   Exercises

▽ **Exercise 1.1 (An Alternative Method for the Computation of Moments)** Let X be any nonnegative random variable.

1. Show that for every $p \geq 1$ such that $\mathbb{E}(X^p)$ is finite,

$$\mathbb{E}(X^p) = \int_{\mathbb{R}_+} pt^{p-1}\mathbb{P}(X > t)\,dt. \tag{1.9}$$

2. Show that

$$\sum_{n \geq 0} \mathbb{P}(X > n + 1) \leq \mathbb{E}\,X \leq \sum_{n \geq 0} \mathbb{P}(X > n).$$

3. Show that

$$\sum_{n \geq 1} \mathbb{P}(X \geq n) \leq \mathbb{E}\,X \leq \sum_{n \geq 0} \mathbb{P}(X \geq n).$$

Solution

1. By the transfer theorem, $\mathbb{P}(X > t) = \int_t^{+\infty} d\mathbb{P}_X(x)$. Hence

$$p \int_{\mathbb{R}_+} t^{p-1} \mathbb{P}(X > t) \, dt = \int_{\mathbb{R}_+} \int_t^{+\infty} p t^{p-1} d\mathbb{P}_X(x) \, dt$$

$$\overset{(1)}{=} \int_{\mathbb{R}_+} \int_0^x p t^{p-1} \, dt \, d\mathbb{P}_X(x) = \int_{\mathbb{R}_+} x^p d\mathbb{P}_X(x) = \mathbb{E}(X^p).$$

(1) by Fubini's theorem.

2. Suppose that X is almost surely finite—otherwise the result is obvious, with all infinite quantities. By 1.,

$$\mathbb{E}\,X = \int_0^{+\infty} \mathbb{P}(X > t) \, dt = \sum_{n \geq 0} \int_n^{n+1} \mathbb{P}(X > t) \, dt.$$

Clearly,

$$\sum_{n \geq 0} \mathbb{P}(X > n + 1) = \sum_{n \geq 0} \int_n^{n+1} \mathbb{P}(X > n + 1) \, dt$$

$$\text{and} \quad \sum_{n \geq 0} \mathbb{P}(X > n) = \sum_{n \geq 0} \int_n^{n+1} \mathbb{P}(X > n) \, dt.$$

Since $\mathbb{P}(X > n + 1) \leq \mathbb{P}(X > t) \leq \mathbb{P}(X > n)$ for all $n \leq t < n + 1$, the result follows.

Note that for an integer valued variable we get $\mathbb{E}\,X = \sum_{n \geq 0} \mathbb{P}(X > n)$, a formula often of use in applications.

3. We compute

$$\sum_{n \geq 1} \mathbb{P}(X \geq n) = \sum_{n \geq 1} \mathbb{P}\left[\bigcup_{i \geq n} (i \leq X < i + 1)\right] = \sum_{n \geq 1} \sum_{i \geq n} \mathbb{P}(i \leq X < i + 1)$$

$$= \sum_{i \geq 1} \sum_{n=1}^i \mathbb{P}(i \leq X < i + 1) = \sum_{i \geq 1} \mathbb{P}(i \leq X < i + 1)i,$$

or

$$\sum_{n \geq 1} \mathbb{P}(X \geq n) \leq \sum_{i \geq 1} \mathbb{E}\,[X \mathbb{1}_{(i \leq X < i+1)}] \leq \mathbb{E}\,X.$$

Finally, $\mathbb{P}(X \geq n) \geq \mathbb{P}(X > n)$ so that 2. yields the second inequality. $\triangle$

▽ **Exercise 1.2 (Superior Limits of Events and Real Numbers)** Let (X_n) be any random sequence.

1. Set $\overline{\lim}\, X_n = \inf_{n\geq 0} \sup_{k\geq n} X_k$. Show that $\overline{\lim}(X_n > x) = (\overline{\lim}\, X_n > x)$ for all $x \in \mathbb{R}$.
2. The sequence is supposed to be i.i.d. Show that the distribution function F of $\overline{\lim}\, X_n$ takes values in $\{0, 1\}$. Show then that $\overline{\lim}\, X_n$ is almost surely constant, equal to $x_0 = \inf\{x \in \mathbb{R} : F(x) = 1\}$.
3. Show that $\overline{\lim}\, X_n$ is measurable for the tail σ-algebra of the natural filtration of (X_n) and prove again that $\overline{\lim}\, X_n$ is almost surely constant.

Solution

1. By definition, $\overline{\lim}(X_n > x) = \cap_{n\geq 0} \cup_{k\geq n} (X_n > x)$, so is a measurable event. We have

$$\omega \in \overline{\lim}(X_n > x) \Leftrightarrow \forall n, \exists k \geq n, X_k(\omega) > x \Leftrightarrow \forall n, \sup_{k\geq n} X_k(\omega) > x$$
$$\Leftrightarrow \inf_n \sup_{k\geq n} X_k(\omega) > x \Leftrightarrow \overline{\lim}\, X_n(\omega) > x$$
$$\Leftrightarrow \omega \in (\overline{\lim}\, X_n > x).$$

2. Borel-Cantelli lemma implies that $\mathbb{P}(\overline{\lim}(X_n > x)) = 0$ or 1 for all real numbers x. According to 1., $F(x) = \mathbb{P}(\overline{\lim}\, X_n > x) = 0$ or 1 too. Further

$$\mathbb{P}(\overline{\lim}\, X_n = x_0) = F(x_0) - \lim_{n\to+\infty} F(x_0 - 1/n) = 1 - 0 = 1.$$

3. Using 1. yields $(\overline{\lim}\, X_n > x) = \cap_{n\geq 0} A_n$, where $A_n = \cup_{k\geq n}(X_k > x)$. Clearly, $A_n \in \sigma(X_n, X_{n+1}, \ldots)$. The sequence (A_n) is decreasing, so $\cap_{n\geq 0} A_n$ is a tail event, and the conclusion follows from Corollary 1.28.   △

▽ **Exercise 1.3 (Indefinitely Divisible Distributions)**   Let (X_n) be an i.i.d. random sequence with distribution P. Let $N \sim \mathcal{P}(\lambda)$ be a random variable independent of (X_n). Set $S_0 = 0$ and $S_n = \sum_{i=1}^n X_i$ for $n \geq 1$.

1. Determine the distribution of S_N, denoted by $\pi_{\lambda, P}$.
2. Show that the Fourier transform of $\pi_{\lambda, P}$ can be written

$$\varphi_{\pi_{\lambda, P}}(t) = \exp\left[\int_{\mathbb{R}} (e^{itx} - 1) dM(x)\right],$$

where M depends on λ and P.
3. Show that $\pi_{\lambda, P}$ is indefinitely divisible for $\lambda \in \mathbb{R}_+^*$.

Solution

1. By Definition 1.74, the random sum S_N has a compound Poisson distribution. Precisely,

$$\pi_{\lambda,P}(B) = \sum_{n \geq 0} \mathbb{P}(S_N \in B, N = n) = \sum_{n \geq 0} \mathbb{P}(S_n \in B)\mathbb{P}(N = n)$$

$$= \sum_{n \geq 0} e^{-\lambda} \frac{\lambda^n}{n!} P^{*(n)}(B),$$

where $P^{*(0)} = \delta_0$, a Dirac distribution at 0.

2. By Proposition 1.73, the characteristic function of S_N is the composition of the generating function of N and of the Fourier transform $\varphi_P(t) = \int_{\mathbb{R}} e^{itx} P(dx)$ of P. We compute

$$\varphi_{\pi_{\lambda,P}}(t) = e^{-\lambda} \sum_{n \geq 0} \frac{[\lambda\varphi_P(t)]^n}{n!} = e^{-\lambda[1-\varphi_P(t)]}$$

$$= \exp\left[-\lambda \int_{\mathbb{R}} (1 - e^{itx}) P(dx)\right] = \exp\left[\int_{\mathbb{R}} (e^{itx} - 1) M(dx)\right],$$

with $M = \lambda P$.

3. Since the Fourier transform characterizes the distribution and transforms convolution product into usual product, we get $\pi_{\lambda,P} = (\pi_{\lambda/n,P})^{*(n)}$, for all $n \in \mathbb{N}$.   △

∇ **Exercise 1.4 (Extension of the Fourier Transform)** Let M be a measure on $(\mathbb{R}, \mathcal{B}(\mathbb{R}))$ such that

$$\int_{\mathbb{R}} (|x| \wedge 1) M(dx) < +\infty.$$

Show that the function

$$\widehat{M}(t) = \exp\left[\int_{\mathbb{R}} (e^{itx} - 1) M(dx)\right]$$

is well-defined for all $t \in \mathbb{R}$—even if M is not a finite measure.

Solution We compute

$$\int_{\mathbb{R}} (e^{itx} - 1) M(dx) = \int_{0 < |x| < 1} (e^{itx} - 1) M(dx) + \int_{|x| \geq 1} (e^{itx} - 1) M(dx)$$

$$= I + J.$$

On the one hand,

$$|I| \leq \int_{0<|x|<1} |e^{itx} - 1| M(dx) = \int_{0<|x|<1} \left| \sum_{k\geq 1} \frac{i^k t^k x^k}{k!} \right| M(dx)$$

$$\leq \int_{0<|x|<1} \sum_{k\geq 1} \frac{|t|^k |x|}{k!} M(dx) = \int_{0<|x|<1} (e^{|t|} - 1)|x| M(dx),$$

or $|I| \leq C \int_{0<|x|<1} |x| M(dx)$.
  On the other hand,

$$|J| \leq \int_{|x|\geq 1} |e^{itx}| M(dx) = \int_{|x|\geq 1} M(dx) = 2M(\mathbb{R}\backslash] - 1, 1[).$$

The conclusion follows.   △

▽ **Exercise 1.5 (Total Cost of a Component)**   A component that is essential
for the continuous running of a machine is automatically and instantaneously
replaced at each failure by a new identical one. The time life of the component
is exponentially distributed with parameter λ.

1. Suppose each replacement costs b Euros. The rate of devaluation of the Euro
   is assumed to be constant and equal to $a > 0$. Compute the mean total cost of
   failures.
2. The updated cost of of each failure at time $t \geq 0$ is now assumed to be given by a
   nonnegative function g defined on $\mathbb{R}_+$. Compute the mean total cost of failures.
3. Deduce 1. from 2.

Solution

1. Let $(T_n)_{n\in\mathbb{N}^*}$ denote the sequence of the failure times of the components. We
   have $T_n = X_1 + \cdots + X_n, n \geq 1$, where (X_n) is the i.i.d. with distribution $\mathcal{E}(\lambda)$
   random sequence of the times between two successive failures.
     The cost of the n-th failure is $b \exp(-aT_n)$ in updated Euros, so the total cost
   of the failures is $C = \sum_{n\geq 1} b \exp(-aT_n)$ and the mean total cost can be written

$$\mathbb{E}C = \mathbb{E}\left(\sum_{n\geq 1} be^{-aT_n}\right) = b\sum_{n\geq 0} \mathbb{E}(e^{-aT_n})$$

$$= b\sum_{n\geq 1} \mathbb{E}(e^{-aX_1})^n = b\sum_{n\geq 1} \left(\frac{\lambda}{\lambda + a}\right)^n = \frac{\lambda b}{a}.$$

2. The cost of the n-th failure is $g(T_n)$, and the mean total cost is $\mathbb{E}\,C = \sum_{n \geq 1} \mathbb{E}\,g(T_n)$. Since $T_n \sim \gamma(n, \lambda)$, we compute

$$\mathbb{E}\,g(T_n) = \int_0^{+\infty} g(t) f_n(t)\,dt = \int_0^{+\infty} g(t) \frac{(\lambda t)^{n-1}}{(n-1)!} \lambda e^{-\lambda t}\,dt,$$

so $\mathbb{E}\,C = \lambda \int_0^{+\infty} g(t)\,dt$.

3. Setting $g(t) = b e^{-at}$, we get $\mathbb{E}\,C = \lambda b \int_0^t e^{-at}\,dt = \lambda b / a$. △

∇ Exercise 1.6 (Maximum of Entropy Under Constraints)

1. Compute the entropy of a non degenerated centered Gaussian vector $\widetilde{X} \sim \mathcal{N}_n(0, \Gamma)$, with variance-covariance matrix Γ such that $\det \Gamma \neq 0$.
2. Show that among the centered n-dimensional random vectors whose variance-covariance matrix is equal to Γ, the maximum of entropy is obtained for the Gaussian vector $\widetilde{X}$.

Solution

1. The density of a non-degenerated Gaussian vector $\widetilde{X} \sim \mathcal{N}_n(0, \Gamma)$ is

$$f_{\widetilde{X}}(x) = (2\pi)^{-n/2} (\det \Gamma)^{-1/2} \exp[-\frac{1}{2} x' \Gamma^{-1} x], \quad x \in \mathbb{R}^n.$$

Thus its entropy is

$$\mathcal{S}(\widetilde{X}) = \frac{1}{2}\Big[n \log(2\pi) + \log(\det \Gamma) + \int_{\mathbb{R}^n} x' \Gamma^{-1} x f_{\widetilde{X}}(x)\,dx \Big].$$

Let $\mathrm{Tr}M = \sum_{i=1}^n M_{ii}$ denote the trace of an n-dimensional matrix. We compute

$$\int_{\mathbb{R}^n} x' \Gamma^{-1} x f_{\widetilde{X}}(x)\,dx = \mathbb{E}\,(\widetilde{X}' \Gamma^{-1} \widetilde{X}) = \mathbb{E}\,[\mathrm{Tr}(\widetilde{X}' \widetilde{X}) \Gamma^{-1})]$$

$$= \mathrm{Tr}[\mathbb{E}\,(\widetilde{X}' \widetilde{X}) \Gamma^{-1}] = \mathrm{Tr}(\Gamma \Gamma^{-1}) = n.$$

Therefore, $\mathcal{S}(\widetilde{X}) = \log[\sqrt{(2\pi e)^n \det \Gamma}]$.

2. For any n-dimensional random vector X with density f, we can write

$$\mathcal{S}(X) = \int_{\mathbb{R}^n} f(x) \log \frac{f_{\widetilde{X}}(x)}{f(x)}\,dx - \int_{\mathbb{R}^n} f(x) \log f_{\widetilde{X}}(x)\,dx = (i) + (ii).$$

On the one hand, since $\log x \leq x - 1$ for all positive x, we obtain

$$+ - *(i) \leq \int_{\mathbb{R}^n} f(x)\Big[\frac{f_{\widetilde{X}}(x)}{f(x)} - 1 \Big] dx = 0.$$

On the other hand,

$$(ii) = \frac{1}{2}\left[n\log(2\pi) + \log(\det\Gamma) + \int_{\mathbb{R}^n} x'\Gamma^{-1}xf(x)\,dx\right].$$

If $\mathbb{E}(X'X) = \Gamma$, we obtain in the same way as in 1.,

$$\int_{\mathbb{R}^n} x'\Gamma^{-1}xf(x)\,dx = \mathrm{Tr}(\Gamma\Gamma^{-1}) = n.$$

Thus, $(ii) = S(\tilde{X})$ and the conclusion follows.   △

▽ Exercise 1.7 (Convergence of Non Integrable Variables)

1. Let (X_n) be an i.i.d. random sequence. Let P denote its distribution and φ its characteristic function. Set $Y_n = (X_n + Y_{n-1})/n$ for $n \geq 1$, with $Y_0 = X_0/2$.
   a. Write Y_n in terms of X_i with $0 \leq i \leq n$.
   b. Write the characteristic function ϕ_n of Y_n in terms of ϕ and n.
2. Suppose P is the Cauchy distribution $\mathcal{C}(1)$.
   a. Let Z be a random variable with a symmetric exponential distribution with parameter $\alpha > 0$, with density defined by $f(z) = \alpha e^{-\alpha|z|}/2$ on $\mathbb{R}$. Determine the characteristic function of Z and then the characteristic function of a random variable with Cauchy distribution $\mathcal{C}(\alpha)$.
   b. Show that (Y_n) converges in distribution to the same distribution.
   c. Study the convergence in distribution of the sequences $(\overline{X}_n)$ and (T_n) defined by $\overline{X}_n = \sum_{i=1}^{n} X_i/n$ and $T_n = \sum_{i=1}^{n} X_i/\sqrt{n}$.

Solution

1. a. Clearly, $Y_n = \sum_{i=0}^{n} X_i/2^{n-i+1}$.
   b. Since according to 1.a., X_n and Y_{n-1} are independent, Theorem 1.68 and properties of characteristic functions jointly imply that

$$\phi_n(t) = \phi_{(Y_{n-1}+X_n)/2}(t) = \phi_{n-1}\left(\frac{t}{2}\right)\phi\left(\frac{t}{2}\right),$$

from which it follows that $\phi_n(t) = \prod_{i=0}^{n}\phi(t/2^{n-i+1}) = \prod_{j=1}^{n+1}\phi(t/2^j)$ for all $t \in \mathbb{R}$.

2. a. We compute

$$\varphi_Z(t) = \frac{\alpha}{2}\int_{\mathbb{R}_-} e^{itz}\frac{\alpha}{2}e^{\alpha z}\,dz + \frac{\alpha}{2}\int_{\mathbb{R}_+} e^{itz}\frac{\alpha}{2}e^{-\alpha z}\,dz$$

$$= \frac{\alpha}{2(t+\alpha)} - \frac{\alpha}{2(t-\alpha)} = \frac{\alpha^2}{\alpha^2 + t^2}.$$

Since the function $t \longrightarrow \alpha^2/(\alpha^2 + t^2)$ is integrable, Fourier's inversion formula yields

$$\frac{\alpha}{2} e^{-\alpha|z|} = \frac{1}{2\pi} \int_{\mathbb{R}} e^{-itz} \frac{\alpha^2}{\alpha^2 + t^2} \, dt.$$

Therefore the characteristic function of a Cauchy distribution $\mathcal{C}(\alpha)$ is $\varphi(t) = \alpha e^{-\alpha|t|}$.

b. We compute

$$\phi_n(t) = \prod_{j=1}^{n+1} \phi\left(\frac{t}{2^j}\right) = \prod_{j=1}^{n+1} e^{-|t|/2^j} = \exp\left(-\sum_{j=1}^{n+1} \frac{|t|}{2^j}\right).$$

Since $\sum_{j=1}^{n+1}(1/2^j)$ is a geometrical series that converges to 1 when n tends to infinity, (Y_n) converges in distribution to a Cauchy distribution $\mathcal{C}(1)$.

c. The sequence (X_n) is not integrable, so neither the strong law of large numbers nor the limit central theorem apply. Nevertheless,

$$\phi_{\overline{X}_n}(t) = \phi\left(\frac{t}{n}\right)^n = \left(e^{-|t/n|}\right)^n = e^{-|t|}, \quad t \in \mathbb{R}.$$

The sequence $(\overline{X}_n)$ converges in distribution to the same Cauchy distribution.
  On the contrary, $\phi_{T_n}(t) = \phi(t/\sqrt{n})^n = e^{-\sqrt{n}|t|}$ for all t. Hence ϕ_{T_n} converges to the function $\mathbb{1}_{\{0\}}$. Since this function is not continuous at zero, it is not a characteristic function and (T_n) cannot converge in distribution. $\triangle$

▽ Exercise 1.8 (The Ballot Theorem—A Random Walk)

1. An urn contains a bowls bearing number 0 and b bowls bearing number 2, where $a > b$. All the bowls are drawn at random and without replacement. Let X_i be the random variable equal to the number of the bowl drawn at the i-th drawing, for $i \in [\![1, n]\!]$, where $n = a + b$.
   a. Compute the expectation of $S_m = \sum_{i=1}^m X_i$ for $m \in [\![1, n]\!]$.
   b. Show that for all $b' \in [\![0, b]\!]$,

   $$\mathbb{P}(S_m < m, \; m = 1, \ldots, n | S_{2b} = 2b') = \mathbb{P}(S_m < m, \; m = 1, \ldots, 2b | S_{2b} = 2b').$$

   c. Show by induction using b. that

   $$\mathbb{P}(S_m < m, \; m = 1, \ldots, n \mid S_n = 2b) = 1 - 2b/n. \tag{1.10}$$

2. Application: two candidates A and B are confronted in a ballot. Candidate A wins. Compute the probability that A has led the ballot all along the counting of the votes.

Solution

1. a. Since $\mathbb{P}(S_n = 2b) = 1$, we have $\mathbb{E}\, S_n = 2b$. Moreover, the variables X_i are exchangeable—that is $(X_1, \ldots, X_n) \sim (X_{i_1}, \ldots, X_{i_n})$ for all $(i_1, \ldots, i_n)$, so $\mathbb{E}\, X_i = \mathbb{E}\, S_n/n = 2b/n$. Therefore, $\mathbb{E}\, S_m = 2bm/n$.
   b. We know that $S_n = 2b$. Hence, if $S_{2b} = 2b'$, then necessarily, $S_m < m$ for all $m \in [\![n, 2b+1]\!]$, from which the desired equality follows.
   c. For (1.10) to hold for $n = 1$, it is necessary that $a = 1$ et $b = 0$ and so $\mathbb{P}(X_1 < 1) = \mathbb{P}(X_1 = 0) = 1$.
   Suppose (1.10) holds for all $m < n$. On the first hand,

$$\mathbb{P}(S_m < m, \ m = 1, \ldots, n) =$$
$$= \sum_{b'=0}^{b} \mathbb{P}(S_m < m, \ m = 1, \ldots, n \mid S_{2b} = 2b')\mathbb{P}(S_{2b} = 2b'),$$

and on the other hand, since $\mathbb{P}(S_n = 2b) = 1$, we have

$$\mathbb{P}(S_n < m, m = 1, \ldots, n) = \mathbb{P}(S_n < m, m = 1, \ldots, n \mid S_n = 2b).$$

Using the induction hypothesis for $n = 2b$, we get

$$\mathbb{P}(S_m < m, \ m = 1, \ldots, 2b \mid S_{2b} = 2b') = 1 - b'/b.$$

Thus,

$$\mathbb{P}(S_m < m, m = 1, \ldots, n) \overset{(1)}{=} \sum_{b'=0}^{b}(1 - b'/b)\mathbb{P}(S_{2b} = 2b') = 1 - \mathbb{E}\, S_{2b}/2b$$
$$\overset{(2)}{=} 1 - \frac{2b}{n}.$$

   (1) by 1.b. and (2) by 1.a. The proof is complete.
2. Representing the votes by the bowls in the urn, and their counting as the drawing of these bowls, the searched probability is $(a-b)/(a+b)$ where a is the number of votes for A and b that for B. An illustration is shown in Fig. 1.3.
   Note that (S_n) is an example of a random walk.   △

▽ **Exercise 1.9 (A Simple Random Walk)**   A coin is tossed indefinitely. Let $p \in]0, 1] \backslash \{1/2\}$ be the probability of obtaining heads at one toss. Let S_n be the random

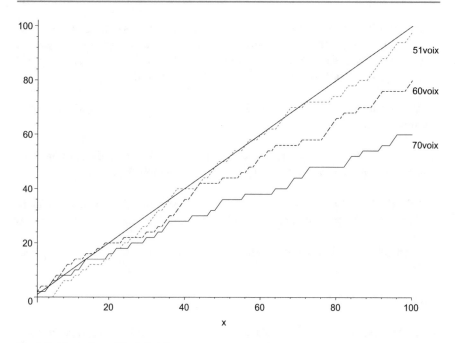

Fig. 1.3 Illustration of the ballot theorem

variable equal to the number of heads less the number of tails in n tosses. We set $S_0 = 0$.

1. Show that S_n is a simple random walk.
2. Compute $\mathbb{P}(S_n = 0)$, depending on the parity of n.
3. Let $N = |\{n \in \mathbb{N}^* : S_n = 0\}|$ be the number of times when the trajectory cuts the time axis. Show that the expected value of N is finite.
4. Compute $\mathbb{P}(\overline{\lim}(S_n = 0))$.

Solution

1. It is enough to take in Definition 1.76 the random variable X_n equal to 1 if the result of the n-th toss is heads and -1 otherwise.
2. Set $A_n = (S_n = 0)$. Clearly, $\mathbb{P}(A_{2n+1}) = 0$, and we can write $A_{2n} = $ "$X_i = 1$ for n indices i among the first n", so $\mathbb{P}(A_{2n}) = \binom{2n}{n} p^n (1 - p)^n$.
3. We can write $N = \sum_{n \geq 0} \mathbb{1}_{A_n}$, so $\mathbb{E}\, N = \sum_{n \geq 0} \mathbb{P}(A_n)$. Moreover,

$$\frac{\mathbb{P}(A_{2n+2})}{\mathbb{P}(A_{2n})} = 4p(1 - p).$$

Since this quantity is less than one, the series is convergent.

4. By Definition 1.19, $\overline{\lim} A_n$ is the set of all the trajectories cutting the time axis
   an infinite number of times. The series with general term $\mathbb{P}(A_n)$ converges, so,
   thanks to Borel-Cantelli lemma, $\mathbb{P}(\overline{\lim} A_n) = 0$. Therefore the trajectory will
   almost surely cut the time axis only a finite number of times.

The above recurrence problem for $p = 1/2$ will be investigated below in
Example 3.36.   △

∇ **Exercise 1.10 (Convergence of the Maximum)** Let F be a distribution function. Set $S = \inf\{x \in \mathbb{R} : F(x) = 1\}$, with the convention $\inf \emptyset = +\infty$.

1. Let (X_n) be an i.i.d. random sequence with distribution $\mathcal{E}(\lambda)$. Compute S and the
   distribution function $F_{(n)}$ of $X_{(n)}$. Show then that $X_{(n)}$ converges almost surely
   to S.
2. Same questions for the convergence in distribution of an i.i.d. random sequence
   (Y_n) with density $f(y) = \alpha(1 - y)^{\alpha-1}\mathbb{1}_{[0,1]}(y)$, for $\alpha \in \mathbb{N}^*$.

Solution

1. Since $F(x) = 1 - e^{-\lambda x}$, we have $S = +\infty$. We compute $F_{(n)}(x) = \mathbb{P}(X_1 \leq x, \dots, X_n \leq x) = F(x)^n$. For all $\varepsilon > 0$,

$$\sum_{n\geq 1} \mathbb{P}[X_{(n)} \leq \varepsilon] = \sum_{n\geq 1} F(\varepsilon)^n = \frac{F(\varepsilon)}{1 - F(\varepsilon)} = \frac{1}{e^{-\lambda\varepsilon}} - 1.$$

   Borel-Cantelli lemma thus yields $\mathbb{P}[\overline{\lim}(X_{(n)} \leq \varepsilon)] = 0$, meaning that $(X_{(n)})$
   converges almost surely to $S = +\infty$.
2. We have $F(y) = [1 - (1 - y)^\alpha]\mathbb{1}_{[0,1]}(y) + \mathbb{1}_{]1,+\infty[}(y)$. Hence $F_{X_{(n)}} = F^n$
   converges to $\mathbb{1}_{[1,+\infty[}$, that is to the distribution function of the Dirac distribution
   δ_1. Finally, we check that $S = 1$.   △

∇ **Exercise 1.11 (Convergence in Probability and in Mean)** Let (X_n) be an
integrable random sequence converging in probability to some random variable X.
Set $X'_n = \min(X_n, X)$.

1. Show that (X'_n) converges in probability to X.
2. Suppose X is integrable. Show that X'_n converges to X in mean.
3. Suppose moreover that $\mathbb{E} X_n$ converges to $\mathbb{E} X$. Show that X_n converges to X in
   mean.

Solution

1. We have

$$\mathbb{P}(|X'_n - X| > \varepsilon) = \mathbb{P}(|X_n - X| > \varepsilon, X_n \le X) \le \mathbb{P}(|X_n - X| > \varepsilon),$$

   and this quantity converges to zero by assumption.
2. We have $0 \le X'_n \le X$, so $|X'_n - X| = X - X'_n$ is integrable.

$$\mathbb{E}\,|X'_n - X| = \underbrace{\mathbb{E}\,[(X - X'_n)\mathbb{1}_{(X-X'_n \le \varepsilon)}]}_{(i)} + \underbrace{\mathbb{E}\,[(X - X'_n)\mathbb{1}_{(X-X'_n \ge \varepsilon)}]}_{(ii)}.$$

   On the one hand, (i) converges to 0. On the other hand, set $A_n = (X - X'_n \ge \varepsilon)$. Then $\mathbb{P}(A_n)$ converges to 0 by 1., so $\mathbb{1}_{A_n}$ converges almost surely to 0, and $X\mathbb{1}_{A_n}$ too. Since $X\mathbb{1}_{A_n} \le X$ and X is integrable, the dominated convergence theorem applies to show that (ii) converges to 0. Finally, X'_n converges to X in mean.
3. Let us set $X''_n = \max(X_n, X)$. We have $\mathbb{E}\,X'_n + \mathbb{E}\,X''_n = \mathbb{E}\,X_n + \mathbb{E}\,X$. Since $\mathbb{E}\,X'_n$ converges to $\mathbb{E}\,X$ by 2., and $\mathbb{E}\,X_n$ converges to $\mathbb{E}\,X$ by assumption, we get that $\mathbb{E}\,X''_n$ converges to $\mathbb{E}\,X$. The conclusion follows because $\mathbb{E}\,|X_n - X| = \mathbb{E}\,(X''_n - X'_n)$.

Note that the dominated convergence theorem does not apply to the sequence $(X_n - X)$, which explains why the result had to be proven first for (X'_n). △

▽ **Exercise 1.12 (Equi-Integrability and Convergence)** Let (X_n) be an integrable random sequence. A random sequence (X_n) is said to be uniformly integrable if

$$\mathbb{E}\,(\mathbb{1}_{A_N}|X_n|) \longrightarrow 0 \text{ uniformly in } n, \quad N \to +\infty,$$

for all sequence of events (A_N) such that $\mathbb{P}(A_N)$ converges to zero when N tends to infinity.

1. a. Show that (X_n) is equi-integrable according to Definition 1.17 if and only if it is bounded in L^1 and uniformly integrable.
   b. *Extended Fatou's lemma.* Show that if (X_n) is equi-integrable, then

$$\mathbb{E}\,(\underline{\lim}\, X_n) \le \underline{\lim}\,\mathbb{E}\,X_n \le \overline{\lim}\,\mathbb{E}\,X_n \le \mathbb{E}\,(\overline{\lim}\, X_n).$$

2. Show that:
   a. if (X_n) is equi-integrable and converges almost surely to X, then X is integrable and $\mathbb{E}\,X_n$ converges to $\mathbb{E}\,X$.
   b. if (X_n) is equi-integrable and converges in probability to X, then X is integrable and $\mathbb{E}\,X_n$ converges to $\mathbb{E}\,X$.

c. if an integrable sequence (X_n) converges in probability to $X \in L^1$, and if some $Y \in L^p$ exists such that $|X_n| < Y$ a.s. for all n, then (X_n) converges to X in L^1.

d. if $(|X_n|^p)$ is equi-integrable and converges in probability to X, then X_n converges to X in L^p-norm for all $p \geq 1$. (Use the inequality $(x + y)^p \leq 2^{p-1}(x^p + y^p)$, true for all nonnegative real numbers x and y).

e. if an L^p random sequence (X_n) converges in probability to $X \in L^p$, and if some $Y \in L^p$ exists such that $|X_n| < Y$ a.s. for all n, then (X_n) converges to X in L^p, for $p > 1$.

Solution

1. a. Suppose that (X_n) is bounded in L^1 and uniformly integrable. Thanks to Markov's inequality, for all $N \in \mathbb{N}^*$,

$$\mathbb{P}(|X_n| > N) \leq \frac{1}{N}\mathbb{E}|X_n| \leq \frac{1}{N}\sup_{n \geq 0}\mathbb{E}|X_n| \longrightarrow 0, \quad N \to +\infty,$$

so, using the uniform continuity, (X_n) is equi-integrable.

Conversely, let N_0 be such that if $N \geq N_0$, then $\mathbb{E}[\mathbb{1}_{(X_n>N)}|X_n|] \leq 1$. We can write

$$\mathbb{E}(|X_n|) = \mathbb{E}[\mathbb{1}_{(X_n>N_0)}|X_n|] + \mathbb{E}[\mathbb{1}_{(X_n \leq N_0)}|X_n|] \leq 1 + N_0,$$

so the sequence $(\mathbb{E}(|X_n|))$ is bounded in L^1. Let $A_N \in \mathcal{F}$ and $\varepsilon > 0$. Then

$$\mathbb{E}[\mathbb{1}_{A_N}|X_n|] = \mathbb{E}[\mathbb{1}_{A_N \cap (X_n>N)}|X_n|] + \mathbb{E}[\mathbb{1}_{A_N \cap (X_n \leq N)}|X_n|]$$
$$\leq \mathbb{E}[\mathbb{1}_{(X_n>N)}|X_n|] + N_1\mathbb{P}(A_N) \leq \varepsilon,$$

if N_1 satisfies $\mathbb{E}[\mathbb{1}_{(X_n>N_1)}|X_n|] \leq \varepsilon/2$ and $\mathbb{P}(A_N) \leq \varepsilon/2N_1$.

b. Fatou's lemma yields

$$\mathbb{E}(\underline{\lim}\, X_n) \leq \mathbb{E}[\underline{\lim}\, \mathbb{1}_{(X_n \geq -N)}X_n] \leq \underline{\lim}\,\mathbb{E}[\mathbb{1}_{(X_n \geq -N)}X_n].$$

We can write

$$\mathbb{E}\,X_n = \mathbb{E}[\mathbb{1}_{(X_n < -N)}X_n] + \mathbb{E}[\mathbb{1}_{(X_n \geq -N)}X_n].$$

Due to equi-integrability, $|\mathbb{E}[\mathbb{1}_{(X_n < -N)}X_n]|$ converges to 0 when N tends to infinity, so $\underline{\lim}_n \mathbb{E}[\mathbb{1}_{(X_n \geq -N)}X_n]$ converges to $\underline{\lim}\,\mathbb{E}\,X_n$.

Therefore, $\mathbb{E}(\underline{\lim}\, X_n) \leq \underline{\lim}\,\mathbb{E}\,X_n$. The second inequality can be proven in the same way.

2. a. According to 1.a., the sequence is bounded in L^1, say by M. Hence, thanks to Fatou's lemma again,

$$\mathbb{E}(|X|) \leq \underline{\lim} \, \mathbb{E}(|X_n|) \leq M < +\infty.$$

The convergence of $\mathbb{E} X_n$ to $\mathbb{E} X$ is an immediate consequence of 1.b.

b. If $(\mathbb{E} X_n)$ did not converge to $\mathbb{E} X$, $\varepsilon > 0$ would exist such that $|EX_n - \mathbb{E} X| > \varepsilon$ for an infinite number of n, that is for a subsequence of (X_n). This subsequence would converge in probability to X, so a subsequence converging almost surely to X would exist. This contradicts 2.a.

c. According to Example 1.18, a sequence that is bounded by an integrable variable is equi-integrable.

d. According to 2.b., X^p is integrable and $\mathbb{E}(|X_n|^p)$ converges to $\mathbb{E}(|X|^p)$. Moreover, the sequence $(|X_n|^p)$ is equi-integrable, and hence, according to 1.a., is bounded in L^1. Since

$$|X_n - X|^p \leq (|X_n| + |X|)^p \leq 2^{p-1}(|X_n|^p + |X|^p),$$

the sequence $(|X_n - X|^p)$ is bounded in L^1. Let $A \in \mathcal{F}$. We have

$$\mathbb{E}(\mathbb{1}_A|X_n - X|^p) \leq 2^{p-1}[\mathbb{E}(\mathbb{1}_A|X_n|^p) + \mathbb{E}(\mathbb{1}_A|X|^p)].$$

Since (X_n) is equi-integrable, it is uniformly integrable. Therefore, $(|X_n - X|^p)$ is also uniformly integrable, hence equi-integrable. Moreover, it converges in probability to 0. Then, according to 2.b., $\mathbb{E}(|X_n - X|^p)$ converges to 0 when n tends to infinity.

e. From 2.d., it is sufficient to prove that $(|X_n|^p)$ is equi-integrable. Set $M = \sup_{n \geq 0} \mathbb{E}[|X_n|^p]$. For all $\varepsilon > 0$, some N exists such that $|x|^{p-1} \geq M/\varepsilon$ for all real number $x \geq N$. For this N, we have

$$\mathbb{E}[\mathbb{1}_{(|X_n|>N)}|X_n|] \leq \mathbb{E}\left[\frac{\varepsilon}{M}\mathbb{1}_{(|X_n|>N)}|X_n|^p\right] \leq \frac{\mathbb{E}(|X_n|^p)}{M}\varepsilon \leq \varepsilon,$$

and the conclusion follows. $\triangle$

Conditioning and Martingales

2

In this chapter, conditional distributions and expectations are presented, following steps of increasing difficulty. Conditional distributions are first supposed to exist, which is true for all random variables taking values in $\mathbb{R}^d$. Still, conditional expectation is defined in the most general case. A section is dedicated to determining practically conditional distributions and expectations, and another to the linear model in which a random phenomenon is assumed to be linearly related to other simultaneously observed phenomena.

Then, main properties of stopping times are given; they are interesting by themselves and will constitute a useful tool in studying martingales thereafter, and Markov chains in the next chapter. The discrete time martingale theory is based on conditional expectation. The fundamental properties of martingales are detailed, among which the stopping theorem, some remarkable inequalities and different convergence theorems, especially for square integrable martingales.

2.1 Conditioning

First, basic probabilities conditional to events are presented, and then they are extended to conditioning with respect to any σ-algebra. Conditional distributions follow. Conditional expectation is first defined as the expectation of the conditional distribution, and then extended in terms of projection. Relations between conditioning and independence are highlighted, and practical methods of computation of conditional distributions are given.

© Springer Nature Switzerland AG 2018
V. Girardin, N. Limnios, *Applied Probability*,
https://doi.org/10.1007/978-3-319-97412-5_2

2.1.1 Conditioning with Respect to an Event

Let $(\Omega, \mathcal{F}, \mathbb{P})$ be a probability space. Let $B \in \mathcal{F}$ be an event with positive probability. The formula

$$\mathbb{P}(A \mid B) = \frac{\mathbb{P}(A \cap B)}{\mathbb{P}(B)} \tag{2.1}$$

is well-known to define a probability $\mathbb{P}(\cdot \mid B)$ on $(\Omega, \mathcal{F})$, called probability conditional on B.

If X is an integrable random variable, by definition, the expectation of X conditional on B is its expectation with respect to the probability conditional on B, that is

$$\mathbb{E}(X \mid B) = \int_{\Omega} X(\omega) \mathbb{P}(d\omega \mid B).$$

Therefore, it shares all properties of ordinary mean. Its practical computation relies on the following formula.

Proposition 2.1 *If X is an integrable random variable and B an event with positive probability, then*

$$\mathbb{E}(X \mid B) = \frac{\mathbb{E}(X \mathbb{1}_B)}{\mathbb{P}(B)} = \frac{1}{\mathbb{P}(B)} \int_B X(\omega) d\mathbb{P}(\omega). \tag{2.2}$$

Proof For $X = \mathbb{1}_A$, where $A \in \mathcal{F}$, (2.2) is satisfied thanks to Definition 2.1.

For an elementary variable $X_n = \sum_{i=1}^{n} a_i \mathbb{1}_{A_i}$, where all the A_i are events, (2.2) is also satisfied, due to the linearity of expectation.

Any nonnegative variable X is the increasing limit of a sequence (X_n) of elementary variables. Hence, the monotone convergence theorem induces that $\mathbb{E}(X_n \mid B) \nearrow \mathbb{E}(X \mid B)$ and $\mathbb{E}(X_n \mathbb{1}_B) \nearrow \mathbb{E}(X \mathbb{1}_B)$.

Finally, for any real random variable X, the conclusion follows by considering the difference of nonnegative variables $X = X^+ - X^-$. $\square$

Conditioning with respect to an event B can be regarded as conditioning with respect to the σ-algebra generated by B, by defining a real random variable, called conditional probability with respect to $\sigma(B)$ (or B), and such that, for any $A \in \mathcal{F}$,

$$\mathbb{P}[A \mid \sigma(B)](\omega) = \begin{cases} \mathbb{P}(A \mid B) & \text{if } \omega \in B, \\ \mathbb{P}(A \mid \overline{B}) & \text{otherwise.} \end{cases}$$

Thus, for any real random variable X, a real random variable called conditional expectation with respect to $\sigma(B)$ (or B) is also defined by setting

$$\mathbb{E}[X \mid \sigma(B)](\omega) = \begin{cases} \mathbb{E}(X \mid B) & \text{if } \omega \in B, \\ \mathbb{E}(X \mid \overline{B}) & \text{otherwise.} \end{cases}$$

We will now extend these notions to conditioning with respect to any σ-algebra included in $\mathcal{F}$.

2.1.2 Conditional Probabilities

First, let us consider a partition $\mathcal{B} = \{B_1, \ldots, B_n\}$ of Ω into events with positive probabilities. If the event B_i was known to have occurred, then the probability of an event A conditional on $\mathcal{B}$ would naturally be $\mathbb{P}(A \cap B_i)/\mathbb{P}(B_i)$. Therefore, the probability of A conditional on $\mathcal{B}$ can be set as

$$\mathbb{P}(A \mid \mathcal{B})(\omega) = \sum_{i=1}^{n} \frac{\mathbb{P}(A \cap B_i)}{\mathbb{P}(B_i)} \mathbb{1}_{B_i}(\omega), \quad \omega \in \Omega.$$

So is defined the conditional probability $\mathbb{P}(A \mid \mathcal{B})$. This random variable is $\sigma(\mathcal{B})$-measurable and satisfies

$$\int_B \mathbb{P}(A \mid \mathcal{B})d\mathbb{P} = \mathbb{P}(A \cap B), \quad B \in \mathcal{B}.$$

Moreover, $\mathbb{E}[\mathbb{P}(A \mid \mathcal{B})] = \mathbb{P}(A)$.

This leads to the following definition of the probability $\mathbb{P}(A \mid \mathcal{G})$ conditional on any given σ-algebra $\mathcal{G}$ included in $\mathcal{F}$.

Definition 2.2 Let $\mathcal{G}$ be a σ-algebra included in $\mathcal{F}$. Let A be any event. The conditional probability of A with respect to $\mathcal{G}$ is a $\mathcal{G}$-measurable random variable, say $\mathbb{P}(A \mid \mathcal{G})$, satisfying

$$\int_B \mathbb{P}(A \mid \mathcal{G})d\mathbb{P} = \mathbb{P}(A \cap B), \quad B \in \mathcal{G}.$$

The conditional probability obviously depends on the probability $\mathbb{P}$. Moreover, since it is a random variable, it is only defined a.s.

$\triangleright$ *Example 2.3* The conditional probability $\mathbb{P}(A \mid \{\emptyset, \Omega\})$ is a constant random variable taking the value $\mathbb{P}(A)$. $\triangleleft$

An alternative definition of the conditional probability, more easy to use in practice, is given by the following result.

Theorem 2.4 *The conditional probability* $\mathbb{P}(A \mid \mathcal{G})$ *is the orthogonal projection of the random variable* $\mathbb{1}_A \in L^2(\Omega, \mathcal{F}, \mathbb{P})$ *on* $L^2(\Omega, \mathcal{G}, \mathbb{P})$, *meaning that* $\mathbb{P}(A \mid \mathcal{G})$ *is the class in* $L^2(\Omega, \mathcal{F}, \mathbb{P})$ *of all the* $\mathcal{G}$-*measurable random variables* $Y \in \mathcal{L}^2(\Omega, \mathcal{F}, \mathbb{P})$ *such that*

$$\mathbb{E}\,(YZ) = \mathbb{E}\,(\mathbb{1}_A Z), \quad Z \in L^2(\Omega, \mathcal{G}, \mathbb{P}).$$

Proof The above property is obviously satisfied for $Z = \sum_{i=1}^n a_i \mathbb{1}_{A_i}$, with all $a_i \in \mathbb{R}$ and $A_i \in \mathcal{G}$. We proceed then as usual for any nonnegative variable Z and conclude for any $Z \in L^2(\Omega, \mathcal{G}, \mathbb{P})$ by considering $Z = Z^+ - Z^-$. $\qquad\square$

▷ *Example 2.5* If $\mathcal{G} = \sigma(B_i, i \in I)$, where $\{B_i : i \in I\}$ is a partition of Ω into events with positive probabilities, then

$$\mathbb{P}(A \mid \mathcal{G}) = \sum_{i \in I} \frac{\mathbb{P}(A \cap B_i)}{\mathbb{P}(B_i)} \mathbb{1}_{B_i}.$$

Indeed, since the random variable $Y = \mathbb{P}(A \mid \mathcal{G})$ is $\mathcal{G}$-measurable, we can write $Y = \sum_{i \in I} a_i \mathbb{1}_{B_i}$, with all $a_i \in \mathbb{R}$. Therefore,

$$\mathbb{P}(A \cap B_i) = \int_{B_i} Y d\mathbb{P} = a_i \mathbb{P}(B_i),$$

and hence,

$$\mathbb{P}(A \mid B_i) = a_i = \frac{1}{\mathbb{P}(B_i)} \int_{B_i} \mathbb{P}(A \mid \mathcal{G}) d\mathbb{P}.$$

Since $\mathbb{P}(A \cap B_i) = \mathbb{P}(A \mid B_i)\mathbb{P}(B_i)$, the conclusion follows. ◁

Theorem 2.6 *Let* $(\Omega, \mathcal{F}, \mathbb{P})$ *be a probability space. If* $\mathcal{G}$ *is a* σ-*algebra included in* $\mathcal{F}$, *then:*

1. $\mathbb{P}(\Omega \mid \mathcal{G}) = 1$ *a.s.;*
2. *for any* $A \in \mathcal{F}$, *we have* $0 \leq \mathbb{P}(A \mid \mathcal{G}) \leq 1$ *a.s.;*
3. *for any sequence* (A_n) *of pairwise disjoint events of* $\mathcal{F}$, *we have*

$$\mathbb{P}\Big(\bigcup_{n \geq 0} A_n \mid \mathcal{G}\Big) = \sum_{n \geq 0} \mathbb{P}(A_n \mid \mathcal{G}) \quad a.s..$$

Proof

1. For all $B \in \mathcal{G}$, we have $\int_B \mathbb{P}(\Omega \mid \mathcal{G})d\mathbb{P} = \mathbb{P}(B \cap \Omega) = \mathbb{P}(B)$, and hence $\mathbb{P}(\Omega \mid \mathcal{G}) = 1$ a.s.
2. Set $C = \{\omega \in \Omega : \mathbb{P}(A \mid \mathcal{G})(\omega) > 1\}$. We have

$$\mathbb{P}(C) \le \int_C \mathbb{P}(A \mid \mathcal{G})d\mathbb{P} = \mathbb{P}(A \cap C),$$

from which a contradiction arises if $\mathbb{P}(C) > 0$.
3. For all $B \in \mathcal{G}$, we have

$$\int_B \sum_{n \ge 0} \mathbb{P}(A_n \mid \mathcal{G})d\mathbb{P} = \sum_{n \ge 0} \int_B \mathbb{P}(A_n \mid \mathcal{G})d\mathbb{P}$$

$$= \sum_{n \ge 0} \mathbb{P}(A_n \cap B) = \mathbb{P}\Big[\Big(\bigcup_{n \ge 0} A_n\Big) \cap B\Big].$$

So, $\sum_{n \ge 0} \mathbb{P}(A_n \mid \mathcal{G})$, which is obviously $\mathcal{G}$-measurable, is indeed the probability of $\cup_{n \ge 0} A_n$ conditional on $\mathcal{G}$. □

As we have already remarked, for any given sequence (A_n), Point 3. is satisfied only a.s., say on $\Omega \setminus N(A_n)$ with $\mathbb{P}[N(A_n)] = 0$, and not for all ω. The function $A \longrightarrow \mathbb{P}(A \mid \mathcal{G})(\omega)$ is not a probability on $\mathcal{F}$ for all ω in general, but only on $\mathcal{F} \setminus N$ where N is the union of all $N(A_n)$, for all the sequences of events of $\mathcal{F}$. Since this union is not countable for general σ-algebras $\mathcal{F}$, the event N may have a positive probability. If $\mathbb{P}(\cdot \mid \mathcal{G})(\omega)$ is a probability for any $\omega \in \Omega$ then it is said to be a regular conditional probability.

Proposition 2.7 *Let $\mathcal{G}$ be a σ-algebra included in $\mathcal{F}$. If $C \in \mathcal{F}$ and if $A \in \mathcal{G}$, then $\mathbb{P}(A \cap C \mid \mathcal{G}) = \mathbb{P}(C \mid \mathcal{G})\mathbb{1}_A$ a.s.*

Setting above $C = \Omega$ and $A \in \mathcal{G}$, we obtain $\mathbb{P}(A \mid \mathcal{G}) = \mathbb{1}_A$ a.s.

Proof For all $B \in \mathcal{G}$, we have on the one hand,

$$\int_B \mathbb{P}(A \cap C \mid \mathcal{G})d\mathbb{P} = \mathbb{P}(A \cap C \cap B),$$

and on the other hand,

$$\int_B \mathbb{1}_A \mathbb{P}(C \mid \mathcal{G})d\mathbb{P} = \mathbb{P}(A \cap C \cap B),$$

from which the result follows. □

It remains to prove that the conditional probability does exist. For all nonnegative random variable X defined on $(\Omega, \mathcal{F}, \mathbb{P})$, the quantity

$$\mu(B) = \int_B X d\mathbb{P}, \quad B \in \mathcal{G}, \tag{2.3}$$

clearly defines a measure μ on $(\Omega, \mathcal{G}, \mathbb{P})$ taking values in $\overline{\mathbb{R}}_+$. This measure is absolutely continuous with respect to $\mathbb{P}$ and its Radon-Nikodym derivative is a $\mathcal{G}$-measurable random variable. Finally, for $X = \mathbb{1}_A$, $A \in \mathcal{F}$, Radon-Nikodym theorem and (2.3) jointly yield that

$$\int_B \mathbb{P}(A \mid \mathcal{G}) d\mathbb{P} = \mu(B) = \int_B \mathbb{1}_A d\mathbb{P} = \mathbb{P}(A \cap B), \quad B \in \mathcal{G},$$

which proves the existence of a random variable satisfying the properties of Definition 2.2.

A conditional probability can also be regarded as a particular case of a transition probability (or Markov kernel) between two measurable spaces.

Definition 2.8 Let $(\Omega, \mathcal{F})$ and $(E, \mathcal{E})$ be two measurable spaces. A function π : $\Omega \times \mathcal{E} \longrightarrow [0, 1]$ such that $\pi(\omega, .)$ is a probability on $(E, \mathcal{E})$ for almost all $\omega \in \Omega$ and $\pi(., A)$ is an $\mathcal{F}$-measurable function for all $A \in \mathcal{E}$, is called a transition probability from Ω to $\mathcal{E}$.

▷ *Example 2.9* Let f : $(\Omega \times E, \mathcal{F} \otimes \mathcal{E}) \longrightarrow (\mathbb{R}, \mathcal{B}(\mathbb{R}))$ be a nonnegative measurable function and let μ be a nonnegative measure on $(E, \mathcal{E})$ such that

$$\int_E f(\omega, \omega') d\mu(\omega') = 1, \quad \omega \in \Omega.$$

Clearly, the formula $\pi(\omega, A) = \int_A f(\omega, \omega') d\mu(\omega')$ defines a transition probability from Ω to $\mathcal{E}$. Moreover, $\pi(\omega, .)$ is absolutely continuous with respect to μ, with density $f(\omega, \cdot)$.   ◁

Suppose now that $(E, \mathcal{E}) = (\Omega, \mathcal{F})$.

Definition 2.10 Let $\mathcal{G}$ be a σ-algebra included in $\mathcal{F}$. A transition probability from Ω to Ω is called a conditional probability with respect to $\mathcal{G}$ if, for any $A \in \mathcal{F}$, the random variable $\pi(\cdot, A)$ is a version of the conditional probability $\mathbb{P}(A \mid \mathcal{G})$, that is if $\pi(\cdot, A) = \mathbb{P}(A \mid \mathcal{G})(\cdot)$ $\mathbb{P}$-a.s.

2.1.3 Conditional Distributions

Let $(\Omega, \mathcal{F})$ be a measurable space. Let **P** denote the set of all probabilities on $(\mathbb{R}^d, \mathcal{B}(\mathbb{R}^d))$. Let $Q : \Omega \longrightarrow \mathbf{P}$ be a function; in other words $Q(\omega)$ is a probability on $(\mathbb{R}^d, \mathcal{B}(\mathbb{R}^d))$ for all $\omega \in \Omega$. We will denote by $Q(\omega, B)$ the measure by $Q(\omega)$ of any Borel set $B \in \mathcal{B}(\mathbb{R}^d)$.

In order to regard Q as a random variable, $\mathbf{P}$ has to be equipped with a σ-algebra. We will consider the measurable space $(\mathbf{P}, \mathcal{H})$, where $\mathcal{H}$ is the σ-algebra generated by the functions defined on $\mathbf{P}$ by $P \longrightarrow P(B)$ for all $B \in \mathcal{B}(\mathbb{R}^d)$.

Definition 2.11 Let $(\Omega, \mathcal{F})$ be a measurable space. Any measurable function from $(\Omega, \mathcal{F})$ into $(\mathbf{P}, \mathcal{H})$ is called a random distribution (or probability) on $(\mathbb{R}^d, \mathcal{B}(\mathbb{R}^d))$.

This notion of random probability allows us to define the conditional distribution of a random variable with respect to a σ-algebra included in $\mathcal{F}$.

Definition 2.12 Let $X : (\Omega, \mathcal{F}, \mathbb{P}) \longrightarrow (\mathbb{R}^d, \mathcal{B}(\mathbb{R}^d))$ be a random variable and $\mathcal{G}$ a σ-algebra included in $\mathcal{F}$. A function $Q : \Omega \times \mathcal{B}(\mathbb{R}^d) \longrightarrow \mathbf{P}$ is called a conditional distribution of X given $\mathcal{G}$ if for all given $B \in \mathcal{B}(\mathbb{R}^d)$, the variable $Q(\cdot, B)$ is equal to the conditional probability given $\mathcal{G}$ of the event $(X \in B)$.

The conditional distribution function and the density of the conditional distribution are defined similarly.

Definition 2.13 Let $X = (X_1, \dots, X_d) : (\Omega, \mathcal{F}, \mathbb{P}) \longrightarrow (\mathbb{R}^d, \mathcal{B}(\mathbb{R}^d))$ be a random vector and let $\mathcal{G}$ be a σ-algebra included in $\mathcal{F}$.

A function $F : \Omega \times \mathbb{R}^d \longrightarrow [0, 1]$ is called a conditional distribution function of X given $\mathcal{G}$ if $F(\omega, \cdot)$ is a distribution function for all $\omega \in \Omega$, and if for all $x = (x_1, \dots, x_d) \in \mathbb{R}^d$,

$$F(\cdot, x) = \mathbb{P}(X_1 \leq x_1, \dots, X_d \leq x_d \mid \mathcal{G}) \quad \text{a.s..}$$

A nonnegative Borel function q defined on $(\Omega \times \mathbb{R}^d, \mathcal{F} \otimes \mathcal{B}(\mathbb{R}^d))$ is called a conditional density of X given $\mathcal{G}$ if the function Q defined on $\Omega \times \mathcal{B}(\mathbb{R}^d)$ by

$$Q(\omega, B) = \int_B q(\omega, x) \, dx$$

is a conditional distribution of X given $\mathcal{G}$.

It can be proven that for all real random variable X, conditional distributions and distribution functions exist with respect to any σ-algebra $\mathcal{G}$ included in $\mathcal{F}$. They are not uniquely defined but are all versions of the same random variable; in other words, if both Q_1 and Q_2 are conditional distributions of X given $\mathcal{G}$, then $Q_1(\cdot, B) = Q_2(\cdot, B)$ a.s. for all $B \in \mathcal{G}$.

When $F(\omega, x)$ is a distribution function on $\mathbb{R}^d$ for any $\omega \in \Omega$ and $F(\omega, x) = \mathbb{P}(X \leq x \mid \mathcal{G})$, then it is said to be a regular distribution function of X given $\mathcal{G}$. Any random variable X taking values in $\mathbb{R}^d$, or more generally in a Borel space, has a regular distribution function with respect to any $\mathcal{G} \subset \mathcal{F}$. The same applies to characterize regular conditional distributions $Q(\omega, B)$.

2.1.4 Conditional Expectation

When an integrable random variable X has a regular conditional distribution function given the σ-algebra $\mathcal{G} \subset \mathcal{F}$, say $F(\omega, x)$, the conditional expectation of X given $\mathcal{G}$ is defined as the usual expectation, by setting

$$\mathbb{E}(X \mid \mathcal{G})(\omega) = \int_{\mathbb{R}} x F(\omega, dx),$$

or, using the regular conditional distribution,

$$\mathbb{E}(X \mid \mathcal{G})(\omega) = \int_{\Omega} X(\omega') Q(\omega, d\omega').$$

When this conditional distribution does not exist, the conditional expectation of a random variable can still be defined, under some conditions. In both cases, the conditional expectation shows valuable interpretation in terms of projection.

Definition 2.14 Let X be a random variable defined on $(\Omega, \mathcal{F}, \mathbb{P})$ and let $\mathcal{G}$ be a σ-algebra included in $\mathcal{F}$.

If X is nonnegative, the conditional expectation of X with respect to $\mathcal{G}$ is the $\mathcal{G}$-measurable random variable denoted by $\mathbb{E}(X \mid \mathcal{G})$, taking values in $\overline{\mathbb{R}}_+$ and defined by

$$\int_{C} \mathbb{E}(X \mid \mathcal{G}) d\mathbb{P} = \int_{C} X d\mathbb{P}, \quad C \in \mathcal{G}. \tag{2.4}$$

If X is real valued, and if

$$\max(\mathbb{E}(X^+ \mid \mathcal{G}), \mathbb{E}(X^- \mid \mathcal{G})) < +\infty \quad \text{a.s.,} \tag{2.5}$$

the conditional expectation of X given $\mathcal{G}$ is defined as

$$\mathbb{E}(X \mid \mathcal{G}) = \mathbb{E}(X^+ \mid \mathcal{G}) - \mathbb{E}(X^- \mid \mathcal{G}).$$

In case $\min(\mathbb{E}(X^+ \mid \mathcal{G}), \mathbb{E}(X^- \mid \mathcal{G})) < +\infty$ a.s., then $\mathbb{E}(X \mid \mathcal{G})$ may be defined with an infinite value.

Note that $\mathbb{E}(X \mid \mathcal{G})$ satisfies (2.4) in both above cases. Also, if X is $\mathcal{G}$-measurable, then $\mathbb{E}(X \mid \mathcal{G}) = X$.

Since this carries over to random vectors coordinate by coordinate without change, we will study only the case of real random variables.

As shown in the next example, the definition of the conditional expectation is not limited to integrable variables, but, if X is integrable, then it satisfies condition (2.5) and

$$\int_\Omega \mathbb{E}\,X d\mathbb{P} = \mathbb{E}\,X = \int_\Omega X d\mathbb{P} \quad \text{and} \quad \int_\varnothing \mathbb{E}\,X d\mathbb{P} = 0 = \int_\varnothing X d\mathbb{P},$$

so that $\mathbb{E}\,(X \mid \{\varnothing, \Omega\}) = \mathbb{E}\,X$.

▷ *Example 2.15* Let X be any real random variable defined on $(\Omega, \mathcal{F}, \mathbb{P})$ such that $\mathbb{E}\,X^+ = +\infty$ and $\mathbb{E}\,X^- = +\infty$. We have

$$\mathbb{E}\,(X \mid \mathcal{F}) = \mathbb{E}\,(X^+ \mid \mathcal{F}) - \mathbb{E}\,(X^- \mid \mathcal{F}) = X^+ - X^- = X.$$

On the contrary, the expectation of X, that is its conditional expectation given the σ-algebra $\{\varnothing, \Omega\}$, is not defined, and X is not integrable. ◁

Proposition 2.16 *Condition (2.4) is equivalent to*

$$\mathbb{E}\,(ZX) = \mathbb{E}\,[Z\mathbb{E}\,(X \mid \mathcal{G})], \tag{2.6}$$

for any $\mathcal{G}$-measurable random variable Z.

In particular, due to Theorem 2.4, we have $\mathbb{E}\,(\mathbb{1}_A \mid \mathcal{G}) = \mathbb{P}(A \mid \mathcal{G})$.

Proof The direct implication is evident for an indicator function, then we follow the usual procedure, passing to general $\mathcal{G}$-measurable variables via $\mathcal{G}$-measurable step functions and nonnegative variables.

The converse is obtained by setting $Z = \mathbb{1}_C$. □

The method used above for proving the existence of the conditional probability works for proving the existence of the conditional expectation satisfying (2.4). We will rather show it through the orthogonal projection in L^2 for square integrable variables. For this, we need a linear algebra lemma, which we state without proof.

Lemma 2.17 *Let H be a Hilbert space and let $G \subset H$ be a closed subset. For all $x \in H$, there exists a unique $y \in G$ (called the orthogonal projection of x on G) such that for all $z \in G$, we have $\langle x - y, z \rangle_H = 0$ and $\|x - y\|_H^2 \leq \|x - z\|_H^2$.*

Theorem 2.18 *The conditional expectation of an integrable random variable exists with respect to any σ-algebra included in $\mathcal{F}$, and is almost surely unique.*

Proof First suppose that $X \in L^2(\Omega, \mathcal{F}, \mathbb{P})$. Let $G \subset L^2(\Omega, \mathcal{F}, \mathbb{P})$ be the set of all $\mathcal{G}$-measurable functions in $L^2(\Omega, \mathcal{F}, \mathbb{P})$. This set is isomorphic to $L^2(\Omega, \mathcal{G}, \mathbb{P})$, so it is a closed set.

Let Y be the orthogonal projection of X on G; thanks to Lemma 2.17, Y satisfies (2.6), so due to Proposition 2.16, $\mathbb{E}(X \mid \mathcal{G}) = Y$.

Suppose now that X is nonnegative. Set $X_n = X \wedge n$. The sequence (X_n) is square integrable. Let Y_n be the projection of X_n on G. Let Z be a nonnegative bounded $\mathcal{G}$-measurable variable. If $n \geq m$, then $X_n - X_m$ is nonnegative too, so that

$$\mathbb{E}[Z(Y_n - Y_m)] = \mathbb{E}[Z(X_n - X_m)] \geq 0.$$

Set $Y = \sup_n Y_n$. We have $\mathbb{E}(ZY_n) = \mathbb{E}(ZX_n)$, and by increasing limit of the sequence $(\mathbb{E}(ZY_n))$, we obtain $\mathbb{E}(ZY) = \mathbb{E}(ZX)$, so that $\mathbb{E}(X \mid \mathcal{G}) = Y$.

If Z is not bounded, taking $Z_n = Z \wedge n$ yields the result.

Finally, the result is proven for a general variable X via the usual decomposition $X = X^+ - X^-$.   □

$\triangleright$ *Example 2.19 (Quadratic Approximation)*   If $X \in L^2$, then $\mathbb{E}(X \mid \mathcal{G})$ is the projection of X on $\mathcal{G}$, that is the closest variable in $\mathcal{G}$ to X in the sense of least squares, in mathematical words

$$\|X - \mathbb{E}(X \mid \mathcal{G})\|_2^2 \leq \|X - Z\|_2^2, \quad Z \in \mathcal{G},$$

also called the quadratic approximation of X by $\mathcal{G}$. See Fig. 2.1 for an illustration. $\triangleleft$

$\triangleright$ *Example 2.20 (Generated Sub σ-Algebras)*   If $\mathcal{B}$ is closed under finite intersections and generates $\mathcal{G}$, then $Y = \mathbb{E}(X \mid \mathcal{G})$ if and only if Y is $\mathcal{G}$-measurable and $\mathbb{E}(X\mathbb{1}_B) = \mathbb{E}(Y\mathbb{1}_B)$, for any $B \in \mathcal{B}$.

In particular, if $\mathcal{G} = \sigma(B)$, then any $\mathcal{G}$-measurable function Z can be written $Z = a_1 \mathbb{1}_B + a_2 \mathbb{1}_{\overline{B}}$. The conditional expectation Y of X given $\mathcal{G}$ can be written $Y = b_1 \mathbb{1}_B + b_2 \mathbb{1}_{\overline{B}}$ with $\mathbb{E}(X\mathbb{1}_B) = b_1 \mathbb{P}(B)$ and $\mathbb{E}(X\mathbb{1}_{\overline{B}}) = b_2 \mathbb{P}(\overline{B})$, so that

$$\mathbb{E}(X \mid \mathcal{G}) = \frac{\mathbb{E}(X\mathbb{1}_B)}{\mathbb{P}(B)} \mathbb{1}_B + \frac{\mathbb{E}(X\mathbb{1}_{\overline{B}})}{1 - \mathbb{P}(B)} \mathbb{1}_{\overline{B}}.$$

Fig. 2.1 Conditional expectation—Example 2.19

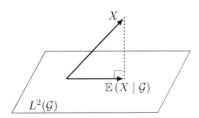

When $\mathcal{G} = \sigma(B_1, \ldots, B_n)$, where $\{B_1, \ldots, B_n\}$ is a partition of Ω into events with positive probabilities, we obtain similarly

$$\mathbb{E}(X \mid \mathcal{G}) = \sum_{i=1}^{n} \frac{\mathbb{E}(X 1_{B_i})}{\mathbb{P}(B_i)} 1_{B_i}.$$

When $\mathcal{G} = \sigma(B_n, n \geq 1)$, where (B_n) is a sequence of pairwise disjoint events with positive probabilities, then

$$\mathbb{E}(X \mid \mathcal{G}) = \sum_{n \geq 1} \frac{\mathbb{E}(X 1_{B_n})}{\mathbb{P}(B_n)} 1_{B_n}.$$

Since for each $\omega \in \Omega$ only one term of this sum is not null, this random variable is well defined. $\triangleleft$

Many properties of the conditional expectation are similar to the properties of expectation.

Properties of Conditional Expectation

Let X and Y be two nonnegative or integrable real random variables and let $\mathcal{G}$ be a σ-algebra included in $\mathcal{F}$.

1. **(Positive)** If $X \geq Y$ a.s., then $\mathbb{E}(X \mid \mathcal{G}) \geq \mathbb{E}(Y \mid \mathcal{G})$ a.s. In particular, if X is nonnegative, then its conditional expectation is nonnegative too—but may take infinite values.
2. **(Linear)** For any $(a, b) \in \mathbb{R}^2$, we have

$$\mathbb{E}(aX + bY \mid \mathcal{G}) = a\mathbb{E}(X \mid \mathcal{G}) + b\mathbb{E}(Y \mid \mathcal{G}) \quad \text{a.s..}$$

3. **(Monotone convergence theorem)** If $0 \leq X_m \nearrow X$, then

$$\mathbb{E}(X_n \mid \mathcal{G}) \nearrow \mathbb{E}(X \mid \mathcal{G}).$$

4. **(Fatou's lemma)** If X_n is nonnegative for $n \in \mathbb{N}$, then

$$\mathbb{E}(\underline{\lim} X_n \mid \mathcal{G}) \leq \underline{\lim} \mathbb{E}(X_n \mid \mathcal{G}).$$

5. **(Dominated convergence theorem)** If (X_n) converges a.s., and if $|X_n| \leq Y$, where Y is integrable, then

$$\mathbb{E}(\lim_{n \to +\infty} X_n \mid \mathcal{G}) = \lim_{n \to +\infty} \mathbb{E}(X_n \mid \mathcal{G}), \quad \text{a.s. and in } L^1.$$

6. **(Jensen's inequality)** If $\varphi : I \longrightarrow \mathbb{R}$, where I is an interval of $\mathbb{R}$, is a convex function such that $\varphi(X)$ is integrable, then

$$\varphi(\mathbb{E}(X \mid \mathcal{G})) \leq \mathbb{E}[\varphi(X) \mid \mathcal{G}] \quad \text{a.s..}$$

For example, if $X \in L^p$, with $1 \leq p \leq +\infty$, we have

$$|\mathbb{E}(X \mid \mathcal{G})|^p \leq \mathbb{E}(|X|^p \mid \mathcal{G}).$$

This yields $|\mathbb{E}(X \mid \mathcal{G})| \leq \mathbb{E}(|X| \mid \mathcal{G})$, inducing that $\mathbb{E}[|\mathbb{E}(X \mid \mathcal{G})|] \leq \mathbb{E}(|X|)$.

The conditional expectation has also some specific properties.

Theorem 2.21 *Let $\mathcal{G}'$ be a σ-algebra included in $\mathcal{G}$. We have*

$$\mathbb{E}[\mathbb{E}(X \mid \mathcal{G}) \mid \mathcal{G}'] = \mathbb{E}(X \mid \mathcal{G}') \quad \text{a.s..}$$

In particular,

$$\mathbb{E}[\mathbb{E}(X \mid \mathcal{G})] = \mathbb{E}X \quad \text{a.s.,} \tag{2.7}$$

which, for $\mathcal{G} = \{\emptyset, \Omega\}$, yields again $\mathbb{E}(X \mid \mathcal{G}) = \mathbb{E}X$ a.s.

Note that, due to the properties of projection, this theorem is obvious if the variables are square integrable.

Proof Suppose that X is nonnegative. Let Z be a variable that is $\mathcal{G}'$-measurable, and hence also $\mathcal{G}$-measurable. Set $V = \mathbb{E}(X \mid \mathcal{G})$ and $U = \mathbb{E}(V \mid \mathcal{G}')$. By definition of the conditional expectation, $\mathbb{E}(ZU) = \mathbb{E}(ZV) = \mathbb{E}[Z\mathbb{E}(X \mid \mathcal{G})] = \mathbb{E}(ZX)$.

If X is an integrable random variable, the conclusion follows as usual through $X = X^+ - X^-$.   □

Proposition 2.22 *If X is $\mathcal{G}$-measurable and if XY satisfies (2.5), then*

$$\mathbb{E}(XY \mid \mathcal{G}) = X\mathbb{E}(Y \mid \mathcal{G}) \quad \text{a.s..}$$

In particular, if X is $\mathcal{G}$-measurable, then $\mathbb{E}(X \mid \mathcal{G}) = X$ a.s.

Proof For nonnegative X and Y.

If Z is $\mathcal{G}$-measurable, then ZX is $\mathcal{G}$-measurable, so

$$\mathbb{E}[(ZX)\mathbb{E}(Y \mid \mathcal{G})] = \mathbb{E}[(ZX)Y] = \mathbb{E}[Z(XY)] = \mathbb{E}[Z\mathbb{E}(XY \mid \mathcal{G})] \quad \text{a.s.,}$$

by definition of the conditional expectation.   □

The definition of conditional variance for variables having a conditional distribution is dimilar to the definition of conditional expectation.

Definition 2.23 Let X be a random variable defined on $(\Omega, \mathcal{F}, \mathbb{P})$ and let Q be its conditional distribution given a σ-algebra $\mathcal{G}$ included in $\mathcal{F}$. The conditional variance of X given $\mathcal{G}$, denoted by $\mathbb{V}\mathrm{ar}\,(X \mid \mathcal{G})$, is the random variable defined by

$$\mathbb{V}\mathrm{ar}\,(X \mid \mathcal{G})(\omega) = \int_{\mathbb{R}} [x - \mathbb{E}\,(X \mid \mathcal{G})(\omega)]^2 Q(\omega,\, dx),$$

for any ω such that $\mathbb{E}\,(X \mid \mathcal{G})(\omega)$ is finite.

The conditional version of the well-known König's formula $\mathbb{V}\mathrm{ar}\,(X) = \mathbb{E}\,(X^2) - [\mathbb{E}\,(X)]^2$ reads

$$\mathbb{V}\mathrm{ar}\,(X \mid \mathcal{G}) = \mathbb{E}\,(X^2 \mid \mathcal{G}) - [\mathbb{E}\,(X \mid \mathcal{G})]^2 = \mathbb{E}\,([X - \mathbb{E}\,(X \mid \mathcal{G})]^2 \mid \mathcal{G}) \quad \text{a.s..}$$

Proposition 2.24 (Total Variance Theorem) *If X is square integrable, then*

$$\mathbb{V}\mathrm{ar}\,(X) = \mathbb{E}\,[\mathbb{V}\mathrm{ar}\,(X \mid \mathcal{G})] + \mathbb{V}\mathrm{ar}\,[\mathbb{E}\,(X \mid \mathcal{G})]. \tag{2.8}$$

Proof We can write

$$\mathbb{V}\mathrm{ar}\,X = \mathbb{E}\,([X - \mathbb{E}\,(X \mid \mathcal{G}) + \mathbb{E}\,(X \mid \mathcal{G}) - \mathbb{E}\,X]^2).$$

Since $\mathbb{E}\,Z = \mathbb{E}\,[\mathbb{E}\,(Z \mid \mathcal{G})]$ for any variable Z, we have

$$\mathbb{E}\,([X - \mathbb{E}\,(X \mid \mathcal{G})]^2) = \mathbb{E}\,[\mathbb{E}\,([X - \mathbb{E}\,(X \mid \mathcal{G})]^2 \mid \mathcal{G})] = \mathbb{E}\,[\mathbb{V}\mathrm{ar}\,(X \mid \mathcal{G})]$$

and

$$\mathbb{E}\,([\mathbb{E}\,(X \mid \mathcal{G}) - \mathbb{E}\,X]^2) = \mathbb{E}\,[(\mathbb{E}\,(X \mid \mathcal{G}) - \mathbb{E}\,[\mathbb{E}\,(X \mid \mathcal{G})])^2] = \mathbb{V}\mathrm{ar}\,[\mathbb{E}\,(X \mid \mathcal{G})].$$

Finally,

$$\mathbb{E}\,([X - \mathbb{E}\,(X \mid \mathcal{G})][\mathbb{E}\,(X \mid \mathcal{G}) - \mathbb{E}\,X])$$
$$= \mathbb{E}\,([X - \mathbb{E}\,(X \mid \mathcal{G})][\mathbb{E}\,(X \mid \mathcal{G}) - \mathbb{E}\,X] \mid \mathcal{G})$$
$$= \mathbb{E}\,[X - \mathbb{E}\,(X \mid \mathcal{G}) \mid \mathcal{G}][\mathbb{E}\,(X \mid \mathcal{G}) - \mathbb{E}\,X],$$

because $\mathbb{E}\,(X \mid \mathcal{G}) - \mathbb{E}\,X$ is $\mathcal{G}$-mesurable. The result is proven since $\mathbb{E}\,[X - \mathbb{E}\,(X \mid \mathcal{G}) \mid \mathcal{G}] = 0$ by definition and linearity of $\mathbb{E}\,(X \mid \mathcal{G})$. □

Until the end of the section, we will study the particular case of the conditioning with respect to a σ-algebra generated by some random variable T, that is $\mathcal{G} = \sigma(T)$. We will set $\mathbb{E}[X \mid \sigma(T)] = \mathbb{E}(X \mid T)$, and more generally $\mathbb{E}[X \mid \sigma(T_1, \ldots, T_n)] = \mathbb{E}(X \mid T_1, \ldots, T_n)$.

Proposition 2.25 *If X and T are two random variables, then $\mathbb{E}(X \mid T) = \varphi(T)$ if and only if the Borel function φ satisfies $\mathbb{E}[Xg(T)] = \mathbb{E}[\varphi(T)g(T)]$ for any Borel function g, or equivalently $\mathbb{E}[X\mathbb{1}_{(T \in B)}] = E[\varphi(T)\mathbb{1}_{(T \in B)}]$ for any Borel set B.*

Proof Any $\sigma(T)$-measurable variable Z can be written $Z = g(T)$ for a Borel function g, from which the result follows by definition of the conditional expectation.
□

Definition 2.26 Let X and T be two random variables and let Q denote the conditional distribution of X given T. The conditional distribution of X given $(T = t)$ is the transition probability π from $\mathbb{R}$ to $\mathbb{R}$ satisfying for any Borel sets A and B,

$$\mathbb{P}(X \in B, T \in A) = \int_A \pi(t, B)d\mathbb{P}_T(t), \qquad (2.9)$$

where $\mathbb{P}_T$ is the distribution of T. We will write $\pi(t, B) = \mathbb{P}(X \in B \mid T = t)$.

Proposition 2.27 *Let X and T be two random variables. A transition probability π from $\mathbb{R}$ to $\mathbb{R}$ is the conditional distribution of X given $(T = t)$ if and only if for any bounded Borel function h,*

$$\mathbb{E}[h(X) \mid T] = \int_{\mathbb{R}} h(x)\pi(T, dx) \quad a.s..$$

Proof Due to (2.9), the equality

$$\mathbb{E}[h(X)g(T)] = \int_{\mathbb{R}^2} g(t)h(x)\pi(t, dx)d\mathbb{P}_T(t)$$

is directly satisfied for indicator functions, and then is proven to hold for a general bounded Borel function g as usual.

The converse is obtained by setting $h(X) = \mathbb{1}_{(X \in B)}$ and $g(T) = \mathbb{1}_{(T \in A)}$.
□

We will denote $\mathbb{E}(X \mid T = t)$ the expectation of the conditional distribution of X given $(T = t)$.

Corollary 2.28 *If X and T are two random variables and if φ is a Borel function, then $\mathbb{E}(X \mid T) = \varphi(T)$ if and only if $\mathbb{E}(X \mid T = t) = \varphi(t)$ for any real number t.*

Proof This is a direct consequence of Proposition 2.27, since $\mathbb{E}\,(X \mid T = t) = \int_{\mathbb{R}} x\pi(t, dx)$. □

The conditional expectation can also be determined independently of the conditional distribution as shown in the next example.

▷ *Example 2.29* Let X and Y be independent random variables with the same distribution $\gamma(1, 1)$. The $T = X + Y \sim \gamma(2, 1)$; see Example 1.66.

Let us determine the conditional expectation of X given T. We are looking for φ such that for all bounded Borel function g, we have $\mathbb{E}\,[g(T)X] = \mathbb{E}\,[g(T)\varphi(T)]$. We compute

$$\mathbb{E}\,[g(T)X] = \iint_{\mathbb{R}_+^2} g(x + y)xe^{-(x+y)}\,dxdy = \int_{\mathbb{R}_+}\int_0^t g(t)e^{-t}u\,dudt$$

$$\overset{(1)}{=} \int_{\mathbb{R}_+} \varphi(t)g(t)te^{-t}\,dt = \mathbb{E}\,[g(T)\varphi(T)].$$

Since (1) is satisfied for $\varphi(t) = \int_0^t u\,du/t = t/2$, we obtain $\mathbb{E}\,(X \mid T) = T/2$ a.s.

This result could have been guessed without knowing the distribution of the variables, as shown in Exercise 2.3 below.

In the same way, for all bounded Borel function h, we have

$$\mathbb{E}\,[h(X) \mid T = t] = \frac{1}{t}\int_0^t h(u)\,du,$$

so that the conditional distribution of X given $(T = t)$ is uniform over $[0, t]$. ◁

2.1.5 Conditioning and Independence

The probabilistic notions of dependence and conditioning are obviously linked.

Theorem 2.30 *If a random variable X defined on $(\Omega, \mathcal{F}, \mathbb{P})$ and a σ-algebra $\mathcal{G}$ included in $\mathcal{F}$ are independent, then $\mathbb{E}\,(X \mid \mathcal{G}) = \mathbb{E}\,X$ a.s.*

Proof Let Z be a $\mathcal{G}$-measurable random variable. We have

$$\mathbb{E}\,(XZ) = (\mathbb{E}\,X)(\mathbb{E}\,Z) = \mathbb{E}\,[(\mathbb{E}\,X)Z],$$

and since $\mathbb{E}\,X$ is a.s. constant, it is $\mathcal{G}$-measurable. □

Even if two random variables are independent, their conditional expectations may not be, as shown in the next example.

▷ *Example 2.31* Suppose that $(S, T) \sim \mathcal{N}_2(0, I_2)$. Set $X = S + T$ and $Y = S - T$. Then (X, Y) is a Gaussian vector and $\text{Cov}\,(X, Y) = 0$, so that X and Y are independent. Still $\mathbb{E}\,(X \mid S) = S = \mathbb{E}\,(Y \mid S)$ a.s.   ◁

Nevertheless, conditional expectation gives a criterion of independence.

Theorem 2.32 *Two random variables X and Y are independent if and only if for all bounded Borel function h,*

$$\mathbb{E}\,[h(X) \mid Y] = \mathbb{E}\,[h(X)] \quad a.s..$$

In particular, if X and Y are independent, then $\mathbb{E}\,(X \mid Y) = \mathbb{E}\,X$ a.s.

Proof The variables X and Y are independent if their generated σ-algebras are independent. Therefore, if Z is $\sigma(Y)$-measurable, then Z is independent of any variable $h(X)$, so

$$\mathbb{E}\,[Zh(X)] = (\mathbb{E}\,Z)(\mathbb{E}\,[h(X)]) = \mathbb{E}\,(Z\mathbb{E}\,[h(X)]).$$

Conversely,

$$\mathbb{E}\,[f(Y)h(X)] = \mathbb{E}\,(f(Y)\mathbb{E}\,[h(X) \mid Y])$$
$$= \mathbb{E}\,(f(Y)\mathbb{E}\,[h(X)]) = \mathbb{E}\,[f(Y)]\mathbb{E}\,[h(X)].$$

Since this holds true for any bounded Borel functions f and h, X and Y are independent.   □

Corollary 2.33 *If $\mathcal{G}_1$ and $\mathcal{G}_2$ are two independent σ-algebras included in $\mathcal{F}$ and if X is an integrable random variable such that $\sigma(X, \mathcal{G}_1)$ and $\mathcal{G}_2$ are independent, then*

$$\mathbb{E}\,[X \mid \sigma(\mathcal{G}_1, \mathcal{G}_2)] = \mathbb{E}\,(X \mid \mathcal{G}_1) \quad a.s..$$

Proof By definition $\sigma(\mathcal{G}_1, \mathcal{G}_2) = \sigma(\{B_1 \cap B_2 \;:\; B_i \in \mathcal{G}_i\})$. So, using Example 2.20, it is sufficient to consider the indicator functions $\mathbb{1}_{B_1 \cap B_2} = \mathbb{1}_{B_1}\mathbb{1}_{B_2}$. We have

$$\mathbb{E}\,[\mathbb{1}_{B_1}\mathbb{1}_{B_2}\mathbb{E}\,(X \mid \mathcal{G}_1)] \overset{(1)}{=} \mathbb{E}\,[\mathbb{1}_{B_2}\mathbb{E}\,(X\mathbb{1}_{B_1} \mid \mathcal{G}_1)]$$

$$\overset{(2)}{=} \mathbb{E}\,(\mathbb{1}_{B_2})\mathbb{E}\,[\mathbb{E}\,(X\mathbb{1}_{B_1} \mid \mathcal{G}_1)]$$

$$\overset{(3)}{=} \mathbb{E}\,(\mathbb{1}_{B_2})\mathbb{E}\,(X\mathbb{1}_{B_1}) = \mathbb{E}\,(\mathbb{1}_{B_1}\mathbb{1}_{B_2}X).$$

(1) because $B_1 \in \mathcal{G}_1$, (2) because $\mathcal{G}_1$ and $\mathcal{G}_2$ are independent, and (3) by (2.7) p. 70. □

This corollary is often of use in applications, as shown in the next example.

▷ *Example 2.34* Let (X_n) be an i.i.d. integrable sequence. Let us determine the conditional expectation of X_1 given $\sigma(S_n, S_{n+1}, \ldots)$, where $S_n = \sum_{i=1}^{n} X_i$. Corollary 2.33 applies to the random variable $X = X_1$ and the σ-algebras $\mathcal{G}_1 = \sigma(S_n)$ and $\mathcal{G}_2 = \sigma(X_{n+1}, X_{n+2}, \ldots)$. Since $\sigma(\mathcal{G}_1, \mathcal{G}_2) = \sigma(S_n, S_{n+1}, \ldots)$, $\sigma(X, \mathcal{G}_1)$ and $\mathcal{G}_2$ are independent, it follows that

$$\mathbb{E}[X_1 \mid \sigma(S_n, S_{n+1}, \ldots)] = \mathbb{E}[X_1 \mid \sigma(S_n)] \quad \text{a.s..}$$

Since the variables X_i are i.i.d., we have

$$\sum_{i=1}^{n} \mathbb{E}(X_i \mid S_n) = \mathbb{E}(S_n \mid S_n) = S_n \quad \text{a.s.}$$

and also $\mathbb{E}(X_i \mid S_n) = \mathbb{E}(X_1 \mid S_n)$ a.s. for $i = 1, \ldots, n$.
  Finally $\mathbb{E}(X_1 \mid S_n) = S_n/n$. ◁

The conditional independence of σ-algebras is defined as follows. The conditional independence of events or of variables follows straightforwardly through the generated σ-algebras.

Definition 2.35 Let $(\Omega, \mathcal{F}, \mathbb{P})$ be a probability space and let $\mathcal{G}$, $\mathcal{G}_1$, and $\mathcal{G}_2$ be σ-algebras included in $\mathcal{F}$. Then $\mathcal{G}_1$ and $\mathcal{G}_2$ are said to be conditionally independent with respect to $\mathcal{G}$ if

$$\mathbb{P}(A \cap B \mid \mathcal{G}) = \mathbb{P}(A \mid \mathcal{G})\mathbb{P}(B \mid \mathcal{G}) \quad \text{a.s.,} \quad A \in \mathcal{G}_1, \ B \in \mathcal{G}_2.$$

2.1.6 Practical Determination

Different methods are here detailed for determining the conditional distribution and expectation of a variable with respect to another.

For Discrete Random Variables
Since elementary techniques are sufficient in the discrete case, we only give one example.

▷ *Example 2.36* Let X and Y be two independent random variables such that $X \sim \mathcal{P}(\lambda)$ and $Y \sim \mathcal{P}(\mu)$. We have $T = X + Y \sim \mathcal{P}(\lambda + \mu)$. Hence, we compute

$$
\mathbb{P}(X = k \mid T = t) = \frac{\mathbb{P}(X = k, Y = t - k)}{\mathbb{P}(X + Y = t)}
$$

$$
= \frac{e^{-\lambda}\lambda^k}{k!} \frac{e^{-\mu}\mu^{t-k}}{(t-k)!} \frac{t!}{e^{-(\lambda+\mu)}(\lambda+\mu)^t}
$$

$$
= \binom{t}{k} \left(\frac{\lambda}{\lambda+\mu}\right)^k \left(1 - \frac{\lambda}{\lambda+\mu}\right)^{t-k}, \quad 0 \le k \le t,
$$

Therefore, the conditional distribution of X given $(T = t)$ is a binomial distribution $\mathcal{B}(t, \lambda/(\lambda + \mu))$.

Hence, $\mathbb{E}(X \mid T = t) = \lambda t/(\lambda + \mu)$ and $\mathbb{V}\mathrm{ar}(X \mid T = t) = \lambda\mu t/(\lambda + \mu)^2$, so that $\mathbb{E}(X \mid T) = \lambda T/(\lambda + \mu)$, and $\mathbb{V}\mathrm{ar}(X \mid T) = \lambda\mu T/(\lambda + \mu)^2$. ◁

For Random Variables with a Joint Density

Proposition 2.37 *If X and Y are two random variables such that (X, Y) has a density f, then the conditional distribution of X given $(Y = y)$ has a density too, given by*

$$
f_{X|Y}(x \mid y) = \frac{f(x, y)}{f_Y(y)} = \frac{f(x, y)}{\int_{\mathbb{R}} f(x, y)\, dx}
$$

if $f_Y(y) \ne 0$ and 0 otherwise.

Proof Let g and h be two bounded Borel functions. We have

$$
\mathbb{E}[h(X)g(Y)] = \iint_{\mathbb{R}^2} h(x)g(y)f(x, y)\, dxdy
$$

$$
= \iint_{\mathbb{R}^2} h(x)g(y)f_{X|Y}(x \mid y)f_Y(y)\, dxdy.
$$

So by Fubini's theorem,

$$
\mathbb{E}[h(X)g(Y)] = \int_{\mathbb{R}} g(y) \left[\int_{\mathbb{R}} h(x)f_{X|Y}(x \mid y)\, dx\right] f_Y(y)dy,
$$

and the result follows from Fubini's theorem and Proposition 2.27.

If $f_Y(y) = 0$ on $E \subset \mathbb{R}$, one can consider $\mathbb{R} \setminus E$. □

If both the conditional distribution and the marginal distribution are known, the distribution of the pair can be determined. This can be generalized to any finite number of random variables. For instance, for three variables, we have

$$f_{X,Y,Z}(x, y, z) = f_X(x) f_{Y|X}(y \mid x) f_{Z|(X,Y)}(z \mid x, y).$$

▷ *Example 2.38* In Example 2.29, let us determine the density of (X, T). For any bounded Borel function h,

$$\mathbb{E}\left[h(X, T)\right] = \iint_{\mathbb{R}^2} h(y, x + y) e^{-(x+y)} \, dx dy = \iint_{\mathbb{R}^2} h(x, t) e^{-t} \mathbb{1}_{[0,t]}(x) \, dx dt,$$

so that $f(x, t) = e^{-t} \mathbb{1}_{[0,t]}(x) \mathbb{1}_{\mathbb{R}_+}(t)$.

Since $T \sim \gamma(2, 1)$, we have $f_T(t) = t e^{-t} \mathbb{1}_{\mathbb{R}_+}(t)$, and hence $f_{X|T}(x \mid t) = (1/t) \mathbb{1}_{[0,t]}(x) \mathbb{1}_{\mathbb{R}_+}(t)$, and hence $\mathbb{E}(X \mid T = t) = \int_0^t (x/t) \, dx$, from which we deduce again that $\mathbb{E}(X \mid T) = T/2$ a.s. ◁

For a Random Variable with a Density and a Discrete Random Variable
Proposition 2.39 *Let X be a random variable with density f_X and let T be a discrete random variable taking values $\{t_k : k \in K\}$.*

1. *If the distribution of X and the conditional distribution of T given $(X = x)$ are known, then the distribution of (T, X) is given by*

$$\mathbb{P}(T = t, X \in B) = \int_B \mathbb{P}(T = t \mid X = x) f_X(x) \, dx.$$

2. *If $\mathbb{P}(T = t, X \in B) = \int_B f(t, x) \, dx$ for any $B \in \mathcal{B}(\mathbb{R})$, then*
   a. *the conditional distribution of T given $(X = x)$ is given by*

$$\mathbb{P}(T = t \mid X = x) = \frac{f(t, x)}{\sum_{k \in K} f(t_k, x)} = \frac{f(t, x)}{f_X(x)};$$

   b. *the conditional distribution of X given $(T = t)$ has the density*

$$f_{X|T}(x \mid t) = \frac{f(t, x)}{\int_{\mathbb{R}} f(t, x) \, dx}.$$

Proof By definition, $\mathbb{P}(T = t \mid X = x) = \mathbb{E}\left[\mathbb{1}_{(T=t)} \mid X = x\right] = \varphi(x)$, with

$$\mathbb{E}\left[\mathbb{1}_{(X \in B)} \mathbb{1}_{(T=t)}\right] = \mathbb{E}\left[\mathbb{1}_{(X \in B)} \varphi(X)\right], \quad B \in \mathcal{B}(\mathbb{R}).$$

It follows that $\mathbb{P}(T = t, X \in B) = \int_B \varphi(x) f_X(x) \, dx$, which proves 1.

Point 2. is a straightforward consequence of Theorem 2.37, by interpreting $f(t, x)$ as the density of (T, X) with respect to $\sum_{k \in K} \delta_{t_k} \otimes \lambda$, where δ_{t_k} is the Dirac measure at t_k and λ is the Lebesgue measure on $\mathbb{R}$. □

▷ *Example 2.40* Let X be a discrete random variable taking integer values and let Y be a nonnegative random variable. Suppose that

$$\mathbb{P}(X = n, Y \leq t) = \int_0^t \beta e^{-(\alpha+\beta)y} \frac{(\alpha y)^n}{n!} \, dy,$$

for some positive real numbers α and β. Let us determine the conditional distribution of X given $(Y = y)$. We have

$$\mathbb{P}(Y \leq t) = \sum_{n \geq 0} \mathbb{P}(X = n, Y \leq t)$$

$$= \int_0^t \sum_{n \geq 0} \frac{(\alpha y)^n}{n!} \beta e^{-(\alpha+\beta)y} dy = \int_0^t \beta e^{-\beta y} \, dy,$$

and hence Y has the distribution $\mathcal{E}(\beta)$. Moreover,

$$\mathbb{P}(X = n \mid Y = y) = \frac{(\alpha y)^n \beta e^{-(\alpha+\beta)y}/n!}{\beta e^{-\beta y}} = e^{-\alpha y} \frac{(\alpha y)^n}{n!},$$

meaning that the conditional distribution of X given $(Y = y)$ is the Poisson distribution $\mathcal{P}(\alpha y)$. Hence, the conditional expectation of X given Y is αY.   ◁

Vector and Gaussian Cases

Proposition 2.41 *Let T and Z be two independent random vectors. Let $X = Z + g(T)$, where g is a Borel function. The conditional distribution of X given $(T = t)$ is that of $Z + g(t)$, or*

$$\mathbb{E}(X \mid T = t) = \mathbb{E}(Z \mid T = t) + g(t).$$

Proof For random variables.

For any bounded Borel function h, $\mathbb{E}[h(X)g(T)] = \mathbb{E}[h(Z + g(T))g(T)]$, that is to say

$$\mathbb{E}[h(X)g(T)] = \iint_{\mathbb{R}^2} h(z + g(t))g(t)d\mathbb{P}_T(t)d\mathbb{P}_Z(z)$$

$$= \int_{\mathbb{R}} g(t) \left[\int_{\mathbb{R}} h(z + g(t))d\mathbb{P}_Z(z) \right] d\mathbb{P}_T(t)$$

$$= \int_{\mathbb{R}} g(t) \left[\int_{\mathbb{R}} h(z)d\mathbb{P}_{Z+g(t)}(z) \right] d\mathbb{P}_T(t),$$

and Proposition 2.27 yields the result.   □

When the variables constitute a Gaussian vector, the conditional distribution follows straightforwardly.

Proposition 2.42 *If $(X, T) \sim \mathcal{N}_n(m, \Gamma)$ with $X : \Omega \longrightarrow \mathbb{R}^p$, $T : \Omega \longrightarrow \mathbb{R}^q$ and $\det \Gamma \neq 0$, then the conditional distribution of X given $(T = t)$ is a Gaussian distribution*

$$
\mathcal{N}_p\Big(\mathbb{E} X + \mathbb{C}\mathrm{ov}\,(X, T)\,\Gamma_T^{-1}(t - \mathbb{E} T), \Gamma_X - \mathbb{C}\mathrm{ov}\,(X, T)\Gamma_T^{-1}\mathbb{C}\mathrm{ov}\,(T, X)\Big).
$$

Proof For $p = q = 1$.

Set $Z = X - AT$, where $A = \mathbb{C}\mathrm{ov}\,(X, T)/\mathbb{V}\mathrm{ar}\,T$. The vector (T, Z) is a Gaussian vector, and hence, since the variables Z and T are uncorrelated, they are independent. Moreover,

$$
\mathbb{E}\,Z = \mathbb{E}\,X - \frac{\mathbb{C}\mathrm{ov}\,(X, T)}{\mathbb{V}\mathrm{ar}\,T}\mathbb{E}\,T \quad \text{and} \quad \mathbb{V}\mathrm{ar}\,Z = \mathbb{V}\mathrm{ar}\,X - \frac{\mathbb{C}\mathrm{ov}\,(X, T)^2}{\mathbb{V}\mathrm{ar}\,T}.
$$

Due to Proposition 2.41, the conditional distribution of X given $(T = t)$ is a Gaussian distribution with expectation $\mathbb{E}\,X + \mathbb{C}\mathrm{ov}\,(X, T)(t - \mathbb{E}\,T)/\mathbb{V}\mathrm{ar}\,T$ and variance $\mathbb{V}\mathrm{ar}\,X - \mathbb{C}\mathrm{ov}\,(X, T)^2/\mathbb{V}\mathrm{ar}\,T$. $\qquad\square$

2.2 The Linear Model

In the linear model, the studied random variable Y is assumed to be linked (up to some additive error) by a linear relation to d other random variables $X_1, \ldots, X_d$ that are simultaneously observed. The model is said to be simple if $d = 1$. If $d > 1$, it is multiple. If the random variables X_j are continuous, the model is called a regression model; a particular case is the Gaussian linear model in which $(X_1, \ldots, X_d, Y)$ is supposed to form a Gaussian vector.

Let Y be a square integrable random variable. First, the expectation of Y is its orthogonal projection in L^2 on the subspace of constant random variables, that is the best approximation of Y by a constant in the sense of least squares. Note that the standard deviation σ_Y is the length of $Y - \mathbb{E}\,Y$ in L^2. Let now $X_1, \ldots, X_d$ also be square integrable random variables, with covariance matrix Γ_X.

The conditional expectation of Y given $(X_1, \ldots, X_d)$, that is $Z = \mathbb{E}\,(Y \mid X_1, \ldots, X_d)$, satisfies

$$
\mathbb{E}\,[(Y - Z)^2] \leq \mathbb{E}\,([Y - g(X_1, \ldots, X_d)]^2),
$$

for any bounded Borel function g. It is the best approximation in the sense of the L^2-norm of Y by some function of $(X_1, \ldots, X_d)$, called the minimum mean square error predictor. It is also the projection of Y on the sub-space $\mathrm{Sp}(X_1, \ldots, X_d)$ in L^2.

Fig. 2.2 Illustration of the
linear model

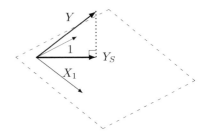

Second, Y is to be approximated by a linear function of $X_1, \ldots, X_d$; see Fig. 2.2, in which $d = 1$. The projection $Y_S = M + \sum_{j=1}^{d} \alpha_j X_j$ has to satisfy the (normal) equations

$$\langle Y - Y_S, 1 \rangle = 0, \tag{2.10}$$

$$\langle Y - Y_S, X_j \rangle = 0, \quad \text{for } j = 1, \ldots, d.$$

Since the scalar product is that of centered variables in L^2, defined by

$$< X, Y >= \mathbb{E}\,(X - \mathbb{E}\,X)(Y - \mathbb{E}\,Y),$$

(2.10) implies $M = \mathbb{E}\,Y - \sum_{j=1}^{d} \alpha_j \mathbb{E}\,X_j$, from which we deduce that

$$\langle Y - Y_S, \; X_j \rangle = \langle Y - \mathbb{E}\,Y - \sum_{i=1}^{d} \alpha_i (X_i - \mathbb{E}\,X_i), X_j \rangle.$$

Due to (2.10), $\langle Y - \mathbb{E}\,Y - \sum_{i=1}^{d} \alpha_i (X_i - \mathbb{E}\,X_i), \; \mathbb{E}\,X_j \rangle = 0$, so we also have

$$\langle Y - \mathbb{E}\,Y - \sum_{i=1}^{d} \alpha_i (X_i - \mathbb{E}\,X_i), \; X_j - \mathbb{E}\,X_j \rangle = 0$$

or, in matrix form,

$$\Gamma_X \begin{pmatrix} \alpha_1 \\ \vdots \\ \alpha_d \end{pmatrix} = \begin{pmatrix} \mathbb{C}\mathrm{ov}\,(Y, X_1) \\ \vdots \\ \mathbb{C}\mathrm{ov}\,(Y, X_d) \end{pmatrix},$$

where Γ_X is the covariance matrix of the vector $X = (X_1, \ldots, X_d)$. If X is not degenerated—that is if Γ_X is invertible, then

$$(\alpha_1, \ldots, \alpha_d) = (\mathbb{C}\mathrm{ov}\,(Y, X_1), \ldots, \mathbb{C}\mathrm{ov}\,(Y, X_d))\, \Gamma_X^{-1}.$$

This defines the best linear approximation of Y by $(X_1, \ldots X_d)$ in the sense of the L^2-norm or least squares.

Moreover, if $(Y, X_1, \ldots, X_d)$ is a Gaussian vector, then $Y - Y_S$ is independent of all the X_j, and

$$\|Y\|_2^2 = \|Y - \sum_{i=1}^d \alpha_i X_i\|_2^2 + \|\sum_{i=1}^d \alpha_i X_i\|_2^2,$$

so the Pythagorean theorem holds true.

▷ *Example 2.43 (Regression Line)* Let Y and X be two square integrable random variables. The best linear approximation of Y by X is $Y_S = \alpha X + m$, where

$$\alpha = \frac{\mathrm{Cov}\,(X, Y)}{\mathrm{Var}\,X} \quad \text{and} \quad m = \mathbb{E}\,Y - a\mathbb{E}\,X. \tag{2.11}$$

The line with equation

$$y = \mathbb{E}\,Y + \frac{\mathrm{Cov}\,(X, Y)}{\mathrm{Var}\,X}(x - \mathbb{E}\,X)$$

is called the regression line of Y given X.

This line is the one around which the distribution of (X, Y) is concentrated; in other words,

$$\int_{\mathbb{R}^2} (y - \alpha x - m)^2 d P_{(X,Y)}(x, y)$$

is minimum on $\mathbb{R}^2$ for α and m given by (2.11).

An example of a regression line is shown on Fig. 2.3.   ◁

The next result is a straightforward consequence of Theorem 2.42. Nevertheless, we give here a direct proof.

Theorem 2.44 *If $(X, Y)'$ is a Gaussian vector, the best quadratic and linear approximations of Y by X are equal.*

In other words, the conditional expectation of Y given X is a linear function of X.

Proof Let α and m be given by (2.11) p. 81. Set $Y = \alpha X + m + \varepsilon$.

We know that $\alpha X + m$ is the orthogonal projection of Y on $\mathrm{Sp}(1, X)$, so $\mathbb{E}\,\varepsilon = 0$ and $\mathbb{E}\,\varepsilon X = 0$. Moreover, (X, Y) is a Gaussian vector, so (X, ε) is a Gaussian vector too. Since ε and X are uncorrelated, they are independent, and hence, $\mathbb{E}\,(\varepsilon \mid X) =$

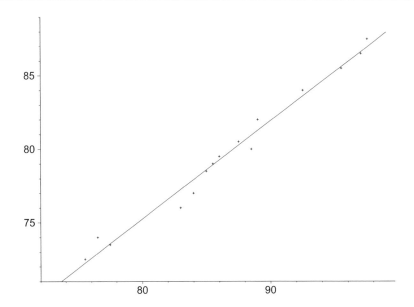

Fig. 2.3 A regression line

$\mathbb{E}\,\varepsilon = 0$ a.s., from which we deduce that

$$\mathbb{E}\,(Y \mid X) = \mathbb{E}\,(\alpha X + m + \varepsilon \mid X) = \alpha X + m.$$

The above equation holds a.s.   □

2.3 Stopping Times

In order to investigate stopping times and then martingales, it is first necessary to extend the Kolmogorov probability space $(\Omega, \mathcal{F}, \mathbb{P})$ by adding to it a filtration $\mathbf{F} = (\mathcal{F}_n)_{n \geq 0}$, an increasing sequence of σ-algebras included in $\mathcal{F}$. This constitutes a stochastic basis $\mathbf{B} = (\Omega, \mathcal{F}, \mathbf{F}, \mathbb{P})$, also called filtered space.

A stopping time is a random variable which does not depend on the "future".

Definition 2.45 Let $\mathbf{F} = (\mathcal{F}_n)$ be a filtration of $\mathcal{F}$. A stopping time for $\mathbf{F}$ (or $\mathbf{F}$-stopping time) is a random variable T taking values in $\overline{\mathbb{N}}$ and satisfying $(T \leq n) \in \mathcal{F}_n$ for all $n \in \mathbb{N}$.

Note that the condition $(T \leq n) \in \mathcal{F}_n$ is equivalent to either $(T = n) \in \mathcal{F}_n$ or $\mathbb{1}_{(T=n)}$ is $\mathcal{F}_n$-measurable, for all $n \in \mathbb{N}$.

▷ *Example 2.46* Any constant random variable T, that is equal to some $m \in \mathbb{N}$, is a stopping time.

Indeed, $(T = m) = \Omega \in \mathcal{F}_m$ and $(T = n) = \emptyset \in \mathcal{F}_n$ for all $n \neq m$.   ◁

Theorem-Definition 2.47 *Let T be an **F**-stopping time. The set $\mathcal{F}_T = \{A \in \mathcal{F} : \forall n \in \overline{\mathbb{N}}, \ A \cap (T = n) \in \mathcal{F}_n\}$ is a σ-algebra, called the σ-algebra of the events previous to T.*

Proof Let $A \in \mathcal{F}_T$. We can write $A = \cup_{n \geq 0} A \cap (T = n)$, so that $\overline{A} = \cap_{n \geq 0} \overline{A} \cap (T = n) \in \mathcal{F}_T$. If moreover $A_m \in \mathcal{F}$ for all m, we get

$$\bigcup_{m \geq 0} A_m = \bigcup_{n \geq 0} \left[\left(\bigcup_{m \geq 0} A_m \right) \cap (T = n) \right] = \bigcup_{m \geq 0} \left[\bigcup_{n \geq 0} A_m \cap (T = n) \right],$$

and hence $\cup_{m \geq 0} A_m \in \mathcal{F}_T$.   □

2.3.1 Properties of Stopping Times

1. If S and T are two stopping times, then $S + T$, $S \wedge T$ and $S \vee T$ are stopping times; indeed:

$$(S + T = n) = \bigcup_{i=0}^{n} (S = i) \cap (T = n - i) \in \mathcal{F}_n;$$

$$(S \wedge T \leq n) = (S \leq n) \cup (T \leq n) \in \mathcal{F}_n;$$

$$(S \vee T \leq n) = (S \leq n) \cap (T \leq n) \in \mathcal{F}_n.$$

In particular, if T is a stopping time and $k \in \mathbb{N}$, then $T + k$ and $T \wedge k$ are stopping times.
2. If (T_n) is a sequence of stopping times, then both $\sup_{n \geq 0} T_n$ and $\inf_{n \geq 0} T_n$ are stopping times, because for instance $(\sup_{n \geq 0} T_n \leq n) = \cap_{n \geq 0}(T_n \leq n)$.
3. If S and T are stopping times such that $S \leq T$, then $\mathcal{F}_S \subset \mathcal{F}_T$; indeed, if $A \in \mathcal{F}_S$, then $A \cap (T \leq n) = [A \cap (S \leq n)] \cap (T \leq n)$ is the intersection of elements of $\mathcal{F}_n$.

A stopping time for the natural filtration $(\sigma(X_k, 0 \leq k \leq n))$ of a random sequence (X_n) is said to ba a stopping time for (X_n).

▷ *Example 2.48 (Hitting and Exit Times)* Let (X_n) be a random sequence taking values in $\mathbb{R}^d$ and let $B \in \mathcal{B}(\mathbb{R}^d)$.

The variable $T = \inf\{n \in \mathbb{N} : X_n \in B\}$ is called the hitting time of B. This is a stopping time for (X_n). Indeed,

$$(T = n) = (X_1 \notin B, \ldots X_{n-1} \notin B, \ X_n \in B),$$

and $(X_i \notin B) \in \mathcal{F}_i$, $(X_n \in B) \in \mathcal{F}_n$, and $\mathcal{F}_i \subset \mathcal{F}_{i+1}$ for $i \in [\![1, n-1]\!]$, so that $(T = n) \in \mathcal{F}_n$. Note that T may also be called first passage time, depending on the kind of process under study.

On the contrary, the exit time from B, that is $S = \sup\{n \in \mathbb{N} : X_n \in B\}$, is not a stopping time for (X_n), because clearly,

$$(S = n) = (X_1 \in B, \ldots, X_n \in B, \ X_{n+1} \notin B)$$

does not belong to $\mathcal{F}_n$. $\triangleleft$

If T is a stopping time for a random sequence (X_n), the random sequence $(X_{T \wedge n})$ is said to be stopped at time T.

Proposition 2.49 *If T is an **F**-stopping time and if (X_n) is a random sequence for the same filtration **F**, then the variable $X_T \mathbb{1}_{(T < +\infty)}$ is $\mathcal{F}_T$-measurable.*

In particular, this induces that $X_T \mathbb{1}_{(T < +\infty)}$ is a real random variable.

Proof Let $B \in \mathcal{B}(\mathbb{R})$. We have

$$(T = n) \cap (X_T \mathbb{1}_{(T < +\infty)} \in B) = (T = n) \cap (X_n \in B) \in \mathcal{F}_n,$$

so $(X_T \mathbb{1}_{(T < +\infty)} \in B) \in \mathcal{F}_T$. $\square$

Proposition 2.50 (Wald's Identity) *If T is a stopping time with finite mean for an i.i.d. integrable random sequence (X_n), then*

$$\mathbb{E}(X_0 + \cdots + X_T) = (\mathbb{E}\, X_0)(\mathbb{E}\, T).$$

Note that this holds true also for any variable T independent of (X_n), as stated in Proposition 1.73.

Proof For nonnegative variables.

Set $S_n = X_0 + \cdots + X_n$. We have

$$\mathbb{E}\, S_T = \sum_{n \geq 0} \mathbb{E}\left[\mathbb{1}_{(T=n)} \sum_{i=0}^{n} X_i\right] \stackrel{(1)}{=} \sum_{i \geq 0} \mathbb{E}\left[X_i \sum_{n \geq i} \mathbb{1}_{(T=n)}\right];$$

(1) by Fubini's theorem. Therefore

$$\mathbb{E}\, S_T = \sum_{i \geq 0} \mathbb{E}[X_i \mathbb{1}_{(T \geq i)}] = \sum_{i \geq 0} \mathbb{P}(T \geq i)\mathbb{E}\, X_i = (\mathbb{E}\, X_0)(\mathbb{E}\, T),$$

because $(T \geq i) = \overline{(T < i)} \in \sigma(X_0, \ldots, X_{i-1})$, and hence $\mathbb{1}_{(T \geq i)}$ is independent of X_i. $\square$

2.4   Discrete-Time Martingales

The casino players call martingales what mathematicians call submartingales. The origin of the mathematical sense dates back to Ville (1939). Their systematic study is due to Doob in the 1950s.

2.4.1   Definitions and Properties

A random sequence is a martingale if the conditional expectation of its future value given its past and present values is equal to its present value. Submartingales (super-martingales) are the stochastic analogous of increasing (decreasing) functions.

Definition 2.51 Let $\mathbf{B} = (\Omega, \mathcal{F}, \mathbf{F}, \mathbb{P})$ be a stochastic basis. A random sequence (X_n) defined on $(\Omega, \mathcal{F}, \mathbb{P})$ is:

1. an **F**-martingale (or martingale adapted to **F**) if all following three conditions are satisfied:
   a. X_n is **F**-adapted (X_n is $\mathcal{F}_n$-measurable for $n \in \mathbb{N}$).
   b. X_n is integrable for $n \in \mathbb{N}$.
   c. $\mathbb{E}(X_{n+1} \mid \mathcal{F}_n) = X_n$ for $n \in \mathbb{N}$.
2. an **F**-submartingale if it satisfies a., b. and $\geq$ in c.
3. an **F**-supermartingale if it satisfies a., b. and $\leq$ in c.

Note that condition c. is equivalent to $\mathbb{E}(X_{n+1} - X_n \mid \mathcal{F}_n) = 0$ for all $n \in \mathbb{N}$, and also to $\mathbb{E}(\mathbb{1}_B X_{n+1}) = \mathbb{E}(\mathbb{1}_B X_n)$, for all $B \in \mathcal{F}_n$. The sequence $(X_n - X_{n-1})$ is called a martingale difference.

A martingale is both a submartingale and a supermartingale. Thanks to the properties of conditional expectation, if (X_n) is a martingale adapted to some filtration **F**, then it is a martingale adapted to its own natural filtration.

If (X_n) is a martingale, then $\mathbb{E} X_n = \mathbb{E} X_0$ for $n \in \mathbb{N}$. Thus, one can write $\mathbb{E} X$ instead of $\mathbb{E} X_n$.

▷ *Example 2.52 (Random Walk)* Let (ξ_n) be an i.i.d. random sequence each of which with expectation m. The random walk $(S_n) = (\sum_{i=1}^{n} \xi_i)$ is a submartingale for the natural filtration of (ξ_n) if $m > 0$, a martingale if $m = 0$ and a supermartingale if $m < 0$. Indeed,

$$\mathbb{E}(S_{n+1} \mid \mathcal{F}_n) = \mathbb{E}(\xi_1 + \cdots + \xi_{n+1} \mid \mathcal{F}_n) \overset{(1)}{=} \xi_1 + \cdots + \xi_n + \mathbb{E}(\xi_{n+1} \mid \mathcal{F}_n),$$

$$\overset{(2)}{=} S_n + \mathbb{E} \xi_{n+1} = S_n + m;$$

(1) because ξ_i is $\mathcal{F}_n$-measurable for $i = 1, \ldots, n$, and (2) because $\mathcal{F}_n$ and ξ_{n+1} are independent.

Note that if $m = 0$ and $\mathbb{E}\,\xi_n^2 = \sigma^2$, then

$$\mathbb{E}\,(S_{n+1}^2 \mid \mathcal{F}_n) = \mathbb{E}\,[(S_n + \xi_{n+1})^2 \mid \mathcal{F}_n] = S_n^2 + \mathbb{E}\,\xi_n^2 = S_n^2 + \sigma^2;$$

therefore, $\mathbb{E}\,[S_n - (n+1)\sigma^2 \mid \mathcal{F}_n] = S_n^2 - n\sigma^2$, and the sequence $(S_n^2 - n\sigma^2)$ is also a martingale.

Interpretation in terms of game: consider that ξ_i denotes the winnings at the i-th game, and S_n the total winnings after the n-th game. The sequence (ξ_n) is i.i.d., with $\mathbb{P}(\xi_n = 1) = p$ and $\mathbb{P}(\xi_n = -1) = 1 - p$, so that $\mathbb{E}\,\xi_n = 2p - 1$.

If (S_n) is a submartingale (that is if $p > 1/2$), then $\mathbb{E}\,(\xi_{n+1} \mid \mathcal{F}_n) \geq 0$; hence, knowing all games up the n-th one, the expectation of winnings at the $(n+1)$-th game is nonnegative.

Note that $(Z_n) = (S_n - n(2p-1))$ is a martingale for any p.   ◁

▷ *Example 2.53 (A Martingale at the Casino)* A player bets 1 euro at the first game; if he loses, he bets 2 euros at the second game, and so on 2^k euros at the k-th game. He stops playing as soon as he wins at one game. At each game, he either loses or wins with probability $1/2$. This strategy leads him to leave the casino with positive winnings. Indeed, when he will stop playing at the random time N—time of first gain, he will have won $2^N - (1 + 2 + 2^2 + \cdots + 2^{N-1}) = 1$ euro.

If X_n denotes the random variable equal to the winnings of the player after the n-th game, we have

$$X_{n+1} = \begin{cases} X_n - 2^n & \text{with probability } 1/2, \\ X_n + 2^n & \text{with probability } 1/2, \end{cases}$$

if he has lost until the n-th game. Hence $\mathbb{E}\,(X_{n+1} \mid \mathcal{F}_n) = X_n$, where $(\mathcal{F}_n)$ is the natural filtration of (X_n), that is a martingale.

We know that $N \sim \mathcal{G}(1/2)$, and $\mathbb{E}\,(|X_n|) \leq 1 + 2 + 2^2 + \cdots + 2^{n-1}$, so the expectation of loss is $\sum_{k\geq 1}(1 + 2 + 2^2 + \cdots + 2^{k-2})(1/2)^k = +\infty$. Therefore, the strategy followed by the player is valuable only if his initial fortune is much larger than that of the casino.   ◁

▷ *Example 2.54 (Product Martingale)* Let (Y_n) be a nonnegative independent random sequence such that $\mathbb{E}\,Y_n = 1$, for all $n \geq 1$. Let (X_n) be defined by $X_0 = 1$, and $X_n = \prod_{k=1}^n Y_k$. We have

$$\mathbb{E}\,(X_n \mid \mathcal{F}_{n-1}) = \mathbb{E}\,(X_{n-1}Y_n \mid \mathcal{F}_{n-1}) = X_{n-1}\mathbb{E}\,(Y_n \mid \mathcal{F}_{n-1}) = X_{n-1},$$

and hence (X_n) is a martingale adapted to the natural filtration $\mathbf{F}$ of (Y_n) (or of (X_n)).   ◁

▷ *Example 2.55 (Conditional Expectation)* Let X be an integrable real random variable defined on $(\Omega, \mathcal{F}, \mathbb{P})$ and let $\mathbf{F}$ be any filtration of $\mathcal{F}$. The sequence (X_n)

defined by $X_n = \mathbb{E}(X \mid \mathcal{F}_n)$ is an **F**-martingale. Indeed, since $(\mathcal{F}_n)$ is an increasing sequence, we have

$$\mathbb{E}(X_{n+1} \mid \mathcal{F}_n) = \mathbb{E}[\mathbb{E}(X \mid \mathcal{F}_{n+1}) \mid \mathcal{F}_n] = \mathbb{E}(X \mid \mathcal{F}_n) = X_n.$$

Such a martingale is said to be regular. ◁

▷ *Example 2.56 (Radon-Nikodym Derivatives)* Let P and Q be two probabilities defined on a measurable space $(\Omega, \mathcal{F})$ and let **F** be a filtration of $\mathcal{F}$. Let P_n and Q_n denote the restrictions of P and Q to $\mathcal{F}_n$, respectively, and suppose that Q_n is absolutely continuous with respect to P_n. Let (X_n) be the sequence of random variables (Radon-Nikodym derivatives) defined on $(\Omega, \mathcal{F}, P)$ by

$$X_n = \frac{dQ_n}{dP_n}, \quad n \geq 1.$$

We have

$$\mathbb{E}(\mathbb{1}_A X_n) = Q_n(A) = Q_m(A) = \mathbb{E}(\mathbb{1}_A X_m), \quad A \in \mathcal{F}_m, \ m < n,$$

so (X_n) is an **F**-martingale. ◁

Properties of Martingales

1. If (X_n) is an **F**-martingale, then $\mathbb{E}(X_m \mid \mathcal{F}_n) = X_n$ for all $n \leq m$, since $\mathbb{E}[\mathbb{E}(X_m \mid \mathcal{F}_{p+1}) \mid \mathcal{F}_p] = \mathbb{E}(X_m \mid \mathcal{F}_p)$ for all $n \leq p \leq m$.
2. The set of all martingales adapted to a given filtration is an $\mathbb{R}$-linear space.
3. If (X_n) is a submartingale, then $(-X_n)$ and (X_n^+) are supermartingales; indeed, if $\mathbb{E}(X_{n+1} \mid \mathcal{F}_n) \geq X_n$, then first $\mathbb{E}(-X_{n+1} \mid \mathcal{F}_n) \leq -X_n$, and second either $\mathbb{E}(X_{n+1}^+ \mid \mathcal{F}_n) = 0 \leq X_n^+$, or $\mathbb{E}(X_{n+1}^+ \mid \mathcal{F}_n) = \mathbb{E}(X_{n+1} \mid \mathcal{F}_n) \leq X_n \leq X_n^+$.
4. If (X_n) is a submartingale and $\varphi : \mathbb{R} \to \mathbb{R}$ is a convex function such that $\varphi(X_n)^+$ is integrable for all n, then $(\varphi(X_n))$ is a submartingale. This property derives from Jensen's inequality through

$$\mathbb{E}[\varphi(X_{n+1}) \mid \mathcal{F}_n] \geq \varphi(\mathbb{E}(X_{n+1} \mid \mathcal{F}_n)) = \varphi(X_n).$$

In particular, if (X_n) is a submartingale, then both (X_n^2) and $(|X_n|)$ are submartingales.
5. If (X_n) and (Y_n) are submartingales, then their maximum $(X_n \vee Y_n)$ is a submartingale. Indeed, for all $A \in \mathcal{F}_n$, both the events $A_1 = A \cap (X_n < Y_n)$ and $A_2 = A \cap (X_n \geq Y_n)$ are also in $\mathcal{F}_n$ and we can write

$$\mathbb{E}[\mathbb{1}_A(X_n \vee Y_n)] = \mathbb{E}(\mathbb{1}_{A_1} Y_n) + \mathbb{E}(\mathbb{1}_{A_2} X_n)$$
$$\leq \mathbb{E}(\mathbb{1}_{A_1} Y_{n+1}) + \mathbb{E}(\mathbb{1}_{A_2} X_{n+1})$$
$$= \mathbb{E}[\mathbb{1}_A(X_{n+1} \vee Y_{n+1})].$$

In the same way, if (X_n) and (Y_n) are supermartingales, then their minimum $(X_n \wedge Y_n)$ is a supermartingale too.

Definition 2.57 Let (X_n) be an **F**-martingale and (Y_n) an **F**-predictable sequence— Y_n is $\mathcal{F}_{n-1}$-measurable for $n \in \mathbb{N}$. The sequence defined by

$$(Y \bullet X)_n = \sum_{k=1}^{n} Y_k(X_k - X_{k-1}), \tag{2.12}$$

is called the discrete stochastic integral (or martingale transform) of (Y_n) by (X_n).

Theorem 2.58 *If (X_n) is an **F**-martingale and (Y_n) a bounded **F**-predictable sequence, then $((Y \bullet X)_n)$ is an **F**-martingale.*

Proof Since $|Y_n(\omega)| \le K$ for all n and ω,

$$\mathbb{E}\left(|(Y \bullet X)_n|\right) \le \sum_{k=1}^{n} \mathbb{E}\left[|Y_k(X_k - X_{k-1})|\right] \le 2K \sum_{k=1}^{n} \mathbb{E}\left(|X_k|\right) < +\infty,$$

meaning that the sequence is integrable, and

$$\mathbb{E}\left[(Y \bullet X)_{n+1} - (Y \bullet X)_n \mid \mathcal{F}_n\right] = Y_n \mathbb{E}\left(X_{n+1} - X_n \mid \mathcal{F}_n\right) = 0,$$

from which the result follows.   □

▷ *Example 2.59 (Coin Tossing)* Let us consider an infinite coin tossing game in which the result of the n-th tossing is associated with a random variable ξ_n with probability $1/2$ of either winning or losing the stake. The random sequence $(\xi_k)_{k \ge 1}$ is i.i.d. and takes values in $\{-1, +1\}$. The stake C_n at the n-th tossing is function of the results $\xi_1, \ldots, \xi_{n-1}$ of the $n - 1$ preceding tossings, so that the sequence (C_n) is predictable.

The random sequence (S_n) defined by $S_0 = 0$ and $S_n = \xi_1 + \cdots + \xi_n$ for $n \ge 1$, is a random walk on $\mathbb{Z}$.

The winnings of the player at the n-th tossing are $C_n \xi_n = C_n(S_n - S_{n-1})$. The cumulated winnings of the first n tossings are the discrete stochastic integral $(C \bullet S)_n$. Moreover, since (S_n) is a martingale by Example 2.52, $((C \bullet S)_n)$ is a martingale too.   ◁

A random sequence is L^p (for a finite $p \ge 1$) if $\mathbb{E}\left(|X_n^p|\right) < +\infty$ for $n \in \mathbb{N}$, and bounded in L^p if $\sup_{n \ge 0} \mathbb{E}\left(|X_n|^p\right) < +\infty$.

Theorem 2.60 (Doob's Decomposition) *Let (X_n) be an integrable F-adapted sequence. Its Doob's decomposition is*

$$X_n = X_0 + Y_n + A_n \tag{2.13}$$

where (Y_n) is an F-martingale and (A_n) an F-predictable sequence, both equal to zero at 0.

The sequence (X_n) is an F-submartingale if and only if (A_n) is nonnegative and increasing.

If (X_n) is a an F-submartingale and is bounded in L^1, then (A_n) converges a.s. and in mean to an a.s. finite variable A_∞.

The sequence (A_n) is called the compensator of (X_n). This decomposition is a.s. unique, in the sense that if $X_n = X_0 + \widetilde{Y}_n + \widetilde{A}_n$, then $Y_n = \widetilde{Y}_n$ and $A_n = \widetilde{A}_n$ a.s. for all $n \in \mathbb{N}$.

Proof If (2.13) holds true, then

$$\mathbb{E}\,(X_n - X_{n-1} \mid \mathcal{F}_{n-1}) = \mathbb{E}\,(Y_n - Y_{n-1} \mid \mathcal{F}_{n-1}) + \mathbb{E}\,(A_n - A_{n-1} \mid \mathcal{F}_{n-1})$$

$$= A_n - A_{n-1}, \tag{2.14}$$

from which we deduce by induction that

$$A_n = \sum_{k=1}^{n} \mathbb{E}\,(X_k - X_{k-1} \mid \mathcal{F}_{k-1}) \quad \text{a.s..}$$

This relation defines (A_n), with $A_0 = 0$, and then the sequence $Y_n = X_n - A_n$ is indeed a martingale.

If moreover (X_n) is a submartingale, then (2.14) says that $A_{n+1} - A_n \geq 0$ a.s., for all n; since $A_0 = 0$, we obtain $A_n \geq 0$ for all n.

Finally, $\mathbb{E}\,A_n = \mathbb{E}\,X_n - \mathbb{E}\,X_0$, and hence $|\mathbb{E}\,A_n| \leq \sup_{n \geq 1} \mathbb{E}\,(|X_n|) + \mathbb{E}\,X_0$. Thus, the bounded increasing sequence (A_n) converges to a variable A_∞. Thanks to the Lebesgue monotone convergence theorem, $\mathbb{E}\,A_n$ converges and A_∞ is integrable. □

▷ *Example 2.61* Let (X_n) be an i.i.d. random sequence with distribution P, taking values in a measurable space $(E, \mathcal{E})$. Let g be an integrable real function defined on $(E, \mathcal{E})$. Then the sequence (Y_n) defined by

$$Y_n = \left(\sum_{i=1}^{n} g(X_i) - n \int_E g\,dP \right)$$

is a martingale adapted to the natural filtration of (X_n).

This open doors to the computation of integrals by means of series.   ◁

Theorem 2.62 (Krickeberg's Decomposition) *Any bounded in L^1 martingale is the difference of two nonnegative martingales.*

Proof For $X_0 = 0$.

By Doob's decomposition of the submartingale $(|X_n|)$, we obtain $|X_n| = Y_n + A_n$, with A_n converging a.s. to a variable A_∞. The two nonnegative martingales are (Z_n) and $(Z_n - X_n)$, where $Z_n = Y_n + \mathbb{E}(A_\infty \mid \mathcal{F}_n)$.

We have $Z_n - |X_n| = \mathbb{E}(A_\infty - A_n \mid \mathcal{F}_n)$ and (A_n) is an increasing sequence converging to A_∞, so that $Z_n - |X_n| \geq 0$. Therefore Z_n is nonnegative and $Z_n \geq X_n$.

Both the sequences $(\mathbb{E}(A_\infty \mid \mathcal{F}_n))$ and (Y_n) are martingales so their sum (Z_n) is a martingale. Since the difference of two martingales is a martingale, $(Z_n - X_n)$ also is a martingale.   □

2.4.2 Classical Inequalities

Several inequalities are useful tools for studying martingales. Let us present some of the most remarkable ones.

Theorem 2.63 (Doob's Maximal Inequality) *If (X_n) is a submartingale, then*

$$\mathbb{P}(\max(X_{n_0}, \ldots, X_{n_1}) \geq a) \leq \frac{1}{a}\mathbb{E} X_{n_1}^+, \quad a > 0, \ n_0 < n_1.$$

Proof For nonnegative variables X_n.

Let us consider the events $F_n = (\inf\{m \in [\![n_0, n_1]\!] : X_m \geq a\} = n)$ for $n \in \mathbb{N}$. We have $F = \cup_{n=n_0}^{n_1} F_n = (\max(X_{n_0}, \ldots, X_{n_1}) \geq a)$. If $\omega \in F_n$, then $X_n(\omega) \geq a$, and hence

$$a\mathbb{P}(F) = a \sum_{n=n_0}^{n_1} \mathbb{P}(F_n) \leq \sum_{n=n_0}^{n_1} \mathbb{E}(\mathbb{1}_{F_n} X_n)$$

$$\overset{(1)}{\leq} \mathbb{E}(\mathbb{1}_F X_{n_1}) \leq \mathbb{E} X_{n_1}.$$

(1) because (X_n) is a submartingale, so that $\mathbb{E}(\mathbb{1}_{F_n} X_n) \leq \mathbb{E}(\mathbb{1}_{F_n} X_{n_1})$, for $n \leq n_1$.   □

Theorem 2.64 (Hoeffding's Inequality) *If (X_n) is an F-martingale and if (u_n) is a sequence of real numbers such that $\mid X_n - X_{n-1} \mid \leq u_n$ a.s. for $n \in \mathbb{N}$, then*

$$\mathbb{P}(\mid X_n - X_0 \mid \geq x) \leq 2\exp\left(-\frac{x^2/2}{\sum_{i=1}^n u_i^2}\right), \quad x > 0.$$

Proof We deduce from Markov's inequality that

$$\mathbb{P}(X_n - X_0 \geq x) \leq e^{\theta x}\mathbb{E}\left(\exp[\theta(X_n - X_0)]\right), \quad \theta > 0.$$

We can write

$$\exp[\theta(X_n - X_0)] = \exp[\theta(X_{n-1} - X_0)]\exp(\theta\Delta X_n),$$

where $\Delta X_n = X_n - X_{n-1}$. Thanks to the convexity of $x \longrightarrow e^{\lambda x}$ for $\lambda > 0$,

$$2e^{\lambda x} \leq (1-x)e^{-\lambda} + (1+x)e^{\lambda}, \quad |x| \leq 1. \tag{2.15}$$

This inequality, applied to any random variable X such that $\mathbb{P}(|X| \leq 1) = 1$ and $\mathbb{E}(X \mid \mathcal{F}_{n-1}) = 0$ for all n, gives

$$\mathbb{E}(e^{\lambda X}) \leq \frac{e^{-\lambda} + e^{\lambda}}{2} < e^{\lambda^2/2},$$

where the right-hand inequality is obtained by comparison of the coefficients of λ^{2n} in the series associated with the exponentials. Applying to $\Delta X_n / u_n$ yields

$$\mathbb{E}\left(\exp[\theta(X_n - X_0)] \mid \mathcal{F}_{n-1}\right) = \exp[\theta(X_{n-1} - X_0)]\mathbb{E}\left[\exp(\theta\Delta X_n) \mid \mathcal{F}_{n-1}\right]$$
$$\leq \exp[\theta(X_{n-1} - X_0)]\exp(\theta^2 u_n^2/2).$$

Taking expectation, we obtain by induction

$$\mathbb{E}\left(\exp[\theta(X_n - X_0)]\right) \leq \mathbb{E}\left(\exp[\theta(X_{n-1} - X_0)]\right)\exp(\theta^2 u_n^2/2)$$
$$\leq \exp\left(\frac{1}{2}\theta^2 \sum_{i=1}^{n} u_i^2\right).$$

Therefore

$$\mathbb{P}(X_n - X_0 \geq x) \leq \exp\left(-\theta x + \frac{1}{2}\theta^2 \sum_{i=1}^{n} u_i^n\right), \quad \theta > 0.$$

The value $\theta = x/\sum_{i=1}^{n} u_i^n$ minimizes the right-hand side of the above inequality, so

$$\mathbb{P}(X_n - X_0 \geq x) \leq \exp\left(-\frac{x^2/2}{\sum_{i=1}^{n} u_i^n}\right), \quad x > 0. \tag{2.16}$$

We could show similarly that

$$\mathbb{P}(X_0 - X_n \geq x) \leq \exp\left(-\frac{x^2/2}{\sum_{i=1}^{n} u_i^n}\right), \quad x > 0. \tag{2.17}$$

The final inequality is obtained by putting together (2.17) to (2.16).   □

We state without proof the next inequality.

Theorem 2.65 (Doob's L^p Inequality) *If (X_n) is an L^p-martingale and $p > 1$, then*

$$\left(\mathbb{E}\left[\max(X_1, \ldots, X_n)^p\right]\right)^{1/p} \geq \frac{p}{p-1} \left[\mathbb{E}\left(|X_n|^p\right)\right]^{1/p}.$$

In order to state the last inequality, first we need to define crossing times. Let $(X_n)_{n \geq 1}$ be a random sequence. Let $a < b$ be two real numbers. Set $T_0 = \inf\{k \geq 0 : X_k \leq a\}$, and then, supposing that T_{2n} for some $n \geq 0$ is defined, set

$$T_{2n+1} = \inf\{k > T_{2n} : X_k \geq b\} \quad \text{and} \quad T_{2n+2} = \inf\{k > T_{2n+1} : X_k \leq a\}.$$

In other words, the times T_n, for $n \geq 0$, are the successive times at which the sequence (X_n) crosses the interval $[a, b]$. They are called the crossing-times at level (a, b), and are stopping times. Note that T_n may take infinite values.

Let us define for $m > 0$, the sequence of the numbers of crossing times at level (a, b) of (X_n), that is

$$U_m(a, b) = |\{n \in \mathbb{N} : n \text{ is odd and } T_n \leq m\}|,$$

meaning that $2U_m(a, b)$ variables of the sequence (T_n) belong to $[0, m]$, precisely the variables $T_0, T_1, \ldots, T_{2U_m(a,b)-1}$.

The sequence $(U_m(a, b))$ is increasing. Setting $\widetilde{X}_n = (X_n - a)^+$, we have $U_m(a, b) = \widetilde{U}_m(0, b - a)$ for all $a < b$, where $(\widetilde{U}_m)$ denotes the sequence of the numbers of crossing times at level $(0, b - a)$ of $(\widetilde{X}_n)$.

Lemma 2.66 (Doob's Crossing Time Theorem) *Let (X_n) be a random sequence and let $a < b$ be two fixed real numbers. For any integer $m > 0$:*

1. if (X_n) is a submartingale, then

$$\mathbb{E} \, U_m(a, b) \leq \frac{\mathbb{E} \, (X_m - a)^+}{b - a} \, ;$$

2. *if (X_n) is a supermartingale, then*

$$\mathbb{E}\, U_m(a, b) \le \frac{\mathbb{E}\,(X_m - a)^-}{b - a}.$$

Proof

1. For a fixed $m > 0$, set $Y_n = X_{T_n \wedge m}$ for $n \ge 0$. Three cases may occur.
   If $k > U_m(a, b)$, then $Y_{2k-1} = Y_{2k} = X_m$, which induces $Y_{2k-1} - Y_{2k} = 0$.
   If $k = U_m(a, b)$, then $Y_{2k-1} \ge b$ and $Y_{2k} = X_m > a$, which induces $b - a \le Y_{2k-1} - Y_{2k} + (X_m - a)^+$.
   If $k < U_m(a, b)$, then $Y_{2k-1} \ge b$ and $Y_{2k} \le a$, which induces $b - a \le Y_{2k-1} - Y_{2k}$.
   Putting all together yields

$$\sum_{k=1}^{U_m(a,b)} (b - a) \le (X_m - a)^+ + \sum_{k=1}^{U_m(a,b)} (Y_{2k-1} - Y_{2k}),$$

or equivalently,

$$(b - a)U_m(a, b) \le (X_m - a)^+ + \sum_{k=1}^{U_m(a,b)} (Y_{2k-1} - Y_{2k}). \tag{2.18}$$

Set $\mathcal{G}_n = \sigma(Y_0, Y_1, \ldots, Y_n)$. We have $\mathbb{E}\,[Y_{n+1} - Y_n \mid \mathcal{G}_n] \ge 0$, for $n \in \mathbb{N}$, meaning that $(Y_n)_{n \ge 0}$ is a submartingale.
Since $\mathbb{E}\,(Y_{2k-1} - Y_{2k}) \le 0$, we obtain by taking expectation of both sides of (2.18),

$$(b - a)\mathbb{E}\, U_m(a, b) \le \mathbb{E}\,(X_m - a)^+,$$

and the conclusion follows.
2. can be proven similarly.   □

2.4.3   Martingales and Stopping Times

A martingale is characterized by the formula $\mathbb{E}\,(X_m \mid \mathcal{F}_n) = X_{m \wedge n}$. Questions arise: does this equality remains valid for two stopping times T and S instead of the two constants n and m? Is the sequence (X_{T_n}) still a martingale for a sequence of stopping-times (T_n)? The answers lie in the stopping and the sampling theorems, under certain conditions.

Theorem 2.67 (Stopping) *Let (X_n) be an **F**-martingale (submartingale). If S and T are two **F**-stopping times such that:*

1. $\max(\mathbb{E}\,(|X_T|), \mathbb{E}\,(|X_S|)) < +\infty$,
2. $\underline{\lim}\,\mathbb{E}\,[|X_n|\mathbb{1}_{(T>n)}] = 0$ *and* $\underline{\lim}\,\mathbb{E}\,[|X_n|\mathbb{1}_{(S>n)}] = 0$,

then

$$X_{S \wedge T} = \mathbb{E}\,(X_T \mid \mathcal{F}_S), \quad (\leq) \quad \mathbb{P} - a.s.. \tag{2.19}$$

Proof We will write for simplification $\mathbb{E}\,(X\mathbb{1}_A) = \mathbb{E}\,(X; A)$ for any variable X and any event A.

For proving (2.19), it if sufficient to show that for any $A \in \mathcal{F}_S$, we have

$$\mathbb{E}\,[X_T; A \cap (S \leq T)] = \mathbb{E}\,[X_S; A \cap (S \leq T)],$$

or, equivalently, that for any $n \geq 0$,

$$\mathbb{E}\,[X_T; A \cap (S \leq T, S = n)] = \mathbb{E}\,[X_S; A \cap (S \leq T, S = n)].$$

Set $B_n = A \cap (S = n) \in \mathcal{F}_n$. We have

$$\mathbb{E}\,[X_n; B_n \cap (T \geq n)] = \mathbb{E}\,[X_n; B_n \cap (T = n)] + \mathbb{E}\,[X_n; B_n \cap (T \geq n+1)],$$

and, since $B_n \cap (T \geq n+1) \in \mathcal{F}_n$,

$$\mathbb{E}\,[X_n; B_n \cap (T \geq n+1)] = \mathbb{E}\,[\mathbb{E}\,(X_{n+1} \mid \mathcal{F}_n); B_n \cap (T \geq n+1)]$$
$$= \mathbb{E}\,[X_{n+1}; B_n \cap (T \geq n+1)].$$

Therefore

$$\mathbb{E}\,[X_n; B_n \cap (T \geq n)] =$$
$$= \mathbb{E}\,[X_n; B_n \cap (T = n)] + \mathbb{E}\,[X_{n+1}; B_n \cap (T = n+1)]$$
$$+ \mathbb{E}\,[X_{n+1}; B_n \cap (T \geq n+2)]$$
$$= \mathbb{E}\,[X_T; B_n \cap (n \leq T \leq n+1)] + \mathbb{E}\,[X_{n+2}; B_n \cap (T \geq n+2)]$$
$$\vdots$$
$$= \mathbb{E}\,[X_T; B_n \cap (n \leq T \leq n+m)] + \mathbb{E}\,[X_{n+m}; B_n \cap (T > n+m)],$$

and hence

$$\mathbb{E}[X_T; B_n \cap (n \le T \le n+m)] = \mathbb{E}[X_n; B_n \cap (n \le T)] - \mathbb{E}[X_m; B_n \cap (T > n+m)].$$

In other words,

$$\mathbb{E}[X_T; B_n \cap (n \le T)] =$$
$$= \varlimsup_m (\mathbb{E}[X_n; B_n \cap (n \le T)] - \mathbb{E}[X_m; B_n \cap (T > n + m)])$$
$$= \mathbb{E}[X_n; B_n \cap (n \le T)] - \varlimsup_m \mathbb{E}[X_m; B_n \cap (T > n + m)]$$
$$\overset{(1)}{=} \mathbb{E}[X_n; B_n \cap (n \le T)].$$

(1) due to Assumption 2., by the decomposition $X_k = 2X_k^+ - |X_k|$. The result follows.   □

Doob's crossing time theorem induces via the sampling theorem—stated without proof—that for an increasing sequence of stopping times (T_n), the sequence (X_{T_n}) is also a martingale under certain conditions.

Definition 2.68 Let (X_n) be a random sequence and let (T_n) be an increasing sequence of finite stopping times. The sequence (X_{T_n}) is called a sampling of (X_n).

Theorem 2.69 (Sampling) *Let (X_n) be a martingale (respectively submartingale, supermartingale) adapted to a filtration $(\mathcal{F}_n)$ and let (X_{T_n}) be a sampling of (X_n). If X_{T_n} is integrable for all n and if*

$$\varlimsup_N \mathbb{E}[|X_N| \mathbb{1}_{(T_n > N)}] = 0, \quad n \in \mathbb{N}, \tag{2.20}$$

then the sequence (X_{T_n}) is a martingale (respectively submartingale, supermartingale) adapted to the filtration $(\mathcal{F}_{T_n})$.

The hypothesis of the sampling theorem are obviously satisfied if $T_n = f(n)$ where f is an increasing deterministic function of n, and also under the following conditions

Proposition 2.70 *The hypothesis of the sampling theorem are satisfied under any of the following two conditions:*

1. $\exists k \in \mathbb{R}_+^*$ such that $|X_n| < k$ a.s., $n \in \mathbb{N}$.
2. $\forall n \in \mathbb{N}, \exists k_n \in \mathbb{N}$ such that $T_n \le k_n$ a.s.

Proof Set $Y_n = X_{T_n}$.

1. We have

$$(|Y_n| > k) = \bigcup_{i \geq 0} (T_n = i, |X_i| > k) \subset \bigcup_{i \geq 0} (|X_i| > k),$$

thus, $\mathbb{P}(|Y_n| > k) = 0$ or $|Y_n| \leq k$ a.s., and hence $\mathbb{E}\,|Y_n| \leq k < +\infty$.

Moreover, $(T_n > N) \searrow (T_n = +\infty)$ so that $\mathbb{P}(T_n > N) \searrow \mathbb{P}(T_n = +\infty) = 0$ because T_n is a.s. finite, and $0 \leq \mathbb{E}\,[\mathbb{1}_{(T_n > N)}|X_n|] \leq k\mathbb{P}(T_n > N)$.

2. We can write $Y_n = \sum_{i=0}^{k_n} \mathbb{1}_{(T_n = i)} X_i$, so $|Y_n| \leq \sum_{i=0}^{k_n} |X_i|$ and hence $\mathbb{E}\,|Y_n| \leq \sum_{i=0}^{k_n} \mathbb{E}\,|X_i| < +\infty$.

Finally, if $N \geq k_n$, we have $\mathbb{E}\,[\mathbb{1}_{(T_n > N)}|X_n|] = 0$. □

Corollary 2.71 *Let T be a stopping time for a random sequence (X_n). If (X_n) is a martingale (respectively submartingale, supermartingale), then the sequence stopped at time T, that is $(X_{T \wedge n})$, is a supermartingale (respectively submartingale, supermartingale).*

Proof Since the variables $T \wedge n$ are bounded stopping times, condition 2. with $k_n = n$ of Proposition 2.70 is satisfied, and the sampling theorem yields the conclusion. □

The stopping theorem also induces the following inequality. The proof uses sequences stopped at time T.

Theorem 2.72 (Doob's Martingale Inequality) *Let (X_n) be an **F**-submartingale and let θ be a positive constant.*

1. *We have*

$$\mathbb{P}[\max(X_1, \ldots, X_n) \geq \theta] \leq \frac{1}{\theta}\mathbb{E}\,[X_n^+ \mathbb{1}_{(\max(X_1, \ldots, X_n) \geq \theta)}] \leq \frac{1}{\theta}\mathbb{E}\,X_n^+, \quad n \in \mathbb{N}.$$

2. *If X_n is nonnegative for all $n \in \mathbb{N}$, then, for all $1 \leq p \leq +\infty$,*

$$\| \max(X_1, \ldots, X_n)\|_p \leq \frac{p}{p-1}\|X_n\|_p.$$

Proof

1. Set $T_\theta = \inf\{n \in \mathbb{N} : X_n \geq \theta\}$ for $\theta > 0$. On the one hand, the stopping theorem induces that $\mathbb{E}\,X_{T_\theta \wedge n} \leq \mathbb{E}\,X_n$. On the other hand,

$$\mathbb{E}\,X_n = \mathbb{E}\,[X_n \mathbb{1}_{(T_\theta \leq n)}] + \mathbb{E}\,[X_n \mathbb{1}_{(T_\theta > n)}] \geq \theta\mathbb{P}(T_\theta \leq n) + \mathbb{E}\,[X_n \mathbb{1}_{(T_\theta > n)}],$$

hence, if we set $X_n^* = \max(X_1, \ldots, X_n)$,

$$\theta \mathbb{P}(X_n^* \geq \theta) \overset{(1)}{\leq} \theta \mathbb{P}(T_\theta \leq n) \leq \mathbb{E}[X_n(1 - \mathbb{1}_{(T_\theta > n)})] = \mathbb{E}[X_n \mathbb{1}_{(T_\theta \leq n)}]$$

$$\overset{(2)}{\leq} \mathbb{E}[X_n \mathbb{1}_{(X_n^* \geq \theta)}] \leq \mathbb{E} X_n^+.$$

(1) and (2) because $(X_n^* \leq \theta) \subset (T_\theta \geq n)$.

2. By the transfer theorem, $\mathbb{P}(X > t) = \int_t^{+\infty} d\mathbb{P}_X(x)$, and hence

$$p \int_{\mathbb{R}_+} t^{p-1} \mathbb{P}(X > t) \, dt = \int_{\mathbb{R}_+} \int_t^{+\infty} pt^{p-1} d\mathbb{P}_X(x) \, dt$$

$$\overset{(1)}{=} \int_{\mathbb{R}_+} \int_0^x pt^{p-1} \, dt \, d\mathbb{P}_X(x) = \int_{\mathbb{R}_+} x^p d\mathbb{P}_X(x) = \mathbb{E}(X^p).$$

(1) by Fubini's theorem.

Therefore

$$\|X_n^*\|_p^p = \mathbb{E}[(X_n^*)^p] = \int_0^{+\infty} px^{p-1} \mathbb{P}(X_n^* \geq x) \, dx,$$

hence

$$\|X_n^*\|_p^p \overset{(1)}{\leq} \int_0^{+\infty} px^{p-2} \mathbb{E}[X_n \mathbb{1}_{(X_n^* \geq x)}] \, dx,$$

$$= \mathbb{E}\left[X_n \int_0^{+\infty} \mathbb{1}_{(X_n^* \geq x)} px^{p-2} \, dx\right] = \frac{p}{p-1} \mathbb{E}[X_n(X_n^*)^{p-1}]$$

$$\overset{(2)}{\leq} \frac{p}{p-1} \|X_n\|_p \|X_n^*\|_p^{p-1},$$

(1) by *1.* and (2) by Hölder's inequality, and the result follows. □

2.4.4 Convergence of Martingales

Different results of convergence of martingales will be stated, under conditions of integrability. The square integrable martingales will be specifically inverstigated in the next section.

The proof of Doob's convergence theorem requires the following criterion: a submartingale (X_n) is bounded in L^1 if and only if $\sup_{n \geq 0} \mathbb{E} X_n^+ < +\infty$. This is equivalent to the definition $\sup_{n \geq 0} \mathbb{E}\,(|X_n|) < +\infty$, because

$$\mathbb{E}\,X_n^+ \leq \mathbb{E}\,|X_n| = 2\mathbb{E}\,X_n^+ - \mathbb{E}\,X_n \leq 2\mathbb{E}\,X_n^+ - \mathbb{E}\,X_0.$$

Theorem 2.73 (Doob's Convergence) *Any bounded in L^1 martingale, super-martingale or submartingale converges a.s. to an integrable variable.*

Proof For a submartingale (X_n).

If (X_n) did not converges a.s., then necessarily $\mathbb{P}\left(\underline{\lim}\, X_n < \overline{\lim}\, X_n\right) > 0$. Since $\mathbb{Q}$ is dense in $\mathbb{R}$, rational numbers a and b would exist such that

$$\mathbb{P}\left(\underline{\lim}\, X_n < a < b < \overline{\lim}\, X_n\right) > 0. \tag{2.21}$$

The sequence $(U_n(a,b))$ of numbers of crossing times at level (a,b) of (X_n) is an increasing sequence. Set $U_\infty(a,b) = \lim_{n \to +\infty} U_n(a,b)$, possibly infinite. On the one hand, if (2.21) is satisfied, the probability that the total number of crossing times at level (a,b) of (X_n) be infinite is positive, so the expectation of $U_\infty(a,b)$ is infinite. On the other hand, since the submartingale (X_n) is bounded in L^1, the submartingale (X_n^+) is bounded too. Thus, thanks to Doob's inequality,

$$\mathbb{E}\,U_n(a,b) \leq \frac{\mathbb{E}\,(X_n - a)^+}{b - a} \leq \frac{\mathbb{E}\,X_n^+ + a}{b - a} < +\infty. \tag{2.22}$$

Since $(U_n(a,b))$ is nonnegative and increasing, the dominated convergence theorem induces that $\mathbb{E}\,U_n(a,b) \nearrow \mathbb{E}\,U_\infty(a,b)$, so

$$\mathbb{E}\,U_\infty(a,b) \leq \frac{\sup_{n \geq 0} \mathbb{E}\,X_n^+ + a}{b - a} < +\infty,$$

and a contradiction arises.

Therefore, (X_n) converges a.s. to some X. Thanks to Fatou's lemma,

$$\mathbb{E}\,|X| \leq \sup_{n \geq 0} \mathbb{E}\,[|X_n|] < +\infty,$$

so X is integrable. $\square$

Theorem 2.74 (L^p Convergence) *Let $p \in \,]1, +\infty[$. Any bounded in L^p martingale (or nonnegative submartingale) converges a.s. and in L^p-norm.*

Proof If the martingale (X_n) is bounded in L^p, then it is bounded in L^1 so converges a.s. by Doob's convergence theorem.

Exercise 1.12 shows that if a random sequence is a.s. bounded and converges a.s., then it converges in L^p. $\square$

Theorem 2.75 *A martingale converges in mean if and only if it is regular.*

Exercise 1.12 shows this condition to be equivalent to equi-integrability of the martingale.

Proof Let (X_n) be a martingale converging in mean. Then $\mathbb{E}\,|X_n| \leq \mathbb{E}\,|X| + \mathbb{E}\,|X - X_n|$, so (X_n) is bounded in L^1 and, thanks to Doob's convergence theorem, converges a.s.

Since (X_n) is a martingale, $\mathbb{E}\,(X_{n+m}|\mathcal{F}_n) = X_n$ for all $m \geq n$. Therefore $\mathbb{E}\,(ZX_{n+m}) = \mathbb{E}\,(ZX_n)$ for all $\mathcal{F}_n$-measurable variable Z, and letting m tend to infinity gives $\mathbb{E}\,(ZX) = \mathbb{E}\,(ZX_n)$, meaning that $X_n = \mathbb{E}\,(X|\mathcal{F}_n)$.

Conversely, let X be an integrable variable defined on $(\Omega, \mathcal{F}, \mathbb{P})$ and let $\mathbf{F}$ be a filtration of $\mathcal{F}$. Set $X_n = \mathbb{E}\,(X \mid \mathcal{F}_n)$. We have $\mathbb{E}\,|X_n| \leq \mathbb{E}\,|X|$, so (X_n) is bounded in L^1 and, thanks to Doob's convergence theorem again, converges in mean. $\square$

2.4.5 Square Integrable Martingales

Square integrable martingales are involved in many applications, as the statistical study of lifetimes or branching processes.

Definition 2.76 A martingale whose every element is square integrable is called a square integrable martingale (or L^2-martingale).

We will denote all square integrable martingales by (M_n).

$\triangleright$ *Example 2.77 (Conditional Expectation)* Let X be a random variable in $L^2(\Omega, \mathcal{F}, \mathbb{P})$ and let $\mathbf{F}$ be a filtration of $\mathcal{F}$.

The sequence $(M_n) = (\mathbb{E}\,(X \mid \mathcal{F}_n))$ is an L^2-martingale; indeed,

$$\mathbb{E}\,(M_{n+1} \mid \mathcal{F}_n) = \mathbb{E}\,[\mathbb{E}\,(X \mid \mathcal{F}_{n+1}) \mid \mathcal{F}_n] = \mathbb{E}\,(X \mid \mathcal{F}_n) = M_n,$$

and $\mathbb{E}\,(X \mid \mathcal{F}_n) \in L^2$. $\triangleleft$

Properties of Square Integrable Martingales

1. If (M_n) is an L^2-martingale, then (M_n^2) is a submartingale; indeed

$$\mathbb{E}\,(M_{n+1}^2 \mid \mathcal{F}_n) - M_n^2 = \mathbb{E}\,[(M_{n+1} - M_n)^2 \mid \mathcal{F}_n] \geq 0.$$

2. The increases $\Delta M_n = M_n - M_{n-1}$ of a square integrable martingale (M_n) are not correlated, so are orthogonal in L^2. Indeed, for any $m > n$,

$$\mathbb{E}\,(\Delta M_n \Delta M_m) = \mathbb{E}\,[\mathbb{E}\,(\Delta M_n \Delta M_m \mid \mathcal{F}_{m-1})]$$
$$= \mathbb{E}\,[\Delta M_n \mathbb{E}\,(\Delta M_m \mid \mathcal{F}_{m-1})],$$

and $\mathbb{E}\,(\Delta M_m \mid \mathcal{F}_{m-1}) = 0$.

Using property 1. and Doob's decomposition theorem, we can write $M_n^2 = X_n + \langle M \rangle_n$, where the sequence (X_n) is an **F**-martingale and $(\langle M \rangle_n)$ is the compensator of (M_n^2). This is an increasing **F**-predictable sequence called the quadratic characteristic (or predictable quadratic variation) of (M_n). Moreover,

$$\langle M \rangle_n = \sum_{k=1}^{n} \mathbb{E}\left[(\Delta M_k)^2 \mid \mathcal{F}_{k-1}\right].$$

▷ *Example 2.78 (Sum of L^2-variables)* Let (Y_n) be an i.i.d. sequence of centered square integrable random variables. Set $M_0 = 0$ and $M_n = Y_1 + \cdots + Y_n$, for $n \geq 1$. The sequence (M_n) is an L^2-martingale for the natural filtration $(\mathcal{F}_n)$ of (Y_n), and

$$\langle M \rangle_n = \sum_{k=1}^{n} \mathbb{E}\left[(\Delta M_k)^2 \mid \mathcal{F}_{k-1}\right]$$

$$= \sum_{k=1}^{n} \mathbb{E}\left(Y_k^2 \mid \mathcal{F}_{k-1}\right) = \mathbb{V}\mathrm{ar}\, Y_1 + \cdots + \mathbb{V}\mathrm{ar}\, Y_n = \mathbb{V}\mathrm{ar}\, M_n.$$

Therefore, the quadratic characteristic of (M_n) is a constant random variable equal to its variance.   ◁

▷ *Example 2.79 (Continuation of Example 2.61)* The compensator of the sequence (Y_n^2) is the sequence defined by

$$\langle Y^2 \rangle_n = n \left(\int_E g^2 dP - \int_E g\, dP \right)^2$$

for all $n \geq 1$.   ◁

Theorem 2.80 (Kolmogorov's L^2-martingale Inequality) *Let (M_n) be an L^2-martingale. Then*

$$\mathbb{P}[\max(|M_1|, \ldots, |M_n|) \geq \theta] \leq \frac{1}{\theta^2} \mathbb{E}\left(|M_n|^2\right), \quad n \in \mathbb{N},\ \theta > 0.$$

Proof The random variable $T = \inf\{k \in \mathbb{N} : |M_k| \geq \theta\}$ is a stopping time. Thanks to Markov's inequality,

$$\mathbb{P}[\max(|M_1|, \ldots, |M_n|) \geq \theta] = \mathbb{P}(|M_{T \wedge n}| \geq \theta) \leq \frac{1}{\theta^2} \mathbb{E}\left(|M_{T \wedge n}|^2\right).$$

Both $T \wedge n$ and n are stopping times, and $T \wedge n \leq n$, so $M_{T \wedge n} = \mathbb{E}\left(M_n \mid \mathcal{F}_{T \wedge n}\right)$, thanks to the stopping theorem.

Moreover, since $M_{T \wedge n}$ is square integrable, because M_n is. Taking conditional expectation reduces L^2-norms, and hence $\mathbb{E}\left(|M_{T \wedge n}|^2\right) \leq \mathbb{E}\left(|M_n|^2\right)$.   □

Proposition 2.81 *Let (M_n) be an L^2-martingale.*

1. *(M_n) is bounded if and only if*

$$\sum_{n \geq 1} \mathbb{E}\left[(\Delta M_n)^2\right] < +\infty.$$

2. *If $M_0 = 0$, then (M_n) is bounded if and only if $(\langle M \rangle_n)$ converges a.s. to an integrable variable.*

Proof

1. Since $M_n = M_0 + \sum_{k=1}^{n} \Delta M_k$, the Pythagorean theorem induces that

$$\mathbb{E}\left(M_n^2\right) = \mathbb{E}\left(M_0^2\right) + \sum_{k=1}^{n} \mathbb{E}\left[(\Delta M_k)^2\right],$$

and the conclusion follows.
2. Doob's decomposition of (M_n^2) gives $\mathbb{E}\left(M_n^2\right) = \mathbb{E}\left(\langle M \rangle_n\right)$, from which the direct implication derives. Point 1. yields the converse.   □

If the square integrable martingale is bounded, the quadratic mean convergence and the regularity again hold true.

Theorem 2.82 (Strong Convergence) *Let (M_n) be a bounded in L^2 martingale adapted to a filtration $(\mathcal{F}_n)$. Then M_n converges a.s. and in quadratic mean to some square integrable $\mathcal{F}_\infty$-measurable random variable M_∞; moreover $M_n = \mathbb{E}\left(M_\infty \mid \mathcal{F}_n\right)$ for all $n \in \mathbb{N}$.*

Proof First, if (M_n) is bounded in L^2, it is bounded in L^1. Thanks to Doob's convergence theorem, $M_\infty = \lim_{n \to +\infty} M_n$ exists a.s.

Second,

$$\mathbb{E}\left[(M_{n+m} - M_n)^2\right] = \sum_{k=n+1}^{n+m} \mathbb{E}\left[(M_k - M_{k-1})^2\right]$$

and Fatou's lemma for m tending to infinity together induce that

$$\mathbb{E}\left[(M_\infty - M_n)^2\right] \leq \sum_{k \geq n+1} \mathbb{E}\left[(M_k - M_{k-1})^2\right].$$

Moreover, $M_n = \mathbb{E}(M_m \mid \mathcal{F}_n)$ for all $m \geq n$, and $\mathbb{E}(\cdot \mid \mathcal{F}_n)$ is a continuous operator on L^2, so taking the limit yields $M_n = \mathbb{E}(M_\infty \mid \mathcal{F}_n)$. $\square$

The last theorem, stated without proof, represent a large numbers law and a central limit theorem for martingales.

Theorem 2.83 *Let (M_n) be an L^2-martingale and (a_n) a sequence of real numbers increasing to infinity. If $\langle M \rangle_n / a_n$ converges in probability to a positive limit σ^2, and if*

$$\frac{1}{a_n} \sum_{k=1}^n \mathbb{E}\left[(M_k - M_{k-1})^2 \mathbb{1}_{\{|M_k - M_{k-1}| \geq \varepsilon \sqrt{a_n}\}}\right] \xrightarrow{a.s.} 0$$

for all $\varepsilon > 0$, then

$$\frac{M_n}{a_n} \xrightarrow{a.s.} 0 \quad and \quad \frac{M_n}{\sqrt{a_n}} \xrightarrow{\mathcal{L}} \mathcal{N}(0, \sigma^2).$$

2.5   Exercises

▽ **Exercise 2.1 (Random Sum of Random Variables)** Let $S_N = \sum_{k=1}^N X_k$, for $n \geq 1$, be a random sum, with the notation of Definition 1.72 in Chap. 1.

1. Determine the conditional expectation of S_N given N and prove again Proposition 1.73.
2. Determine the conditional distribution of N given $(S_N = s)$.

Solution

1. We have $\mathbb{E}(S_N \mid N = n) = \mathbb{E}(S_n) = n\mathbb{E} X_1$, and hence $\mathbb{E}(S_N \mid N) = N\mathbb{E} X_1$. Therefore

$$\mathbb{E} S_N = \mathbb{E}[\mathbb{E}(S_N \mid N)] = \mathbb{E}[N\mathbb{E} X_1] = (\mathbb{E} N)(\mathbb{E} X_1).$$

2. We know from Theorem 2.4 that $\mathbb{P}(N = n \mid S_N = s) = \varphi(s)$ if

$$\mathbb{E}[\mathbb{1}_{(S_N \in B)} \mathbb{1}_{(N=n)}] = \mathbb{E}[\mathbb{1}_{(S_N \in B)} \varphi(S_N)], \quad B \in \mathcal{B}(\mathbb{R}).$$

On the one hand,

$$\mathbb{E}[\mathbb{1}_{(S_N \in B)} \mathbb{1}_{(N=n)}] = \mathbb{P}(N = n, S_n \in B) = \mathbb{P}(N = n) \int_B f_{S_n}(s) ds,$$

and on the other hand,

$$\mathbb{E}\,[\mathbb{1}_{(S_N \in B)}\varphi(S_N)] = \int_B \varphi(s)\,f_{S_N}(s)ds.$$

Therefore, $\mathbb{P}(N = n \mid S_N = s) = \mathbb{P}(N = n)\,f_{S_n}(s)/f_{S_N}(s)$. △

▽ **Exercise 2.2 (Compound Poisson Distribution in Insurance)** Let S_T be the total amount of money paid by an insurance firm in an interval of time $I = [0, T]$. Suppose that all the accidents occur independently and cannot be simultaneous. Suppose also that the number of accidents in I is finite.

Let us cut I into n regular intervals I_j^n. Let N_j^n be the random variable equal to zero if no accident occurs in I_j^n and to one otherwise; let X_j^n be the total amount of money paid by the firm in I_j^n and let p_j^n be the probability of accident in I_j^n. Let P denote the conditional distribution of X_j^n given $(N_j^n = 1)$ for any large n.

1. Write S_T as a function of the other variables.
2. Suppose that $\sum_{j=1}^n p_j^n$ converges to $\lambda \in \mathbb{R}_+^*$ and that $\sum_{j=1}^n (p_j^n)^2$ converges to zero when n tends to infinity. Show that S_T has a compound Poisson distribution.
3. Application: $p_j^n = 1/n$, with P a Bernoulli distribution $\mathcal{B}(p)$ with $p \in]0, 1[$.

Solution

1. Let $S_n = \sum_{j=1}^n X_j^n$. We will suppose that n is large enough for the probability that two accidents occur in the same I_j^n to be zero, and hence $S_T = S_n$.
2. Let us determine the Laplace transform of S_n. We know by Theorem 1.68 that $\psi_{S_n}(t) = \prod_{j=1}^n \psi_{X_j^n}(t)$. We compute

$$\psi_{X_j^n}(t) = \mathbb{E}\,(e^{-tX_j^n}) = 1 - p_j^n + \mathbb{E}\,(e^{-tX_j^n} \mid N_j^n = 1)p_j^n = 1 - p_j^n[1 - \psi_P(t)].$$

For $x > 0$, we have $-\log(1 - x) = x + x^2[1 + \varepsilon(x)]$, where $\varepsilon(x) < 1$ for $0 < x < x_0$. By assumption, p_j^n converges to zero when n tends to infinity, so $\max_{1 \le j \le n} p_j^n < x_0/2$ for n large enough. By definition of the Laplace transform, $|1 - \psi_P(t)| \le 2$, thus,

$$-\log \psi_{S_n}(t) = [1 - \psi_P(t)] \sum_{j=1}^n p_j^n + [1 - \psi_P(t)]^2 \sum_{j=1}^n (p_j^n)^2 [1 + \varepsilon([1 - \psi_P(t)]p_j^n)]$$

and

$$[1 - \psi_P(t)]^2 \sum_{j=1}^n (p_j^n)^2 [1 + \varepsilon([1 - \psi_P(t)]p_j^n)] \longrightarrow 0, \quad n \to +\infty,$$

and $\psi_{S_n}(t)$ converges to $\exp(\lambda[\psi_P(t) - 1])$, which is known by Proposition 1.73 to be the Laplace transform of the compound Poisson distribution $\mathcal{CP}(\lambda, P)$.

3. As in Example 1.75, we obtain $S_T \sim \mathcal{P}(p)$.   △

∇ **Exercise 2.3 (Conditional Expectation)** Let X and Y be two independent variables with the same distribution P.

1. Determine the conditional expectation of X given $X + Y$.
2. Suppose that P is the standard normal distribution. Determine the conditional expectation of $|X|$ given $X^2 + Y^2$, and then that of X given $X^2 + Y^2$.

Solution

1. We have $\mathbb{E}(X \mid X + Y) = (X + Y)/2$, because

$$X + Y = \mathbb{E}(X + Y \mid X + Y) = \mathbb{E}(X \mid X + Y) + \mathbb{E}(Y \mid X + Y).$$

2. We know from Proposition 2.25 that $\mathbb{E}(|X| \mid X^2 + Y^2) = \varphi(X^2 + Y^2)$ if for any Borel function $h : \mathbb{R} \longrightarrow \mathbb{R}$,

$$\mathbb{E}[h(X^2 + Y^2)\varphi(X^2 + Y^2)] = \mathbb{E}[h(X^2 + Y^2)|X|].$$

On the one hand,

$$\mathbb{E}[h(X^2 + Y^2)\varphi(X^2 + Y^2)] =$$

$$= \iint_{\mathbb{R}^2} h(x^2 + y^2)\varphi(x^2 + y^2)\frac{e^{-(x^2+y^2)/2}}{2\pi}\, dx dy$$

$$= \int_{\mathbb{R}_+}\int_0^{2\pi} h(r^2)\varphi(r^2)\frac{e^{-r^2/2}}{2\pi} r dr d\theta = \int_{\mathbb{R}_+} h(r^2)\varphi(r^2)e^{-r^2/2} r dr,$$

because $\int_0^{2\pi} d\theta = 2\pi$. On the other hand,

$$\mathbb{E}[h(X^2 + Y^2)|X|] = \iint_{\mathbb{R}^2} h(x^2 + y^2)|x|\frac{e^{-(x^2+y^2)/2}}{2\pi}\, dx dy$$

$$= \int_{\mathbb{R}_+}\int_0^{2\pi} h(r^2)r|\cos\theta|\frac{e^{-r^2/2}}{2\pi} r dr d\theta$$

$$\stackrel{(1)}{=} \int_{\mathbb{R}_+} h(r^2)r^2\frac{2e^{-r^2/2}}{\pi}\, dr.$$

(1) because $\int_0^{2\pi} |\cos\theta| d\theta = 4$. Hence $\varphi(r^2) = 2r/\pi$. This leads to

$$\mathbb{E}(|X| \mid X^2 + Y^2) = \frac{2}{\pi}\sqrt{X^2 + Y^2}.$$

On the contrary, $\mathbb{E}(X \mid X^2 + Y^2) = 0$ because $\int_0^{2\pi} \cos\theta d\theta = 0$. $\triangle$

▽ **Exercise 2.4 (Characteristic Function and Conditioning)** Let X and Y be two random variables such that $Y \sim \mathcal{N}(m, \alpha^2)$ and $\mathbb{E}(e^{itX} \mid Y) = e^{-\sigma^2 t^2/2} e^{itY}$.

1. a. Determine the characteristic function of X and then its distribution.
   b. Same question for (X, Y).
2. Determine $\mathbb{E}(X|Y)$ and show that $X - Y$ and Y are independent.
3. a. Determine the conditional distribution of X given $(Y = y)$.
   b. Same question for Y given $(X = x)$.
   c. Give the conditional expectation of Y given X.

Solution

1. a. We compute

$$\phi_X(t) = \mathbb{E}[\mathbb{E}(e^{itX} \mid Y)] = e^{-\sigma^2 t^2/2} \phi_Y(t) = e^{imt} e^{-(\sigma^2+\alpha^2)t^2/2},$$

so $X \sim \mathcal{N}(m, \sigma^2 + \alpha^2)$.
   b. In the same way,

$$\phi_{(X,Y)}(t, s) = \mathbb{E}[\mathbb{E}(e^{itX} e^{isY} \mid Y)] = e^{-\sigma^2 t^2/2} \phi_Y(s + t)$$
$$= e^{im(s+t)} e^{-(\sigma^2+\alpha^2)t^2/2} e^{-\alpha^2 s^2/2} e^{-\alpha^2 st},$$

so $(X, Y)'$ is a Gaussian vector with expectation $(m, m)'$ and covariance matrix

$$\Gamma = \begin{pmatrix} \sigma^2 + \alpha^2 & \alpha^2 \\ \alpha^2 & \alpha^2 \end{pmatrix}.$$

2. Thanks to Theorem 2.42,

$$\mathbb{E}(X \mid Y) = \frac{\mathbb{C}\text{ov}(X, Y)Y + \mathbb{E} X \mathbb{V}\text{ar } Y - \mathbb{C}\text{ov}(X, Y)\mathbb{E} Y}{\mathbb{V}\text{ar } Y} = Y.$$

Since $\mathbb{E} X = \mathbb{E} Y = m$, we have $\mathbb{E}(X - Y) = 0$ and

$$\mathbb{C}\text{ov}(X - Y, Y) = \mathbb{E}[(X - Y)Y] = \mathbb{E}(XY) - \mathbb{E}(Y^2) = \mathbb{E}[Y\mathbb{E}(X \mid Y)] - \mathbb{E}(Y^2),$$

and hence $\mathbb{C}\text{ov}\,(X - Y, Y) = 0$. Since $(X - Y, Y)'$ is the linear transform of the Gaussian vector $(X, Y)'$, it is a Gaussian vector too, so $X - Y$ and Y are independent.

3. a. Thanks to Theorem 2.42, the conditional distribution of X given $(Y = y)$ is a Gaussian distribution

$$\mathcal{N} \left(\frac{\mathbb{E}\,X - \mathbb{C}\text{ov}\,(X, Y)(y - \mathbb{E}\,X)}{\mathbb{V}\text{ar}\,X}, \mathbb{V}\text{ar}\,X - \frac{\mathbb{C}\text{ov}\,(X, Y)^2}{\mathbb{V}\text{ar}\,X} \right).$$

Since $\mathbb{C}\text{ov}\,(X, Y) = \alpha^2$, it follows that the distribution of X given $(Y = y)$ is $\mathcal{N}(y, \sigma^2)$.

b. Using again Theorem 2.42, the distribution of Y given $(X = x)$ is

$$\mathcal{N} \left(\frac{m\sigma^2 + x\alpha^2}{\sigma^2 + \alpha^2}, \frac{\alpha^2\sigma^2}{\sigma^2 + \alpha^2} \right).$$

c. This gives straightforwardly

$$\mathbb{E}\,(Y \mid X) = \frac{m\sigma^2}{\sigma^2 + \alpha^2} + \frac{\alpha^2}{\sigma^2 + \alpha^2} X,$$

the desired conditional expectation. $\triangle$

∇ **Exercise 2.5 (Time of First Success or of First Failure)** Let (X_n) be an i.i.d. random sequence with distribution $\mathcal{B}(1/2)$. Set $N_n = \inf\{k \in \mathbb{N} : X_{n+k} = 0\}$.

1. Show that for all fixed $n \in \mathbb{N}$, the variable N_n is a stopping time for the filtration $(\sigma(X_0, \ldots, X_{n+m}))_{m \in \mathbb{N}}$.
2. Compute the probability that N_n will be infinitely often equal to zero.
3. Same question for the value 1—consider the sequence (N_{2n}).

Solution

1. The variable N_n counts the number of ones before the first zero in the sequence (X_k) after time n. Since

$$(N_n = m) = (X_0 = 1, \ldots, X_{n+m-1} = 1, X_{n+m} = 0),$$

N_n is a stopping time.

2. The event "N_n is infinitely often equal to zero" is $\overline{\lim}(N_n = 0)$, according to definition 1.19. We have $(N_n = 0) = (X_n = 0)$ and the sequence of these events is independent. Moreover, $\mathbb{P}(X_n = 0) = 1/2$, so the series with general term $\mathbb{P}(N_n = 0)$ diverges. We deduce from Borel-Cantelli lemma that the searched probability is equal to 1.

3. Since the sequence pf events $(N_n = 1) = (X_n = 0, X_{n+1} = 1)$is not independent, the procedure of the above question does not apply.

On the contrary, the sequence of events $(N_{2n} = 1)$ is independent. Since $\mathbb{P}(N_{2n} = 1) = 1/4$, the series with general term $\mathbb{P}(N_{2n} = 1)$ diverges and $\mathbb{P}[\overline{\lim}(N_{2n} = 1)] = 1$. Finally $\overline{\lim}(N_{2n} = 1) \subset \overline{\lim}(N_n = 1)$, hence the searched probability is 1 too.

Thus, N_n takes a.s. infinitely often both values zero and one. △

▽ Exercise 2.6 (Stochastic Version of the Travelling Salesman Problem) Let (X_n) and (Y_n) be two i.i.d. random sequences with uniform distribution on $[0,1]$. Consider the n points $P_1 = (X_1, Y_1), \ldots, P_n = (X_n, Y_n)$ distributed at random into the unit rectangle $[0, 1]^2$. Let D_n be the minimum distance necessary to join these points.

1. Write D_n in terms of the permutations $\sigma \in \mathcal{S}_n$ of $[\![1, n]\!]$.
2. Set $\mathcal{F}_k = \sigma(X_1, \ldots, X_k; Y_1, \ldots, Y_k)$ for $k \geq 1$ and $\mathcal{F}_0 = \{\emptyset, \Omega\}$. Let $Z_k = \mathbb{E}[D_n \mid \mathcal{F}_k]$ for $0 \leq k \leq n$. Show that (Z_k) is a martingale adapted to $(\mathcal{F}_k)$.
3. Deduce from above questions above an upper bound for the difference between D_n and its mean value—use Hoeffding's inequality.

Solution

1. The path linked to a permutation $\sigma \in \mathcal{S}_n$ is $(P_{\sigma(1)}, \ldots, P_{\sigma(n)})$, and the associated distance is

$$d(\sigma) = \sum_{i=1}^{n-1} \|P_{\sigma(i+1)} - P_{\sigma(i)}\| + \|P_{\sigma(n)} - P_{\sigma(1)}\|,$$

where $\|\cdot\|$ denotes the Euclidean norm. Thus, $D_n = \inf_{\mathcal{S}_n} d(\sigma)$.
2. Clearly, $Z_0 = \mathbb{E} D_0$ and $Z_n = D_n$. By definition, Z_k is $\mathcal{F}_k$-measurable, with $\mathbb{E} Z_k = \mathbb{E} D_n \leq 2\sqrt{2n}$ and

$$\mathbb{E}(Z_k \mid \mathcal{F}_{k-1}) = \mathbb{E}[\mathbb{E}(D_n \mid \mathcal{F}_k) \mid \mathcal{F}_{k-1}] = \mathbb{E}(D_n \mid \mathcal{F}_{k-1}) = Z_{k-1}.$$

3. Let D_n^k be the random variable equal to the minimum length of the path connecting all the n points but the k-th. Clearly,

$$D_n^k \leq D_n \leq D_n^k + 2L_k, \quad k \leq n,$$

where L_k is the shortest distance from the point P_k to one of the points $P_{k+1}, \ldots, P_n$. Thus,

$$\mathbb{E}(D_n^k \mid \mathcal{F}_{k-1}) \leq Z_{k-1} \leq \mathbb{E}(D_n^k \mid \mathcal{F}_{k-1}) + 2\mathbb{E}(L_k \mid \mathcal{F}_{k-1})$$

and

$$\mathbb{E}\left(D_n^k \mid \mathcal{F}_k\right) \leq Z_k \leq \mathbb{E}\left(D_n^k \mid \mathcal{F}_k\right) + 2\mathbb{E}\left(L_k \mid \mathcal{F}_k\right).$$

Since $\mathbb{E}\left(D_n^k \mid \mathcal{F}_k\right) = \mathbb{E}\left(D_n^k \mid \mathcal{F}_{k-1}\right)$, we get from above inequalities that $|Z_k - Z_{k-1}| \leq 2\max\{\mathbb{E}\left(L_k \mid \mathcal{F}_{k-1}\right), \mathbb{E}\left(L_k \mid \mathcal{F}_k\right)\} \leq 2\sqrt{2}$. Finally, Hoeffding's inequality gives

$$\mathbb{P}(|D_n - \mathbb{E}\,D_n| \geq x) \leq 2\exp(-x^2/(2\sqrt{2}n)).$$

A much better bound, of the order of $2\exp(-Cx^2/\log n)$, can be obtained by studying the distribution of the minimum distance. Note that the problem of identifying the shortest path through a set of points is a classical one in operational research. △

▽ **Exercise 2.7 (Bin Packing)** Let us consider objects whose sizes are represented by an i.i.d. sequence (X_n) with distribution supported in $[0, 1]$. Bins with size 1 are available. Let B_n be the minimum number of bins necessary to pack the first n objects. Set $Y_k = \mathbb{E}\left(B_n \mid \mathcal{F}_k\right)$ where $(\mathcal{F}_k)$ is the natural filtration of (X_k).

1. Show that (Y_n) is a martingale.
2. Show that $\mathbb{P}(|B_n - \mathbb{E}\,B_n| > x) \leq 2e^{-x^2/2n}$.

Solution

1. Clearly, $Y_0 = \mathbb{E}\,B_n$ and $Y_n = B_n$. By definition, Y_k is $\mathcal{F}_k$-measurable, $\mathbb{E}\,Y_k = \mathbb{E}\,B_n \leq n$ and

$$\mathbb{E}\left(Y_k \mid \mathcal{F}_{k-1}\right) = \mathbb{E}\left[\mathbb{E}\left(B_n \mid \mathcal{F}_k\right) \mid \mathcal{F}_{k-1}\right] = \mathbb{E}\left(B_n \mid \mathcal{F}_{k-1}\right) = Y_{k-1}.$$

2. Let B_n^k be the random variable equal to the minimum number of boxes necessary to pack all objects but the k-th one. Obviously,

$$B_n^k \leq B_n \leq B_n^k + 1, \quad k \leq n.$$

Hence

$$\mathbb{E}\left(B_n^k \mid \mathcal{F}_{k-1}\right) \leq Y_{k-1} \leq \mathbb{E}\left(B_n^k \mid \mathcal{F}_{k-1}\right) + 1$$

and

$$\mathbb{E}\left(B_n^k \mid \mathcal{F}_k\right) \leq Y_k \leq \mathbb{E}\left(B_n^k \mid \mathcal{F}_k\right) + 1.$$

Since $\mathbb{E}(B_n^k \mid \mathcal{F}_k) = \mathbb{E}(B_n^k \mid \mathcal{F}_{k-1})$, we deduce from above inequalities that $|Y_k - Y_{k-1}| \leq 1$. Then Hoeffding's inequality gives the searched inequality. $\triangle$

$\triangledown$ **Exercise 2.8 (About the Stopping Theorem)** Let (X_n) be an **F**-martingale and let T be an **F**-stopping time.

1. Suppose for this question that $T \leq t$. Show that $\mathbb{E}(X_t \mid \mathcal{F}_T) = X_T$ without using the stopping theorem.
2. Suppose for this question that T is a.s. finite, that X_T is integrable, and that $\mathbb{E}[|X_n|\mathbb{1}_{(T \geq n)}]$ converges to zero. Show that $\mathbb{E}\,X_T = \mathbb{E}\,X_0$.
3. Suppose that X_n converges in mean to some variable X. Set

$$X_T = \begin{cases} X, \text{ on } (T = n), \\ X \text{ on } (T = +\infty). \end{cases}$$

Suppose that X_T is integrable. Show that $\mathbb{E}\,X_T = \mathbb{E}\,X_0$.
4. Let (Y_n) be a random sequence adapted to **F**. Show that (Y_n) is a martingale if and only if $EY_T = \mathbb{E}\,Y_0$ for all bounded stopping time T.

Solution

1. Let $A \in \mathcal{F}_T$. We have

$$\mathbb{E}(X_t \mathbb{1}_A) = \sum_{p \leq t} \mathbb{E}[X_t \mathbb{1}_{A \cap (T=p)}] = \sum_{p \leq t} \mathbb{E}(\mathbb{E}[X_t \mathbb{1}_{A \cap (T=p)} \mid \mathcal{F}_p])$$

$$= \sum_{p \leq t} \mathbb{E}[X_p \mathbb{1}_{A \cap (T=p)}] = \mathbb{E}\left[\sum_{p \leq t} X_p \mathbb{1}_{A \cap (T=p)}\right] = \mathbb{E}(X_T \mathbb{1}_A),$$

so $\mathbb{E}(X_t \mid \mathcal{F}_T) = X_T$.
2. Since T is a.s. finite, $X_T \mathbb{1}_{(T<n)}$ converges a.s. to X_T. Moreover, $|X_T \mathbb{1}_{(T<n)}| \leq |X_T|$ which is integrable. The dominated convergence theorem yields that $\mathbb{E}[X_T \mathbb{1}_{(T<n)}]$ converges to $\mathbb{E}\,X_T$. We compute

$$\mathbb{E}(X_{T \wedge n}) = \mathbb{E}[X_T \mathbb{1}_{(T<n)}] + \mathbb{E}[X_n \mathbb{1}_{(T \geq n)}], \quad n \in \mathbb{N},$$

so $\mathbb{E}\,X_{T \wedge n}$ converges to EX_T. Since $T \wedge n$ is a bounded stopping time for any fixed n, the stopping theorem yields $\mathbb{E}\,X_{T \wedge n} = \mathbb{E}\,X_0$, and the conclusion follows.
3. We can write $\mathbb{E}\,X_T = \mathbb{E}[X_T \mathbb{1}_{(T=+\infty)}] + \mathbb{E}[X_T \mathbb{1}_{(T<+\infty)}]$. We have

$$\mathbb{E}\,X_0 = \mathbb{E}(X_{T \wedge n}) = \mathbb{E}[X_T \mathbb{1}_{(T<n)}] + \mathbb{E}[X_n \mathbb{1}_{(T \geq n)}], \quad n \in \mathbb{N},$$

and we can show as in 2. that $\mathbb{E}[X_T \mathbb{1}_{(T<n)}]$ converges to $\mathbb{E}[X_T \mathbb{1}_{(T<+\infty)}]$. Moreover,

$$X_T \mathbb{1}_{(T \geq n)} \xrightarrow{L^1} X_T \mathbb{1}_{(T=+\infty)} = X \mathbb{1}_{(T=+\infty)},$$

so $\mathbb{E}[X_T \mathbb{1}_{(T \geq n)}]$ converges to $\mathbb{E}[X \mathbb{1}_{(T=+\infty)}]$, and the conclusion follows.
4. The direct implication is given by the stopping theorem. For showing the converse, let us consider $A \in \mathcal{F}_n$ and the stopping times $T_1 = n\mathbb{1}_A$ and $T_2 = n\mathbb{1}_{\overline{A}} + (n+1)\mathbb{1}_A$, for any fixed n. By assumption, $\mathbb{E}\, Y_{T_1} = \mathbb{E}\,(Y_n \mathbb{1}_A) = \mathbb{E}\, Y_0$ and $\mathbb{E}\, Y_{T_2} = \mathbb{E}\,(Y_{n+1}\mathbb{1}_A) + \mathbb{E}\,(Y_n \mathbb{1}_{\overline{A}}) = \mathbb{E}\, Y_0$. Therefore, $\mathbb{E}\,(Y_{n+1}\mathbb{1}_A) = \mathbb{E}\,(Y_n \mathbb{1}_A)$, or $\mathbb{E}\,(Y_{n+1} \mid \mathcal{F}_n) = Y_n$, by definition of the conditional expectation. $\qquad \triangle$

▽ **Exercise 2.9 (Martingales and a.s. Convergence)** Let (X_n) be an i.i.d. random sequence with exponential distribution with parameter 1. Let (ε_n) be an i.i.d. random sequence, independent of (X_n), and such that $\mathbb{P}(\varepsilon_1 = 1) = \mathbb{P}(\varepsilon_1 = -1) = 1/2$. Let $\mathcal{F}_n$ be the σ-algebra generated by $(\varepsilon_1, \ldots, \varepsilon_n, X_1, \ldots, X_n)$ for $n \geq 1$, with $\mathcal{F}_0 = \{\emptyset, \Omega\}$. Let $p \in]1, 2[$. Set

$$M_n = \sum_{i=1}^n \varepsilon_i (X_1 + \cdots + X_i)^{-1/p}, \quad n \in \mathbb{N}^*.$$

1. Show that (M_n) is a martingale adapted to $(\mathcal{F}_n)$.
2. Show that

$$\mathbb{E}\,(M_n - M_1)^2 \leq \int_{\mathbb{R}_+} x^{-2/p}(1 - e^{-x})\, dx, \quad n \in \mathbb{N}^*,$$

and then that (M_n) converges a.s.

Solution Set $S_n = X_1 + \cdots + X_n$.

1. We have $|\varepsilon_i| = 1$ for any i, so $\mathbb{E}\,|M_n| \leq \sum_{i=1}^n \mathbb{E}\,(S_n^{-1/p})$. Since $S_n \sim \gamma(n, 1)$, the variable M_n is integrable and it is clearly $\mathcal{F}_n$-measurable. Finally,

$$\mathbb{E}\,(M_{n+1} \mid \mathcal{F}_n) = M_n + \mathbb{E}\,(S_{n+1}^{-1/p}\varepsilon_{n+1} \mid \mathcal{F}_n) = M_n + \mathbb{E}\,(S_{n+1}^{-1/p} \mid \mathcal{F}_n)(\mathbb{E}\,\varepsilon_{n+1}),$$

and $\mathbb{E}\,\varepsilon_{n+1} = 0$, so (M_n) is indeed a martingale.
2. Since $\mathbb{E}\,(\varepsilon_i^2) = 1/2 + 1/2 = 1$, we have

$$\mathbb{E}\,(M_n - M_1)^2 = \mathbb{E}\left(\sum_{i=2}^n \frac{\varepsilon_i}{S_i^{1/p}}\right)^2 = \sum_{i=2}^n \mathbb{E}\,(\varepsilon_i^2)\mathbb{E}\,(S_i^{-2/p})$$

$$= \sum_{i=2}^n \int_{\mathbb{R}_+} x^{-2/p} \frac{x^{i-1}}{(i-1)!}e^{-x}\, dx$$

$$= \int_{\mathbb{R}_+} x^{-2/p} \Big(\sum_{i=2}^{n} \frac{x^{i-1}}{(i-1)!} \Big) e^{-x} \, dx$$

$$\leq \int_{\mathbb{R}_+} x^{-2/p} (1 - e^{-x}) \, dx.$$

Moreover, $\int_{\mathbb{R}_+} x^{-2/p}(1 - e^{-x}) \, dx$ is finite for $1 < p < 2$, hence the sequence $(M_n - M_1)$ is bounded in L^2, so in L^1. Since M_1 is integrable, (M_n) is bounded in L^1 too. Theorem 2.73 induces that (M_n) converges a.s. △

▽ **Exercise 2.10 (Martingales and Characteristic Functions)** Let (X_n) be an i.i.d. random sequence with distribution $\mathcal{N}(0, 1)$ and let (a_n) be a sequence of real numbers. Set $M_n = a_1 X_1 + \cdots + a_n X_n$ for $n \in \mathbb{N}^*$.

1. Show that (M_n) is a martingale adapted to the natural filtration $(\mathcal{F}_n)$ of the sequence (X_n).
2. Determine the characteristic function of M_n, and its limit when n tends to infinity.
3. Show that if $\sum_{n \geq 1} a_n^2 = \sigma^2 < +\infty$, then (M_n) converges a.s. to a random variable M. Determine the distribution of M.
4. Show that if $\sum_{n \geq 1} a_n^2 = +\infty$, then (M_n) is not a.s. convergent.

Solution

1. For any $n \in \mathbb{N}$, M_n is $\mathcal{F}_n$-measurable and is integrable. Moreover,

$$\mathbb{E}(M_{n+1} \mid \mathcal{F}_n) = \mathbb{E}(M_n \mid \mathcal{F}_n) + a_{n+1}\mathbb{E}(X_{n+1} \mid \mathcal{F}_n)$$

$$= \mathbb{E}(M_n \mid \mathcal{F}_n) + a_{n+1}\mathbb{E}(X_{n+1}) = M_n.$$

2. We have

$$\phi_{M_n}(t) = \prod_{j=1}^{n} \mathbb{E}(e^{-ita_j X_j}) = \exp(-t^2 \sum_{j=1}^{n} a_j^2/2).$$

If $\sum_{n \geq 1} a_n^2 = \sigma^2$, then $\phi_{M_n}(t)$ converges to $\exp(-t^2\sigma^2/2)$. Otherwise, it converges to zero for all $t \neq 0$ and $\phi_{M_n}(0)$ converges to 1.
3. In this case,

$$\mathbb{E}\, M_n^2 = \mathbb{E}\Big(\sum_{i=1}^{n} a_i X_i \Big)^2 \leq \sum_{i=1}^{n} a_i^2 \mathbb{E}\, X_i^2 < \sigma^2,$$

and hence (M_n) is a bounded in L^2 martingale. Hence, (M_n) converges a.s. to a random variable M, and by 2., we obtain that $M \sim \mathcal{N}(0, \sigma^2)$.
4. If (M_n) converged a.s. to a variable M, then ϕ_{M_n} would converge to ϕ_M, which would be continuous. Hence a contradiction with 2. △

Markov Chains

3

This chapter investigates the homogeneous discrete-time Markov chains with countable—finite or denumerable—state spaces, also called discrete. Markov chains generalize sequences of independent random variables to variables linked by a simple dependence relation. They model for example, phase transitions of substances between solid, liquid and gaseous states, passages of systems between up and down states, etc. A random sequence $(X_n)_{n \in \mathbb{N}}$ is a Markov chain if its future values depend on its previous values only through its present value, the so-called Markov property. The index n is interpreted as a time, even when it is the n-th trial or step in a process.

First we present basic results and typical examples of Markov chains. Then we define and compute their entropy rate, and finally we present many applications to reliability and to the theory of branching processes.

3.1 General Properties

Markov chains are characterized by their transition functions. Basics on these functions are presented together with numerous examples. Relations between Markov chains and martingales or stopping times are then detailed. Finally, classification of states and definition of stationary distributions lead to study the asymptotic behavior of the chains.

3.1.1 Transition Functions with Examples

Thereafter, $(X_n)_{n \in \mathbb{N}}$ will be a sequence of random variables defined on a probability space $(\Omega, \mathcal{F}, \mathbb{P})$ and taking values in a countable set $E \subset \mathbb{R}^d$, referred to as a—discrete—state space.

The future of a Markov chain depends on its past only through its present, in other words, a Markov chain satisfies the Markov property.

© Springer Nature Switzerland AG 2018
V. Girardin, N. Limnios, *Applied Probability*,
https://doi.org/10.1007/978-3-319-97412-5_3

Definition 3.1 (Markov Property) A random sequence (X_n) is a Markov chain if

$$\mathbb{P}(X_{n+1} = j \mid X_0 = i_0, X_1 = i_1, \ldots, X_{n-1} = i_{n-1}, X_n = i) =$$
$$= \mathbb{P}(X_{n+1} = j \mid X_n = i),$$

for all $i, j, i_0, i_1, \ldots, i_{n-1}$ in E, and all $n \in \mathbb{N}$, provided that $\mathbb{P}(X_0 = i_0, X_1 = i_1, \ldots, X_{n-1} = i_{n-1}, X_n = i) \neq 0$.

If the transition probability $\mathbb{P}(X_{n+1} = j \mid X_n = i) = P(i, j)$ does not depend on the index n, the chain is said to be homogeneous or to have stationary transition probabilities.

We will study only homogeneous Markov chains.

The function $(i, j) \longrightarrow P(i, j)$, defined on $E \times E$, is referred to as the transition function of the chain. Indeed,

$$\mathbb{P}(X_{n+1} \in A \mid X_0 = i_0, \ldots, X_{n-1} = i_{n-1}, X_n = i) = \sum_{j \in A} P(i, j),$$

and the function $P : (E, \mathcal{P}(E)) \longrightarrow [0, 1]$ defined by $P(i, A) = \sum_{j \in A} P(i, j)$ is a transition probability from E to E. The quantity $P(i, j)$ is the transition probability from i to j and $P = (P(i, j))_{(i,j) \in E^2}$ is the—possibly infinite—transition matrix. The distribution μ of X_0 is the initial distribution of the chain. We set $\mu(i) = \mu(\{i\})$, define $\mathbb{P}_\mu(A) = \sum_{i \in E} \mu(i)\mathbb{P}(A \mid X_0 = i)$, and denote by $\mathbb{E}_\mu$ the mean associated with the probability $\mathbb{P}_\mu$. If $\mu = \delta_i$, with $i \in E$, we set $\mathbb{P}_i$ instead of $\mathbb{P}_{\delta_i}$; specifically $\mathbb{P}_i(A) = \mathbb{P}(A \mid X_0 = i)$.

Definition 3.2 The n-step transition function P^n of a Markov chain is given by the probabilities that the chain goes from one state to another in n steps, namely,

$$P^n(i, j) = \mathbb{P}(X_{m+n} = j \mid X_m = i) = \mathbb{P}(X_n = j \mid X_0 = i), \quad m \geq 0, \; n \geq 1,$$

where $P^1(i, j) = P(i, j)$ and, by convention, $P^0(i, i) = 1$ and $P^0(i, j) = 0$ for $i \neq j$.

For any function h defined on E, we denote by Ph the function defined on E by $Ph(i) = \sum_{j \in E} P(i, j)h(j)$. If $Ph = h$, then h is said to be harmonic; in this case, $P^n h = h$ for all n.

The Markov property can be stated as

$$\mathbb{E}[h(X_{n+1}) \mid \mathcal{F}_n] = \mathbb{E}[h(X_{n+1}) \mid X_n] = Ph(X_n), \quad \text{a.s.}$$

for all bounded functions $h : \mathbb{E} \longrightarrow \mathbb{R}$.

Properties of Transition Functions

1. The function $j \longrightarrow P(i, j)$ defines a probability on $(E, \mathcal{P}(E))$ for all $i \in E$; indeed, $0 \leq P(i, j) \leq 1$ and $\sum_{j \in E} P(i, j) = 1$, because the chain remains in E. Remaining in i is considered as a virtual transition. Precisely, a jump time is a time when a change of state occurs.
2. (**Chapman-Kolmogorov equation**) For all i and j in E and all integers m and n, we have

$$\sum_{k \in E} P^n(i, k) P^m(k, j) = P^{n+m}(i, j).$$

Indeed,

$$
\begin{aligned}
P^{n+m}(i, j) &= \mathbb{P}(X_{n+m} = j \mid X_0 = i) \\
&\stackrel{(1)}{=} \sum_{k \in E} \mathbb{P}(X_{n+m} = j, X_n = k \mid X_0 = i) \\
&\stackrel{(2)}{=} \sum_{k \in E} \mathbb{P}(X_{n+m} = j \mid X_n = k, X_0 = i) \mathbb{P}(X_n = k \mid X_0 = i) \\
&\stackrel{(3)}{=} \sum_{k \in E} P^m(k, j) P^n(i, k).
\end{aligned}
$$

(1) par the formula of total probability, (2) by the compound probabilities theorem and (3) by the Markov property.
3. When E is a finite set, Point 2. shows that P^n is the n-th power of the matrix P for the ordinary matrix product; more generally, $P^n P^m = P^{n+m} = P^m P^n$.

Random sequences satisfying Chapman-Kolmogorov equation are not all Markov chains, as shown in Exercise 3.3.

▷ *Example 3.3 (Transmission of a Binary Message Among a Population)* At each step, the received message is well transmitted with probability $1 - p$, with $0 < p < 1$. The state space is $E = \{0, 1\}$, coding the binary message. Suppose that the original message is 0, that is $\mathbb{P}(X_0 = 0) = 1$. Then

$$\mathbb{P}(X_n = 1 \mid X_{n-1} = 1) = \mathbb{P}(X_n = 0 \mid X_{n-1} = 0) = 1 - p,$$

$$\mathbb{P}(X_n = 0 \mid X_{n-1} = 1) = \mathbb{P}(X_n = 1 \mid X_{n-1} = 0) = p,$$

and hence

$$P = \begin{pmatrix} 1 - p & p \\ p & 1 - p \end{pmatrix}.$$

By induction, we get

$$P^n = \frac{1}{2} \begin{pmatrix} 1 & 1 \\ 1 & 1 \end{pmatrix} + \frac{(1-2p)^n}{2} \begin{pmatrix} 1 & -1 \\ -1 & 1 \end{pmatrix},$$

which yields

$$\mathbb{P}(X_n = 0) = \frac{1}{2} + \frac{(1-2p)^n}{2}.$$

Regardless of the original message and probability p, the probability of receiving after a large number of steps either this message or its opposite is the same, precisely,

$$P^n \longrightarrow \frac{1}{2} \begin{pmatrix} 1 & 1 \\ 1 & 1 \end{pmatrix}, \quad n \to +\infty,$$

as shown in Fig. 3.1.   ◁

The above simple example illustrates some of the issues linked to Markov chains theory: convergence, limit distribution (meaning the distribution of X_n when n tends to infinity), lack of memory—especially of the initial distribution. The following one illustrates modeling by Markov chains.

▷ *Example 3.4 (Moran's Reservoir)*   A reservoir with capacity $c \in \mathbb{N}^*$ units of volume is observed at integer times n. In the time interval $[n, n+1[$, a random quantity of water, equal to Z_n unities of volume, enters the reservoir. When the capacity of this reservoir is reached, the exceeding quantity is lost by overflowing. At the end of $[n, n+1[$, one unit of water, if available, leaves the reservoir. Clearly, if X_n denotes the level of water in the reservoir at time n, then the level at time $n+1$ is

$$X_{n+1} = (X_n + Z_n - 1)^+ \wedge (c - 1).$$

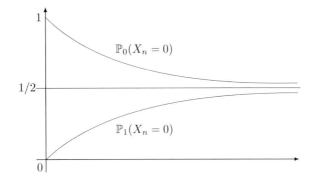

Fig. 3.1   Transmission of a message—Example 3.3, $p = 1/2$

If the variables Z_n are assumed to be nonnegative and i.i.d. with common distribution $\mathbb{P}(Z_n = k) = p_k$, then (X_n) is a Markov chain with state space $E = \{0, 1, \ldots, c - 1\}$ and transition matrix

$$
\begin{array}{c}
\\
0 \\
1 \\
2 \\
\vdots \\
c-2 \\
c-1
\end{array}
\begin{array}{cccccc}
0 & 1 & 2 & \cdots & c-2 & c-1 \\
\left(\begin{array}{cccccc}
p_0 + p_1 & p_2 & p_3 & \cdots & p_{c-1} & h_c \\
p_0 & p_1 & p_2 & \cdots & p_{c-2} & h_{c-1} \\
0 & p_0 & p_1 & \cdots & p_{c-3} & h_{c-2} \\
\vdots & \vdots & \vdots & \ddots & \vdots & \vdots \\
0 & 0 & 0 & \cdots & p_1 & h_2 \\
0 & 0 & 0 & \cdots & p_0 & h_1
\end{array}\right)
\end{array},
$$

where $h_k = \sum_{i \geq k} p_i$ for $1 \leq k \leq c$. The graph of (X_n) for $c = 5$ is given in Fig. 3.2. Note that c could be considered as a state of the chain, but then only X_0 could take this value. $\triangleleft$

Let us now detail some classical types of Markov chains.

$\triangleright$ *Example 3.5 (Random Walk)* A random walk (see Definition 1.76) is a Markov chain with state space $E = \mathbb{Z}$ and transition function P given by

$$
P(i, j) = p(j - i), \quad (i, j) \in \mathbb{Z}^2,
$$

for some function $p : \mathbb{Z} \longrightarrow [0, 1]$.

When $p(1) = p = 1 - p(-1)$, with $p \in [0, 1]$, the random walk is said to be simple and we have

$$
P(i, j) = \begin{cases}
p & \text{if } j = i + 1, \\
1 - p & \text{if } j = i - 1, \\
0 & \text{otherwise.}
\end{cases}
$$

Its graph is given in Fig. 3.3. A trajectory of a simple random walk is shown in Fig. 1.2 in Chap. 1.

Fig. 3.2 Moran's reservoir—Example 3.4, $c = 5$

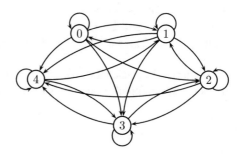

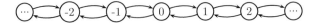

Fig. 3.3 Graph of a simple random walk—Example 3.5

The probability of return to i in n steps is

$$P^{2n+1}(i, i) = 0 \quad \text{and} \quad P^{2n}(i, i) = \binom{2n}{n} p^n (1 - p)^n, \quad n \geq 1. \tag{3.1}$$

One can show by induction and using the properties of the binomial coefficients that the probability of going from i to j in n steps is

$$P^n(i, j) = \binom{n}{(n + j - i)/2} p^{(n+j-i)/2} (1 - p)^{(n-j+i)/2} \quad \text{if } n + j - i = 2m > 0,$$

and is null otherwise.   ◁

▷ *Example 3.6 (Birth and Death Chain)*   A Markov chain is a birth and death chain if its state space is $E = \mathbb{N}$ and its transition function is tridiagonal, that is

$$P(i, j) = \begin{cases} p_i \text{ if } j = i + 1, \\ q_i \text{ if } j = i - 1, \\ r_i \text{ if } j = i, \\ 0 \text{ otherwise,} \end{cases}$$

for $i \geq 0$ and $j \geq 1$. When $q_i = 0$ ($p_i = 0$) for all i, the chain is a pure birth (death) chain.

This chain models various physical and biological phenomena, such as growing of populations, evolution of identical systems with failures and repairs, decay of radio-active substances, etc.

If $E = \mathbb{N}$, with $r_i = 0$, and $P(0, 1) = 1$, the chain is called a Harris Markov chain. If $E = \mathbb{Z}$, with $r_i = 0$, $p_i = p$ and $q_i = 1 - p$ for all i, the chain is a simple random walk.

Birth and death chains are often considered with finite state spaces. If $E = \{1, \ldots, N\}$, with $p_i = 1 - i/N$, $q_i = i/N$ and $r_i = 0$, the chain is called an Ehrenfest model and is used in thermodynamics; if $r_i = 0$, $p_i = p$, $q_i = q$, for all i but $p_0 = 0$ and $q_N = 0$, the chain is the gambler's ruin chain.   ◁

The finite dimensional distributions of a Markov chain are completely specified by its transition function and initial distribution.

Proposition 3.7 *Let (X_n) be a Markov chain with transition function P and initial distribution μ. We have*

$$\mathbb{P}(X_0 \in A_0, X_1 \in A_1, \ldots, X_n \in A_n) =$$

$$= \sum_{i_0 \in A_0} \sum_{i_1 \in A_1} \cdots \sum_{i_n \in A_n} \mu(i_0) P(i_0, i_1) \cdots P(i_{n-1}, i_n). \qquad (3.2)$$

for all $n \geq 1$ and subsets $A_0, A_1, \ldots, A_n$ of E.

If we set $A_0 = A_1 = \cdots = A_{n-1} = E$ and $A_n = A$ in (3.2), we obtain the distribution of X_n, namely

$$\mathbb{P}(X_n \in A) = \mu P^n(A) = \sum_{i \in E} \sum_{j \in A} \mu(i) P^n(i, j). \qquad (3.3)$$

Proof Applying the compound probabilities theorem and the Markov property yields the result. For example,

$$\mathbb{P}(X_0 = i_0, X_1 = i_1, X_2 = i_2) =$$

$$= \mathbb{P}(X_2 = i_2 \mid X_0 = i_0, X_1 = i_1)\mathbb{P}(X_1 = x_1 \mid X_0 = i_0)P(X_0 = i_0)$$

$$= \mathbb{P}(X_2 = i_2 \mid X_1 = i_1)\mathbb{P}(X_1 = i_1 \mid X_0 = i_0)P(X_0 = i_0)$$

$$= \mu(i_0)P(i_0, i_1)P(i_1, i_2),$$

gives the distribution of (X_0, X_1, X_2). □

Conversely, any sequence of random variables taking values in a discrete set E and satisfying (3.2) for some distribution μ and function P is a Markov chain, with initial distribution μ and transition function P. The next result is even more general.

Theorem 3.8 (Canonical Construction) *If E is a countable set, μ a probability on E and P a transition function on E, then a probability space can be defined on which a discrete Markov chain exists taking values in E, with initial distribution μ and transition matrix P.*

Proof Set $\Omega = E^{\mathbb{N}}$ and $\mathcal{F} = \mathcal{P}(E)^{\otimes \mathbb{N}}$. Relation (3.2) defines a probability $\mathbb{P}$ on the infinite cylindrical sets $A_0 \times A_1 \times \cdots \times A_n \times E \times \cdots$ of $(\Omega, \mathcal{F})$.

Let X_n denote the projection on the n-th element, defined by

$$X_n(\omega) = X_n(\omega_0, \omega_1, \ldots, \omega_n, \ldots) = \omega_n.$$

The sequence (X_n) is a Markov chain on $(\Omega, \mathcal{F}, \mathbb{P})$ that fulfills the required conditions. □

The above constructed chain is said to be canonical and $(E^{\mathbb{N}}, \mathcal{P}(E)^{\otimes \mathbb{N}}, \mathbb{P})$ is referred to as the canonical space.

The periodicity is a major characteristic of states of Markov chains.

Definition 3.9 Set $d_i = \text{g.c.d.}\{n \in \mathbb{N}^* : P^n(i, i) > 0\}$ for $i \in E$, with the convention $d_i = +\infty$ if $P^n(i, i) = 0$ for all $n \geq 1$.

If $1 < d_i < +\infty$, then i is said to be periodic with period d_i. If $d_i = 1$, then i is said to be aperiodic.

A Markov chain with only aperiodic states is said to be aperiodic. A state $i \in E$ such that $P(i, i) = 1$ is said to be absorbing. When the chain reaches an absorbing state, it remains there forever.

Proposition 3.10 *If the initial distribution of the chain is δ_i, for a non absorbing state $i \in E$, then:*

1. *the sojourn time in state i—that is the time spent by the chain in i before leaving it—has a geometric distribution with parameter $1 - P(i, i)$;*
2. *the probability for the chain to jump to state j when it leaves state i is $P(i, j)/[1 - P(i, i)]$.*

Proof

1. Let η_i denote the sojourn time in state i. For any $k \geq 1$, we have

$$\mathbb{P}_i(\eta_i = k) = \mathbb{P}_i(X_{m+1} = i, \ldots, X_{m+k-1} = i, X_{m+k} = I/X_{m-1} = i, X_m = i)$$
$$= \sum_{j \in E\setminus\{i\}} P(i, i) \ldots P(i, i) P(i, j) = P(i, i)^{k-1}[1 - P(i, i)].$$

2. For $i \neq j$,

$$\mathbb{P}_i(X_{n+1} = j \mid X_n = i, X_{n+1} \neq i) =$$
$$= \frac{\mathbb{P}_i(X_{n+1} = j, X_n = i)}{\mathbb{P}_i(X_n = i, X_{n+1} \neq i)} = \frac{\mathbb{P}_i(X_{n+1} = j \mid X_n = i)}{\mathbb{P}_i(X_{n+1} \neq i \mid X_n = i)} = \frac{P(i, j)}{1 - P(i, i)},$$

by definition of conditional probabilities. $\qquad\qquad\qquad\qquad\qquad\qquad\square$

The matrix with entries $Q(i, j) = P(i, j)/[1 - P(i, i)]$, for $i \neq j$ is the transition matrix of the embedded Markov chain defined at the jump times of (X_n); see Exercise 3.8.

$\triangleright$ *Example 3.11* The sojourn time in i of a birth and death chain has the distribution $\mathcal{G}(1 - r_i)$ if $r_i > 0$. It is constant and equal to one unit of time if $r_i = 0$. $\qquad \triangleleft$

Trajectories of a Markov chain in a time interval $[0, t]$ can be generated by jump times and visited states, using Proposition 3.10, through the algorithm:

1. Let i be the initial state of the chain: $X_0(\omega) = i$.
   $n := 0$. $j := i$. $S_0(\omega) := W_0(\omega) = 0$.
2. $n := n + 1$.
3. Let $W_n(\omega)$ be the realization of a $\mathcal{G}(1 - P(j, j))$-distributed random variable.
   $S_n(\omega) := S_{n-1}(\omega) + W_n(\omega)$.
4. Let $K(\omega)$ be the realization of a variable taking values in $E \setminus \{j\}$, with distribution
   $\mathbb{P}(K = k) = P(j, k)/[1 - P(j, j)]$. $j := K(\omega)$.
5. If $S_n \geq t$, then end.
6. Continue at Step 2.

Note that if the initial distribution of the chain is not δ_i, Step 1 consists in its realization following some simulation method.

Multidimensional chains are defined and investigated in a similar way. Let us give two typical examples of two-dimensional chains.

▷ *Example 3.12 (Product Markov Chain)* Let (X_n) and (Y_n) be two independent Markov chains, with respective state spaces E_X and E_Y, initial distributions μ^X and μ^Y, and transition functions P^X and P^Y. The sequence $(Z_n) = ((X_n, Y_n))$ is a Markov chain with state space $E = E_X \times E_Y$, initial distribution $\mu = \mu^X \otimes \mu^Y$ and transition function given by

$$P((i, j), (k, l)) = P^X(i, k)P^Y(j, l), \quad (i, j) \in E_X \times E_Y, \ (k, l) \in E_X \times E_Y.$$

Since $\mathbb{P}[Z_0 = (i, j)] = \mathbb{P}(X_0 = i)\mathbb{P}(Y_0 = j)$, the Markov property is clearly satisfied and

$$P((i, j), (k, l)) = \mathbb{P}[Z_1 = (k, l) \mid Z_0 = (i, j)]$$

$$= \mathbb{P}(X_1 = k, Y_1 = l \mid X_0 = i, Y_0 = j)$$

$$= \mathbb{P}(X_1 = k \mid X_0 = i)\mathbb{P}(Y_1 = l \mid Y_0 = j).$$

The product can be generalized to any number of independent Markov chains. ◁

▷ *Example 3.13* Let (X_n) be a Markov chain and set $Z_n = (X_n, X_{n-1})$. Let us prove that (Z_n) satisfies the Markov property. Set $j = (j^2, j^1)$ and $i_k = (i_k^2, i_k^1)$ for $k = 0, \ldots, n$, with necessarily $i_k^1 = i_{k-1}^2$ for $k = 1, \ldots, n$. Then

$$\mathbb{P}(Z_{n+1} = j \mid Z_0 = i_0, \ldots, Z_{n-1} = i_{n-1}, Z_n = i_n) =$$

$$= \mathbb{P}(X_{n+1} = j^2 \mid X_0 = i_0^1, X_1 = i_0^2, \ldots, X_{n-1} = i_n^1, X_n = i_n^2)\delta_{i_n^2 j^1}$$

$$\overset{(1)}{=} \mathbb{P}(X_{n+1} = j^2 \mid X_{n-1} = i_n^1, X_n = i_n^2)\delta_{i_n^2 j^1}$$

$$= \mathbb{P}(X_{n+1} = j^2, X_n = j^1 \mid X_{n-1} = i_n^1, X_n = i_n^2) = \mathbb{P}(Z_{n+1} = j \mid Z_n = i_n).$$

(1) by the Markov property applied to (X_n)—keeping $X_{n-1} = i_n^1$ in order to get Z_n back.

The transition function $\widetilde{P}$ of this two-dimensional chain (X_n, X_{n-1}) is given by $\widetilde{P}((k, l), (i, j)) = P(j, i)\delta_{li}$ for i, j, k, l in E.

Note that the transition function $\widehat{P}$ of the other two-dimensional chain (X_{n-1}, X_n) is given by $\widehat{P}((k, l), (i, j)) = P(i, j)\delta_{li}$ for i, j, k, l in E.

This extends easily to the k-dimensional chain $Z_n = (X_n, \ldots, X_{n-k})$ for any fixed $k \geq 1$.   ◁

3.1.2 Martingales and Markov Chains

The Markov property can be expressed as a martingale property.

Theorem 3.14 *A random sequence (X_n) is a Markov chain with transition function P if and only if for any bounded function $h : E \longrightarrow \mathbb{R}$, the random sequence (M_n) defined by*

$$M_n = h(X_n) - h(X_0) - \sum_{k=0}^{n-1}[Ph(X_k) - h(X_k)] \qquad (3.4)$$

is a martingale for the natural filtration $(\mathcal{F}_n)$ of (X_n).

Proof Taking conditional expectation of both sides of (3.4) yields

$$\mathbb{E}(M_n \mid \mathcal{F}_{n-1}) = \mathbb{E}[h(X_n) \mid \mathcal{F}_{n-1}] - h(X_0) - \sum_{k=0}^{n-1}[Ph(X_k) - h(X_k)].$$

Moreover,

$$-h(X_0) - \sum_{k=0}^{n-1}[Ph(X_k) - h(X_k)] =$$

$$= -h(X_0) - Ph(X_{n-1}) + h(X_{n-1}) - \sum_{k=0}^{n-2}[Ph(X_k) - h(X_k)]$$

$$= -Ph(X_{n-1}) + M_{n-1}.$$

Therefore, $\mathbb{E}[h(X_n) \mid \mathcal{F}_{n-1}] = Ph(X_{n-1})$ is equivalent to the martingale property $\mathbb{E}[M_n \mid \mathcal{F}_{n-1}] = M_{n-1}$.   □

Clearly, (M_n) is square integrable, from which we deduce the following result on its quadratic characteristic.

Proposition 3.15 *If (X_n) is a Markov chain with transition function P, then the quadratic characteristic of the martingale (M_n) defined by (3.4) is the compensatory of the sub-martingale (M_n^2), namely*

$$\langle M \rangle_n = \sum_{k=0}^{n-1} [Ph^2(X_k) - (Ph(X_k))^2].$$

Proof The quadratic characteristic of (M_n) is

$$\langle M \rangle_n = \sum_{k=1}^{n} \mathbb{E}[(\Delta M_k)^2 \mid \mathcal{F}_{k-1}].$$

We have $\Delta M_k = M_k - M_{k-1} = h(X_k) - Ph(X_{k-1})$ and

$$\begin{aligned}
\mathbb{E}[(\Delta M_k)^2 \mid \mathcal{F}_{k-1}] &= \\
&= \mathbb{E}[h^2(X_k) - 2h(X_k)Ph(X_{k-1}) + (Ph(X_{k-1}))^2 \mid \mathcal{F}_{k-1}] \\
&= \mathbb{E}[h^2(X_k) \mid \mathcal{F}_{k-1}] - 2Ph(X_{k-1})\mathbb{E}[h(X_k) \mid \mathcal{F}_{k-1}] + [Ph(X_{k-1})]^2 \\
&= Ph^2(X_{k-1}) - [Ph(X_{k-1})]^2,
\end{aligned}$$

from which the result follows.   □

The harmonic functions of a Markov chain are also characterized by a martingale property.

Proposition 3.16 *Let (X_n) be a Markov chain with state space E and transition matrix P. A function $h \colon E \longrightarrow \mathbb{R}_+$ is harmonic for P if and only if $(h(X_n))$ is a martingale for the natural filtration $(\mathcal{F}_n)$ of (X_n).*

Proof We have $\mathbb{E}[h(X_n) \mid \mathcal{F}_{n-1}] = \mathbb{E}[h(X_n) \mid X_{n-1}]$. If h is harmonic, then

$$\mathbb{E}[h(X_n) \mid X_{n-1} = i] = \sum_{j \in E} \mathbb{P}(X_n = j \mid X_{n-1} = i)h(j)$$

$$= \sum_{j \in E} P(i, j)h(j) = Ph(i) = h(i),$$

or $\mathbb{E}[h(X_n) \mid \mathcal{F}_{n-1}] = h(X_{n-1})$.

Conversely, if $(h(X_n))$ is a martingale, then $\mathbb{E}[h(X_1) \mid X_0] = h(X_0)$. We obtain $\mathbb{E}[h(X_1) \mid X_0] = Ph(X_0)$ as above, from which it follows that $\mathbb{E}[h(X_1) \mid X_0 = i] = h(i) = Ph(i)$ for all $i \in E$.   □

3.1.3   Stopping Times and Markov Chains

Markov chains and stopping times are linked especially through the strong Markov property. In other words, the strong Markov property is an extension to stopping times of the Markov property for integers. Moreover, all discrete Markov chains satisfy the strong Markov property.

Theorem 3.17 (Strong Markov Property) *Let (X_n) be a Markov chain. Let T be an a.s. finite stopping time of (X_n). For all $A \in \mathcal{F}$,*

$$\mathbb{P}[(X_T, X_{T+1}, \ldots) \in A \mid \mathcal{F}_T] = \mathbb{P}_{X_T}[(X_0, X_1, \ldots) \in A] \quad a.s..$$

Note that this property holds true for any stopping time T on the event $(T < +\infty)$.

Proof It is sufficient to prove that for any $C \in \mathcal{F}_T$,

$$\mathbb{E}\left[\mathbb{1}_C \mathbb{P}(X_T = x_0, X_{T+1} = x_1, \ldots \mid \mathcal{F}_T)\right] = \mathbb{E}\left[\mathbb{1}_C \mathbb{P}_{X_T}(X_0 = x_0, X_1 = x_1, \ldots)\right].$$

We compute

$$\mathbb{E}\left[\mathbb{1}_C \mathbb{P}(X_T = x_0, X_{T+1} = x_1, \ldots \mid \mathcal{F}_T)\right] =$$
$$= \sum_{n \geq 0} \mathbb{E}\left[\mathbb{1}_{C \cap (T=n)} \mathbb{P}(X_n = x_0, X_{n+1} = x_1, \ldots \mid \mathcal{F}_n)\right]$$

and

$$\mathbb{E}\left[\mathbb{1}_{C \cap (T=n)} \mathbb{P}(X_n = x_0, X_{n+1} = x_1, \ldots \mid \mathcal{F}_n)\right] =$$
$$\overset{(1)}{=} \mathbb{E}\left[\mathbb{1}_{C \cap (T=n)} \mathbb{P}(X_n = x_0, X_{n+1} = x_1, \ldots \mid X_n)\right]$$
$$\overset{(2)}{=} \mathbb{E}\left[\mathbb{1}_{C \cap (T=n)} \mathbb{P}_{X_n}(X_0 = x_0, X_1 = x_1, \ldots)\right].$$

(1) by the weak Markov property and (2) by homogeneity.   □

The strong Markov property is often used for either hitting or return times.

Definition 3.18 Let (X_n) be a Markov chain with state space E. Let $i \in E$. The random variable $T_i = \inf\{n \in \mathbb{N}^* : X_n = i\}$ is called the (first) return time to state i. If $\mathbb{P}(X_0 = i) = 1$, then T_i is called the recurrence time to i. The random variable $T_i' = \inf\{n \in \mathbb{N} : X_n = i\}$ is called the hitting time of i. By convention, $\inf \emptyset = +\infty$.

Proposition 3.19 *The return time T_i is a stopping time of (X_n) for any $i \in E$.*

Proof We have $T_i \geq 1$. Moreover, $(T_i = 1) = (X_1 = i)$,

$$(T_i = n) = (X_1 \neq i, \ldots, X_{n-1} \neq i, X_n = i) \quad n \geq 2,$$

and $(T_i = +\infty) = (X_n \neq i, n \geq 1)$. Thus T_i is a stopping time of (X_n). $\qquad \square$

Corollary 3.20 *If $Y_n = X_{T_i+n}$, then, conditional to the event $(T_i < +\infty)$, (Y_n) is a Markov chain with initial distribution δ_i and transition function P. Moreover, (Y_n) is independent of X_n for all $n < T_i$.*

The above corollary of the strong Markov property is often used under the form

$$\mathbb{P}_j[T_i < +\infty, h(X_{T_i+n}) \in B] = \mathbb{P}_j(T_i < +\infty)\mathbb{P}_i[h(X_{T_i+n}) \in B],$$

where $(i, j) \in E \times E$, $B \in \mathcal{B}(\mathbb{R})$, $n \in \mathbb{N}$ and $h : E \longrightarrow \mathbb{R}$ is some function.

An alternative stating of the Markov property holds in terms of operators.

Definition 3.21 The function $\theta : \Omega^{\mathbb{N}} \longrightarrow \Omega^{\mathbb{N}}$ defined by

$$\theta(\omega) = \theta(\omega_0, \omega_1, \ldots) = (\omega_1, \omega_2, \ldots)$$

is called the one step translation (or shift) operator.

We denote by θ_k the k-th iterate of θ for $k \geq 1$: $\theta_k = \theta \circ \theta_{k-1}$, where $\theta_1 = \theta$ and θ_0 is the identity on $\Omega^{\mathbb{N}}$. For instance, for any random sequence (X_n), we have $X_n(\theta(\omega)) = X_{n+1}(\omega)$ for all $\omega \in \Omega$, and also $\mathbb{1}_{(X_n=i)} \circ \theta_k = \mathbb{1}_{(X_{n+k}=i)}$, for all $k \geq 1$ and all $n \geq 0$.

For any stopping time T, set

$$\theta_T(\omega) = \begin{cases} \theta_n(\omega) & \text{on } (T = n), \\ \widetilde{\omega} & \text{on } (T = +\infty), \end{cases}$$

where $\widetilde{\omega}$ denotes some point added to Ω. The Markov property can be written

$$\mathbb{E}_i(f \circ X \circ \theta_n \mid \mathcal{F}_n) = \mathbb{E}_{X_n}(f \circ X),$$

and the strong Markov property becomes

$$\mathbb{E}_i(f \circ X \circ \theta_T \mid \mathcal{F}_T) = \mathbb{E}_{X_T}(f \circ X),$$

for all bounded functions $f : E^{\mathbb{N}} \longrightarrow \mathbb{R}$.

3.2   Classification of States

Any state of a Markov chain can be characterized either as recurrent or as transient. The distinction is fundamental for investigating the asymptotic behavior of the chain.

Let $i \in E$ and $j \in E$ with $j \neq i$. The time spent by the chain at i in $[\![1, n]\!]$ is $N_i^n = \sum_{k=1}^n \mathbb{1}_{(X_k=i)}$. The total time spent by the chain at i is $N_i = \sum_{k \geq 1} \mathbb{1}_{(X_k=i)}$. The number of transitions from i to j in one step in $[\![1, n]\!]$ is $N_{ij}^n = \sum_{k=1}^n \mathbb{1}_{(X_{k-1}=i, X_k=j)}$, with $N_{ij} = \sum_{k \geq 1} \mathbb{1}_{(X_{k-1}=i, X_k=j)}$. Both random variables N_i and N_{ij} take values in $\overline{\mathbb{N}}$. Let $\varrho_{ij} = \mathbb{P}_i(T_j < +\infty) = \mathbb{P}_i(N_j \geq 1)$ denote the probability of visiting state j, starting from state i.

Definition 3.22   If $\varrho_{ii} = 1$, then i is said to be recurrent (or persistent). Otherwise— if $\varrho_{ii} < 1$, then i is said to be transient.

If i is recurrent then either $m_i = \mathbb{E}_i T_i < +\infty$ and i is said to be positive recurrent, or $m_i = +\infty$ and i is said to be null recurrent.

A state i is recurrent if, assuming the chain starts at i, it will eventually return to i with probability one. An absorbing state is clearly a positive recurrent state.

$\triangleright$ *Example 3.23 (Gambler's Ruin Problem)*   Suppose the initial fortune of a gambler is k and the initial fortune of the casino is $N - k$. The gambler either wins or loses one euro at each bet with probability $p \in]0, 1[$ and $q = 1 - p$, respectively. This experiment is modelled by a birth and death chain taking values in $[\![0, N]\!]$, with transition matrix given by $P(i, i + 1) = p = 1 - P(i, i - 1) = 1 - q$ and $P(0, 0) = P(N, N) = 1$.

The states 0 and N are absorbing. All other states are transient, because the probability of passage from i to either 0 or N is positive. The time before absorption is $\tau = \inf\{n \in \mathbb{N}^* : X_n = 0 \text{ or } N\}$. The event "The gambler is ruined" corresponds to the absorption of the chain in 0. Thus, the probability of the gambler's ruin is $u_k = \mathbb{P}(X_\tau = 0 \mid X_0 = k)$ and the probability of the casino's ruin is $1 - u_k$. We have $u_0 = 1$, $u_N = 0$ and $u_k = pu_{k+1} + qu_{k-1}$ for all $1 \leq k \leq N - 1$. Therefore, (u_k) is a Fibonacci's sequence. Setting $r = q/p$, we get

$$u_k = \begin{cases} \dfrac{r^k - r^N}{1 - r^N} & \text{if } p \neq q, \\ \dfrac{N - k}{N} & \text{if } p = 1/2. \end{cases}$$

The mean duration of the game can be computed similarly by determining $\mathbb{E}(\tau \mid X_0 = k) = v_k$, that satisfies $1 + p v_{k+1} + q v_{k-1} = v_k$. ◁

Lemma 3.24 *For all $(i, j) \in E \times E$ and $m \in \mathbb{N}^*$, we have*

$$\mathbb{P}_i(N_j \geq m) = \varrho_{ij} \varrho_{jj}^{m-1}. \tag{3.5}$$

Proof Let us prove the result by induction on m.

For $m = 1$, by definition, $\mathbb{P}_i(N_j \geq 1) = \varrho_{ij}$. Let us suppose that (3.5) holds true for some $m > 1$ and let $n_1 < n_2 < \cdots < n_m < n_{m+1}$ denote the $m + 1$ first times of visit to state j. Setting $F_m = \{[\![n_1, n_m]\!] : 1 \leq n_1 < \cdots < n_m < +\infty\}$, we get

$$\mathbb{P}_i(N_j \geq m + 1) = \sum_{(n_1, \dots, n_{m+1}) \in F_{m+1}} \mathbb{P}_i(X_{n_1} = j, \dots, X_{n_{m+1}} = j).$$

Since

$$\mathbb{P}_i(X_{n_1} = j, \dots, X_{n_{m+1}} = j) = \mathbb{P}_i(X_{n_1} = j, \dots, X_{n_m} = j)\mathbb{P}_j(X_{n_{m+1}-n_m} = j),$$

by the Markov property and the compound probabilities theorem, it follows that

$$\mathbb{P}_i(N_j \geq m + 1) = \sum_{(n_1, \dots, n_m) \in F_m} \mathbb{P}_i(X_{n_1} = j, \dots, X_{n_m} = j) \sum_{k \geq 1} \mathbb{P}_j(X_k = j)$$

$$= \mathbb{P}_i(N_j \geq m)\mathbb{P}_j(T_j < +\infty)$$

$$\overset{(1)}{=} \varrho_{ij} \varrho_{jj}^{m-1} \varrho_{jj}.$$

(1) by the induction hypothesis. ∎

Definition 3.25 Let (X_n) be a Markov chain with state space E and transition matrix P. Let $\alpha \in]0, 1]$. The function

$$U^\alpha = I + \alpha P + \alpha^2 P^2 + \cdots + \alpha^n P^n + \cdots$$

from $E \times E$ to $\overline{\mathbb{R}}$ is called the α-potential of (X_n), or P.

The function U^1 is referred to as the potential of the chain and is typically denoted by U.

Proposition 3.26 *We have*

$$U(i, j) = \begin{cases} \mathbb{E}_i N_j & \text{if } i \neq j, \\ 1 + \mathbb{E}_i N_i & \text{if } i = j, \end{cases} \tag{3.6}$$

and for $i \neq j$ we also have $U(i, j) = \varrho_{ij} U(j, j)$.

Proof By definition,

$$U(i, j) = \sum_{n \geq 0} P^n(i, j) = \sum_{n \geq 0} \mathbb{P}_i(X_n = j) = \sum_{n \geq 0} \mathbb{E}_i[\mathbb{1}_{(X_n = j)}],$$

from which (3.6) derives. Hence, for $i \neq j$,

$$U(i, j) = \mathbb{E}_i \left[\sum_{n \geq 1} \mathbb{1}_{(X_n = j)} \right] = \mathbb{E}_i \left[\sum_{n \geq T_j} \mathbb{1}_{(X_n = j)} \right]$$

$$\overset{(1)}{=} \mathbb{E}_i[\mathbb{1}_{(T_j < +\infty)}] \mathbb{E}_j \left[\sum_{n \geq T_j} \mathbb{1}_{(X_n = j)} \right] = \varrho_{ij} U(j, j).$$

(1) by Markov property.   □

Recurrence and transience are equivalent to the following simple conditions.

Theorem 3.27

A. *The three following propositions are equivalent:*
   1. *The state $i \in E$ is recurrent.*
   2. $\mathbb{P}_i(N_i = +\infty) = 1$.
   3. $\sum_{n \geq 1} P^n(i, i) = +\infty$.
B. *The three following propositions are equivalent:*
   1. *The state $i \in E$ is transient.*
   2. $\mathbb{P}_i(N_i = +\infty) = 0$.
   3. $\sum_{n \geq 1} P^n(i, i) < +\infty$.

If i is transient and $\mu = \delta_i$, then N_i is $\mathcal{G}(1 - \varrho_{ii})$ distributed, with respect to $\mathbb{P}_i$.

Proof Since $(N_i \geq m)$ is a decreasing sequence of events, we deduce from Lemma 3.24 that

$$\mathbb{P}_i(N_i = +\infty) = \lim_{m \to +\infty} \mathbb{P}_i(N_i \geq m) = \lim_{m \to +\infty} \varrho_{ii}^m$$

$$= \begin{cases} 0 \text{ if } \varrho_{ii} < 1, \text{ transient state}, \\ 1 \text{ if } \varrho_{ii} = 1, \text{ recurrent state}. \end{cases}$$

Moreover,

$$\mathbb{E}_i N_i = \sum_{n\geq 0} \mathbb{P}_i(N_i \geq n) = \sum_{n\geq 0} \varrho_{ii}^n$$

$$= \begin{cases} (1 - \varrho_{ii})^{-1} < +\infty & \text{if } \varrho_{ii} < 1, \text{ transient state} \\ +\infty & \text{if } \varrho_{ii} = 1, \text{ recurrent state,} \end{cases}$$

and A and B follow.

Suppose now that i is transient. Using Lemma 3.24, we get

$$\mathbb{P}_i(N_i = k) = \mathbb{P}_i(N_i \geq k) - \mathbb{P}_i(N_i \geq k + 1) = \varrho_{ii}^k - \varrho_{ii}^{k+1} = \varrho_{ii}^k(1 - \varrho_{ii}).$$

In other words, N_i is geometrically distributed with parameter $1 - \varrho_{ii}$. $\qquad\square$

Theorem 3.28 *Let $j \in E$ be a recurrent state. Let $i \neq j$ belong to E. Then, either $\varrho_{ji} = 0$ and then $\mathbb{P}_j(N_i = 0) = 1$, or $\varrho_{ji} = 1$ and then $\mathbb{P}_j(N_i = +\infty) = 1$ and i is recurrent too.*

Proof Suppose that $\varrho_{ji} = \mathbb{P}_j(T_i < +\infty) > 0$. Since j is recurrent, the chain visits j infinitely often and the number $N_j^n(i)$ of visits to i, between the $(n-1)$-th and n-th return to j, is well-defined for $n \geq 1$. Due to the strong Markov property, the sequence $(N_j^n(i))$ is i.i.d. with respect to the probability P_j.

Moreover $N_i = \sum_{n\geq 0} N_j^n(i)$, with $N_i^0(j) = 0$. Since $\mathbb{P}_j(T_i < +\infty) > 0$, we have $\mathbb{P}_j(N_i^1 \geq 1) > 0$ and hence $\mathbb{P}_j[N_j^n(i) \geq 1] > 0$. Using Borel-Cantelli Lemma, we obtain $\mathbb{P}_j[\overline{\lim}(N_j^n(i) \geq 1)] = 1$, so that the chain returns infinitely often to i from j, or $\mathbb{P}_j(N_i = +\infty) = 1$, and finally $\mathbb{P}_j(T_i < +\infty) = 1$.

In other words, $\mathbb{P}_j(T_i < +\infty)$ is equal either to zero or to one. If $\mathbb{P}_j(T_i < +\infty) = 0$, clearly, $\mathbb{P}_j(N_i = 0) = 1$.

It only remains to show that i is recurrent when $\mathbb{P}_j(T_i < +\infty) = 1$. Indeed, $\mathbb{P}_j(N_i = +\infty) = \mathbb{P}_j(T_i < +\infty, N_{Y,i} = +\infty)$, where $Y_n = X_{T_i+n}$. Using the strong Markov property, this can be written $1 = \mathbb{P}_j(T_i < +\infty) \mathbb{P}_i(N_i = +\infty)$, thus $\mathbb{P}_i(N_i = +\infty) = 1$. $\qquad\square$

In addition to the recurrent or transient nature of the states, the possibility to pass from one state to another in a finite number of steps is a fundamental property of Markov chains.

Definition 3.29 Let i and j belong to E. If $P^n(i, j) > 0$ for some n, then j is said to be attainable from i, or $i \rightarrow j$. If both $i \rightarrow j$ and $j \rightarrow i$, the states i and j are said to communicate, or $i \leftrightarrow j$.

Note that n cannot be null if $i \neq j$.

Proposition 3.30 *Distinct states i and j communicate if and only if they satisfy $\varrho_{ij}\varrho_{ji} > 0$.*

Proof We have

$$\mathbb{P}_i(T_j < +\infty) = \mathbb{P}_i\left[\bigcup_{n\geq 0}(T_j = n)\right]$$

$$= \sum_{n\geq 0}\mathbb{P}_i(T_j = n) \tag{3.7}$$

$$= \sum_{n\geq 0}\mathbb{P}_i(X_k \neq j,\ 1 \leq k \leq n - 1,\ X_n = j). \tag{3.8}$$

If $\varrho_{ij} > 0$, we deduce from (3.7) that some $n \in \mathbb{N}$ exists such that $\mathbb{P}_i(T_j = n) > 0$. But $\mathbb{P}_i(T_j = n) \leq \mathbb{P}_i(X_n = j)$ and $\mathbb{P}_i(X_n = j) = P^n(i, j)$, and hence $P^n(i, j) > 0$.

Conversely, if $P^n(i, j) = 0$ for all $n \in \mathbb{N}$, then

$$\mathbb{P}_i(X_k \neq j, 1 \leq k \leq n - 1, X_n = j) = 0.$$

We deduce from (3.8) that $\mathbb{P}_i(T_j < +\infty) = 0$; in other words, i and j do not communicate. $\square$

The relation of communication $\leftrightarrow$ is clearly an equivalence relation on E. It is reflexive, $i \leftrightarrow i$, since $P^0(i, i) = 1$ and symmetric, $i \leftrightarrow j \iff j \leftrightarrow i$, by symmetry of the definition of the relation $\leftrightarrow$. Finally it is transitive, $i \leftrightarrow j$ and $j \leftrightarrow k$ imply $i \leftrightarrow k$, because $P^{n+m}(i, k) \geq P^n(i, j)P^m(j, k) > 0$ and $P^{n'+m'}(k, i) \geq P^{n'}(k, j)P^{m'}(j, i) > 0$).

Therefore, E is the disjoint union of the equivalence classes of the relation. All the elements of a given class are communicating and communicate with no state out of the class. Recurrence and transience are class properties.

Proposition 3.31 *All the states of a given class of the relation of communication are of the same nature, either transient, or positive recurrent, or null recurrent.*

The class itself is correspondingly said to be either transient, or positive recurrent, or null recurrent.

Proof Let i and j be two elements of the same class. Some integers m and n exist such that $P^m(j, i) > 0$ and $P^n(i, j) > 0$. For all $r \in \mathbb{N}$,

$$P^{n+r+m}(i, i) \geq P^n(i, j)P^r(j, j)P^m(j, i),$$

thus

$$\sum_{r \geq 0} P^{n+r+m}(i, i) \geq P^n(i, j) P^m(j, i) \sum_{r \geq 0} P^r(j, j). \tag{3.9}$$

We have $\mathbb{E}_i N_i \geq \sum_{r \geq 0} P^{n+r+m}(i, i)$ and $\mathbb{E}_j N_j = \sum_{r \geq 0} P^r(j, j)$. If j is recurrent, then $\mathbb{E}_j N_j = +\infty$, and hence $\mathbb{E}_i N_i = +\infty$, meaning that i is recurrent too. If i is transient, then $\mathbb{E}_i N_i < +\infty$, and hence (3.9) yields $\mathbb{E}_j N_j < +\infty$, meaning that j is transient too.   □

A Markov chain with only recurrent (transient) states is said to be recurrent (transient). A Markov chain with only communicating states is said to be irreducible; it has exactly one communication class and is either recurrent or transient.

▷ *Example 3.32 (A Simple Random Walk—Continuation of Example 3.5)* By (3.1) p. 117 and Stirling's formula, we compute

$$P^{2n}(0, 0) \overset{\sim}{=} \frac{[4p(1-p)]^n}{\sqrt{\pi n}}.$$

Clearly $4p(1 - p) \leq 1$ for all $p \in [0, 1]$ with equality only if $p = 1/2$. From Theorem 3.27, the state 0 is recurrent if $\sum_{n \geq 0} P^n(0, 0)$ is finite—if p=1/2, and is transient otherwise—if $p \neq 1/2$.

The chain is irreducible, and hence is be either recurrent or transient, depending on the nature of 0.   ◁

Definition 3.33 A subset C is said to be closed if $\mathbb{P}_i(X_1 \in C) = 1$ for all $i \in C$.

A closed set is also said to be final since once the chain reaches such a set, it never exits. If a closed set contains only one state, this is necessarily an absorbing state.

Proposition 3.34 *All recurrent classes are closed.*

Proof Let C be a class that is not closed. Some state $i \in C$ exists leading to some $j \notin C$, that does not lead to i.

Hence some $m \geq 1$ exists such that $\mathbb{P}_i(X_m = j) > 0$. Since j does not lead to i and $\mathbb{P}_i(T_i < +\infty) < 1$, the probability to visit i infinitely often is less than 1: i is necessarily transient. Therefore, by Proposition 3.31, the class C is transient itself.   □

We state the next result without proof.

Theorem 3.35 *An irreducible Markov chain is positive recurrent if and only if the system of equations*

$$\sum_{j \in E} P(i, j)x_j = x_i, \quad i \in E, \tag{3.10}$$

has a solution $x = (x_i)_{i \in E}$ *that is absolutely summable—such that* $\sum_{i \in E} |x_i| < +\infty$—*and is not null. Moreover, this solution is unique up to a multiplicative constant, and all the* x_i *have the same sign.*

This system can be written in matrix form $Px = x$. If E is finite, then the column vector $x = \mathbf{1}$ is an eigenvector of the matrix P, it is absolutely summable and the chain is positive recurrent. Finite chains will be specifically studied below in Sect. 3.5.

▷ *Example 3.36 (A Simple Random Walk—Continuation of Example 3.5)* For $p = 1/2$ this Markov chain is irreducible and recurrent.

Solving the system (3.10) yields $x_i = 1$ for all $i \in \mathbb{Z}$. This solution is not summable, and hence the simple random walk is null recurrent.   ◁

3.3   Stationary Distribution and Asymptotic Behavior

Stationary distributions are an essential tool in the investigation of the asymptotic behavior of Markov chains.

Definition 3.37 A positive measure λ defined on E is said to be stationary or invariant for the Markov chain (X_n)—or for its transition function P—if

$$\sum_{i \in E} \lambda(i)P(i, j) = \lambda(j), \quad j \in E.$$

Moreover, if λ is a probability, it is referred to as a stationary distribution, typically denoted by π.

The above relation can be written $\lambda P = \lambda$, where $\boldsymbol{\lambda} = (\lambda(i))_{i \in E}$ is a line vector.

Clearly, any measure stationary for P is stationary for P^k, for any $k \in \mathbb{N}^*$. If the initial distribution of the chain is π, a stationary distribution for P, then X_n has the distribution π for any n: the sequence is stationary in terms of distribution.

▷ *Example 3.38 (A Binary Markov Chain)* Let (X_n) be a Markov chain with two states, with transition matrix

$$P = \begin{pmatrix} 1 - \alpha & \alpha \\ \beta & 1 - \beta \end{pmatrix},$$

where $0 < \alpha < 1$ and $0 < \beta < 1$. The stationary distribution of (X_n) is characterized by the relations

$$\begin{cases} (1 - \alpha)\pi(0) + \beta\pi(1) = \pi(0) \\ \pi(0) + \pi(1) = 1. \end{cases}$$

Therefore, $\pi(0) = \beta/(\alpha + \beta)$ and $\pi(1) = \alpha/(\alpha + \beta)$. ◁

▷ *Example 3.39 (A Stationary Distribution for the Chain of Example 3.13)* If $\pi = (\pi(i))$ is a stationary distribution for (X_n), then $\widetilde{\pi}$, defined by $\widetilde{\pi}(i, j) = \pi(i)P(i, j)$, for $(i, j) \in E^2$, is a stationary distribution for (X_{n-1}, X_n). Indeed, π satisfies

$$\sum_{k \in E} \pi(k)P(k, i)P(i, j) = \pi(i)P(i, j), \quad i, j \in E,$$

or $\sum_{(k,l) \in E^2} \widetilde{\pi}(k, l)\delta_{li}P(i, j) = \widetilde{\pi}(i, j)$; in other words $\widetilde{\pi}\widetilde{P} = \widetilde{\pi}$.
Moreover, $\sum_{(i,j) \in E^2} \widetilde{\pi}(i, j) = 1$. ◁

Definition 3.40 A measure λ on E is said to be reversible for the Markov chain (X_n)—or for its transition function P—if

$$\lambda(i)P(i, j) = \lambda(j)P(j, i), \quad i, j \in E.$$

A reversible measure is clearly stationary, because it satisfies

$$\sum_{i \in E} \lambda(i)P(i, j) = \sum_{i \in E} \lambda(j)P(j, i) = \lambda(j) \sum_{i \in E} P(j, i) = \lambda(j), \quad j \in E.$$

Moreover, it may be easier to determine, as shown in the next example.

▷ *Example 3.41 (Birth and Death Chain—Continuation of Example 3.6)* If π is a reversible distribution for this chain, then

$$\pi(i)p_i = \pi(i + 1)q_{i+1}, \quad i \geq 0. \tag{3.11}$$

Setting

$$\gamma_i = \frac{p_0 \cdots p_{i-1}}{q_1 \cdots q_i}, \quad i \geq 1, \quad \text{and} \quad \gamma_0 = 1,$$

Relation (3.11) yields $\pi(i) = \pi(0)\gamma_i$. Suppose that $\sum_{i \geq 0} \gamma_i < +\infty$. Summing both sides of the above equality on i yields $1/\pi(0) = \sum_{i \geq 0} \gamma_i$. Therefore

$$\pi(i) = \frac{\gamma_i}{\sum_{i \geq 0} \gamma_i}, \quad i \geq 0,$$

and π is the stationary distribution of the chain. $\triangleleft$

When n tends to infinity, a Markov chain visits only its recurrent states. In other words, $P^n(i, j)$ tends to 0 for all transient j.

Theorem 3.42 *If i and j are recurrent states, then,*

$$N_i^n/n \xrightarrow{\text{a.s.}} 1/m_i \quad \text{and} \quad N_{ij}^n/n \xrightarrow{\text{a.s.}} P(i, j)/m_i, \quad n \to +\infty,$$

where $m_i = \mathbb{E}_i T_i$ is the mean recurrence time to i, with the convention that $1/+\infty = 0$.

Proof Let T_i^n denote the time of the n-th visit to i, for $n \geq 1$. Let $u_n = T_i^n - T_i^{n-1}$ denote the time between two successive visits to i for $n \geq 1$, with $u_0 = 0$. We have $T_i^{N_i^n} \leq n < T_i^{N_i^n+1}$, or

$$\frac{T_i^{N_i^n}}{N_i^n} \leq \frac{n}{N_i^n} < \frac{T_i^{N_i^n+1}}{N_i^n}.$$

Due to the strong Markov property, (u_n) is an i.i.d. sequence. Since (N_i^n) tends to infinity a.s. when n tends to infinity, Theorem 1.93 induces that

$$\frac{T_i^{N_i^n}}{N_i^n} = \frac{u_1 + \cdots + u_{N_i^n}}{N_i^n} \xrightarrow{\text{a.s.}} \mathbb{E} u_1 = m_i.$$

In the same way,

$$\frac{T_i^{N_i^n+1}}{N_i^n} = \frac{u_1 + \cdots + u_{N_i^n+1}}{N_i^n} = \frac{u_1 + \cdots + u_{N_i^n+1}}{N_i^n + 1} \cdot \frac{N_i^n + 1}{N_i^n} \xrightarrow{\text{a.s.}} \mathbb{E} T_i = m_i,$$

and hence N_i^n/n converges a.s. to $1/m_i$.

The same argument applied to the two-dimensional Markov chain (X_{n-1}, X_n) yields that N_{ij}^n/n converges a.s. to $P(i, j)/m_i$; see Example 3.13. $\square$

Theorem 3.43 *If π is stationary for (X_n), then $\pi(i) = 1/m_i$ for any recurrent state i.*

Proof Since $N_i^n \le n$ for all i and all n, Theorem 3.42 and the dominated convergence theorem jointly yield for any recurrent j,

$$\frac{\mathbb{E}_j N_i^n}{n} \longrightarrow \frac{1}{m_i}, \quad n \to +\infty.$$

Since π is stationary,

$$\pi(i) = \sum_{j \in E} \pi(j) P^k(j, i), \quad k \ge 1,$$

and hence

$$n\pi(i) = \sum_{j \in E} \pi(j) \sum_{k=1}^{n} P^k(j, i) = \sum_{j \in E} \pi(j) \mathbb{E}_j N_i^n.$$

Therefore

$$\pi(i) = \sum_{j \in E} \pi(j) \frac{\mathbb{E}_j N_i^n}{n} \longrightarrow \sum_{j \in E} \pi(j) \frac{1}{m_i} = \frac{1}{m_i},$$

the searched result. $\qquad\square$

The expectation of the number of visits to any state before return to a fixed state, that is $\mathbb{E}_j N_i^{T_j}$ for all states i and j, yields a stationary measure for any recurrent chain and a stationary distribution for any irreducible positive recurrent chain.

Theorem 3.44 *Let (X_n) be a recurrent Markov chain with state space E. For any $j \in E$, the measure λ^j defined on E by $\lambda^j(i) = \mathbb{E}_j N_i^{T_j}$ is stationary for (X_n).*
If, moreover, (X_n) is irreducible, then $0 < \lambda^j(i) < +\infty$ for all $i \in E$.

Proof Let $j \in E$ and $i \in E$. We have

$$\lambda^j(i) = \mathbb{E}_j N_i^{T_j} = \sum_{n \ge 1} \mathbb{P}_j(X_n = i, n \le T_j).$$

We compute

$$\mathbb{P}_j(X_n = i, n \le T_j) = \sum_{k \in E} \mathbb{P}_j(X_n = i, X_{n-1} = k, n \le T_j)$$

$$= \sum_{k \in E} \mathbb{P}_j(X_{n-1} = k, n \le T_j) P(k, j),$$

and since $\lambda^j(i) = \mathbb{E}_j N_i^{T_j} = \mathbb{E}_j N_i^{T_j-1}$ for $i \neq j$ and $\lambda^j(j) = 1$, we get

$$\lambda^j(i) = \sum_{k \in E} P(k, j) \sum_{n \geq 1} \mathbb{P}_j(X_{n-1} = k, n \leq T_j)$$

$$= \sum_{k \in E} P(k, j) \mathbb{E}_j N_i^{T_j-1} = \sum_{k \in E} P(k, j) \lambda^j(k),$$

and λ^j is stationary for P.

If the chain is irreducible, for all i, there exist some $n > 0$ and $m > 0$ such that $P^n(i, j) > 0$ and $P^m(j, i) > 0$. The measure λ^j is stationary for P^n so $\lambda^j(j) \geq \lambda^j(i) P^n(i, j)$ and $\lambda^j(i)$ is finite.

Finally, for $m_0 = \inf\{m \in \mathbb{N}^* : P^m(j, i) > 0\}$, the probability of visiting i before returning to j is positive, that is to say $\lambda^j(i) > 0$.   □

The next two results are stated without proofs.

Proposition 3.45 *If λ is a stationary measure for an irreducible recurrent Markov chain (X_n), then*

$$\frac{\lambda(i)}{\lambda(j)} = \lambda^j(i) = \mathbb{E}_j N_i^{T_j}, \quad (i, j) \in E \times E.$$

Corollary 3.46 *If λ is a stationary measure for a recurrent Markov chain (X_n), then:*

either $\lambda(E) = +\infty$ and $m_i = +\infty$,

or $\lambda(E) < +\infty$ and (X_n) has a unique stationary distribution π, with $m_i = 1/\pi(i)$, for all $i \in E$.

Therefore, only three types of state spaces exist for irreducible chains: either all the states are transient, or all the states are null recurrent, or all the states are positive recurrent and then the chain has a unique stationary probability.

If E is not irreducible, then it can be divided into two disjoint subspaces, E_t containing transient states and E_r containing recurrent states. Moreover, the recurrent classes (C_k) of the chain constitute a partition of E_r, as shown in Fig. 3.4.

Fig. 3.4 State space decomposition

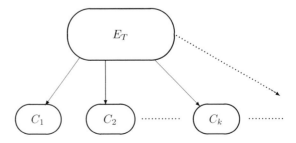

If the states of C_k for $k > 1$ are numbered from $|C_1| + \cdots + |C_{k-1}| + 1$ to $|C_1| + \cdots + |C_k|$, the transition matrix P of (X_n) can be written

$$P = \begin{pmatrix} B & Q_1 & Q_2 & Q_3 & \cdots \\ 0 & P_1 & 0 & 0 & \cdots \\ 0 & 0 & P_2 & 0 & \cdots \\ 0 & 0 & 0 & P_3 & \cdots \\ \vdots & \vdots & \vdots & \vdots & \ddots \end{pmatrix},$$

where P_k is the transition matrix of the restriction of (X_n) to C_k, B the matrix of the transitions from E_t to E_t and Q_k the matrix of the transitions from E_t to C_k.

Moreover, let π_k denotes the stationary distribution for the restriction of (X_n) to C_k and set

$$\Pi = (\, 0 \mid \alpha_1 \pi_1 \mid \alpha_2 \pi_2 \mid \ldots)$$

where the partition is associated with E_t, C_1, C_2, ... Then Π is a stationary distribution for (X_n) for all (α_k) such that $\alpha_k \geq 0$ and $\sum_{k \geq 1} \alpha_k = 1$, as illustrated in the next example.

▷ *Example 3.47* Let (X_n) be a Markov chain with transition matrix

$$P = \begin{pmatrix} 0.5 & 0.4 & 0.1 & 0 & 0 & 0 & 0 & 0 & 0 \\ 0.3 & 0.6 & 0.1 & 0 & 0 & 0 & 0 & 0 & 0 \\ 0 & 0 & 0.6 & 0.1 & 0.1 & 0.1 & 0.1 & 0 & 0 \\ 0 & 0 & 0 & 0.6 & 0.4 & 0 & 0 & 0 & 0 \\ 0 & 0 & 0 & 0.4 & 0.6 & 0 & 0 & 0 & 0 \\ 0 & 0 & 0 & 0 & 0 & 0.6 & 0.2 & 0 & 0.2 \\ 0 & 0 & 0 & 0 & 0 & 0.3 & 0.6 & 0.1 & 0 \\ 0 & 0 & 0 & 0 & 0 & 0 & 0 & 0.5 & 0.5 \\ 0 & 0 & 0 & 0 & 0 & 0.3 & 0 & 0 & 0.7 \end{pmatrix},$$

as shown in Fig. 3.5. Then $E = \{1, \ldots, 9\}$, with $E_t = \{1, 2, 3\}$, $C_1 = \{4, 5\}$ and $C_2 = \{6, 7, 8, 9\}$, from which we deduce that $\pi(1) = (1/2, 1/2)$ and $\pi(2) = (30/73, 15/73, 3/73, 25/73)$. One may check that

$$\Pi = \left(0, 0, 0, \frac{\alpha}{2}, \frac{\alpha}{2}, \frac{30}{73}(1-\alpha), \frac{15}{73}(1-\alpha), \frac{3}{73}(1-\alpha), \frac{25}{73}(1-\alpha) \right)$$

is indeed a stationary distribution for the chain for any $\alpha \in [0, 1]$.   ◁

Fig. 3.5 Graph of the chain
of Example 3.47

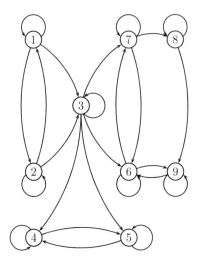

Definition 3.48 A positive recurrent aperiodic state is said to be ergodic.

An irreducible Markov chain with at least one ergodic state—and hence, with only ergodic states—is said to be ergodic.

Definition 3.49 Let (X_n) be a Markov chain with transition matrix P. Let i belong to E. If

$$P^n(i, j) \longrightarrow \pi^i(j), \quad n \to +\infty, \ j \in E,$$

for some probability π^i on E, this probability is called a limit distribution of the chain.

Note that a Markov chain can have several stationary distributions and several limit distributions, as shown in Example 3.57 below. On the contrary, an ergodic chain has a unique limit distribution, equal to its stationary distribution.

Theorem 3.50 (Ergodic) *If (X_n) is an ergodic Markov chain with transition function P and state space E, then its stationary distribution is the unique limit distribution of the chain, and*

$$P^n(i, j) \longrightarrow \pi(j), \quad n \to +\infty, \ i, j \in E.$$

Proof Let μ denote the initial distribution of (X_n). Let (Y_n) be an ergodic Markov chain, independent of (X_n), with the same state space E and transition function P—and hence the same stationary distribution π, but with initial distribution ν.

The two-dimensional sequence $(Z_n) = (X_n, Y_n)$ is also an ergodic Markov chain. Clearly, its initial distribution is $\mu \otimes \nu$, its transition function Q is given by $Q((i, j), (i', j')) = P(i, i')P(j, j')$ and its stationary distribution is $\pi \otimes \pi$.

Let us set $T = \inf\{n \geq 0 : X_n = Y_n\} = \inf\{n \geq 0 : Z_n = (j, j), j \in E\}$ and $\overline{P} = P \otimes P$. We have $\overline{\mathbb{P}}_{\mu \otimes \nu}(X_n = j, T \leq n) = \overline{\mathbb{P}}_{\mu \otimes \nu}(Y_n = j, T \leq n)$, thus

$$\mathbb{P}_\mu(X_n = j) = \overline{\mathbb{P}}_{\mu \otimes \nu}(X_n = j, T \leq n) + \overline{\mathbb{P}}_{\mu \otimes \nu}(X_n = j, T > n)$$

$$\leq \mathbb{P}_\nu(Y_n = j) + \overline{\mathbb{P}}_{\mu \otimes \nu}(T > n).$$

In the same way,

$$\mathbb{P}_\nu(Y_n = j) \leq \mathbb{P}_\mu(X_n = j) + \overline{\mathbb{P}}_{\mu \otimes \nu}(T > n).$$

The two above inequalities jointly imply that

$$|\mathbb{P}_\mu(X_n = j) - \mathbb{P}_\nu(Y_n = j)| \leq \overline{\mathbb{P}}_{\mu \otimes \nu}(T > n).$$

Since $\overline{\mathbb{P}}_{\mu \otimes \nu}(T > n)$ converges to 0 when n tends to infinity, we get for $\mu = \delta_i$ and $\nu = \pi$

$$|\mathbb{P}_i(X_n = j) - \pi(j)| \longrightarrow 0, \quad n \to +\infty,$$

and the result follows. □

The ergodic theorem can also be stated in the following way.

Theorem 3.51 *If (X_n) is an ergodic Markov chain, with stationary distribution π, and if $g : E \to \mathbb{R}$ is such that $\sum_{i \in E} \pi(i)|g(i)|$ is finite, then*

$$\frac{1}{n} \sum_{k=1}^{n} g(X_k) \xrightarrow{a.s.} \sum_{i \in E} \pi(i)g(i), \quad n \to +\infty.$$

Finally, the following result states that the Shannon entropy rate of an ergodic Markov chain is the sum of the entropies of the transition probabilities $(P(i, j))_{j \in E}$ weighted by the probability of occurrence of each state according to the stationary distribution of the chain.

Proposition 3.52 *The entropy rate of an ergodic Markov chain $X = (X_n)$ with transition matrix P and stationary distribution π is*

$$\mathbb{H}(X) = -\sum_{i \in E} \pi(i) \sum_{j \in E} P(i, j) \log P(i, j),$$

if this quantity is finite.

Proof Using (3.2) p. 119, the marginal distribution of $(X_0, \ldots, X_n)$ is

$$f_n^X(x_0, \ldots, x_n) = \mu(x_0) P(x_0, x_1) \ldots P(x_{n-1}, x_n),$$

so

$$-\frac{1}{n} \log f_n^X(X_0, \ldots, X_n) = -\frac{1}{n} \log \mu(X_0) - \frac{1}{n} \sum_{k=1}^{n} \log P(X_{k-1}, X_k)$$

$$= \quad (i) \quad + \quad (ii).$$

When n tends to infinity, (i) tends to 0. Since the two-dimensional chain (X_{n-1}, X_n) is ergodic, with stationary distribution $(\pi(i) P(i, j))$—see Example 3.39, we deduce from Theorem 3.51 that

$$(ii) \xrightarrow{a.s.} -\sum_{i \in E} \pi(i) \sum_{j \in E} P(i, j) \log P(i, j).$$

The expression of the entropy rate follows from Definition 1.59.   □

▷ *Example 3.53 (Entropy Rate of a Two-State Chain)* The entropy rate of the chain of Example 3.38 is

$$\mathbb{H}(\mathbf{X}) = \frac{q}{p+q}[-p \log p - (1-p) \log(1-p)]$$

$$+ \frac{p}{p+q}[-q \log q - (1-q) \log(1-q)].$$

This entropy rate is shown on Fig. 3.6 as a function of p and q.   ◁

3.4   Periodic Markov Chains

We will assume throughout this section that the Markov chain is irreducible. We will denote by d_i the period of state $i \in E$; see Definition 3.9.

Theorem 3.54 *If i and j are communicating states, then $d_i = d_j$.*

Proof By definition, $P^{d_i}(i, i) > 0$. Since $i \leftrightarrow j$, two integers $n > 0$ and $m > 0$ exist such that $P^n(i, j) > 0$ and $P^m(j, i) > 0$. Hence

$$P^{m+d_i+n}(j, j) \geq P^m(j, i) P^{d_i}(i, i) P^n(i, j) > 0.$$

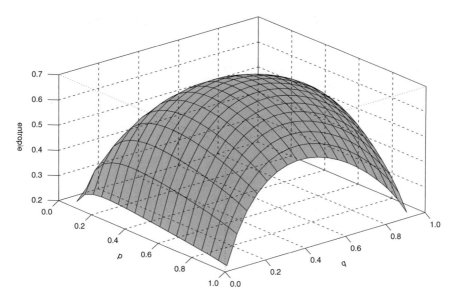

Fig. 3.6 Entropy rate of a two-state chain

In the same way, $P^{n+m+2d_i}(j, j) > 0$ so that d_j is a divisor of $n + m + 2d_i - (n + m + d_i) = d_i$. Therefore, $d_j \leq d_i$. By a symmetric argument, $d_i \leq d_j$, and the conclusion follows.   □

Thus, periodicity is a class property; in other words, if i belongs to a class C and is periodic with period d_i, then all other states $j \in C$ are periodic, with the same period $d_j = d_i = d$. Therefore, if an irreducible Markov chain has one periodic state with period d, then all its states are periodic with period d; the chain is then said to be d-periodic.

The state space E of a d-periodic chain can be divided into d disjoint subsets, say $E_0, E_1, \ldots, E_{d-1}$, in such a way that each transition leads from a state of E_p to a state of E_{p+1}, for $p = 0, 1, \ldots, d - 1$ (with $E_d \equiv E_0$). These sets are referred to as the cyclic classes of the chain. Using these classes, investigating a d-periodic Markov chain (X_n) reduces to investigating an aperiodic chain, as shown in the following proposition. It amounts to observing (X_n) at times $n = md$ for $m \in \mathbb{N}$.

Proposition 3.55 *Let (X_n) be a d-periodic Markov chain with transition function P and cyclic classes $E_0, \ldots, E_{d-1}$.*

1. The sequence (Y_m) of random variables defined by

$$Y_m = X_{md}, \quad m \in \mathbb{N},$$

is an aperiodic Markov chain with transition function $Q = P^d$.

2. *The classes $E_0, \ldots, E_{d-1}$ are closed sets for Q. Moreover, they are irreducible for Q—or (Y_m).*

Proof

1. We have

$$\mathbb{P}(Y_{m+1} = j \mid Y_0 = i_0, Y_1 = i_1, \ldots, Y_{m-1} = i_{m-1}, Y_m = i) =$$
$$= \mathbb{P}(X_{(m+1)d} = j \mid X_0 = i_0, X_d = i_1, \ldots, X_{(m-1)d} = i_{m-1}, X_{md} = i)$$
$$= \mathbb{P}(X_{(m+1)d} = j \mid X_{md} = i) = \mathbb{P}(Y_{m+1} = j \mid Y_m = i),$$

and hence (Y_m) is a Markov chain. Since

$$\mathbb{P}(X_{(m+1)d} = j \mid X_{md} = i) = P^d(i, j),$$

its transition function is Q. For any state i,

$$\text{g.c.d.}\{m \in \mathbb{N} \,:\, Q^m(i, i) > 0\} = \text{g.c.d.}\{m \in \mathbb{N} \,:\, P^{md}(i, i) > 0\}$$
$$= \frac{1}{d}\,\text{g.c.d.}\{n \in \mathbb{N} \,:\, P^n(i, i) > 0\} = 1,$$

so (Y_m) is aperiodic.
2. If $i \in E_p$ and $Q(i, j) \neq 0$, then, by definition of the cyclic classes, $j \in E_{p+md} \equiv E_p$, so E_p is closed.

Let both i and j belong to E_p. Since (X_n) is irreducible, some $n \in \mathbb{N}^*$ exists such that $P^n(i, j) > 0$; since i and j belong to E_p, necessarily $n = md$, with $m > 0$. Thus, $P^{md}(i, j) > 0$, or $Q^m(i, j) > 0$, and hence $i \to j$. In a symmetric way, $j \to i$. Therefore E_p is irreducible. □

Theorem 3.56 *Let (X_n) be a positive recurrent irreducible d-periodic Markov chain, with stationary distribution π. Let $E_0, E_1, \ldots, E_{d-1}$ denote its cyclic classes. For all $i \in E_p$ and $j \in E_q$, with $0 \le p, q \le d - 1$, set*

$$r = r_{ij} = \begin{cases} q - p & \text{if } q \ge p, \\ d - p + q & \text{if } q < p. \end{cases}$$

Then

$$P^{nd+r}(i, j) \longrightarrow d\pi(j), \quad n \to +\infty.$$

Proof We have

$$P^{nd+r}(i, j) = \sum_{k \in E_q} P^r(i, k) P^{nd}(k, j), \quad i \in E_p, \ j \in E_q.$$

Since $\sum_{k \in E_q} P^r(i, k) = 1$ for all $i \in E_p$ by definition of r, it is sufficient to prove that

$$P^{nd}(k, j) = Q^n(k, j) \longrightarrow d\pi(j), \quad k, j \in E_q, \quad n \to +\infty, \tag{3.12}$$

and then applying the dominated convergence theorem will prove the result.

Every transition is from a state of E_q to a state of E_{q+1}. Therefore $\sum_{l \in E_{q+1}} P(k, l) = 1$ for all $k \in E_q$, and thus, by Fubini's theorem,

$$\sum_{l \in E_{q+1}} \sum_{k \in E_q} \pi(k) P(k, l) = \sum_{k \in E_q} \pi(k).$$

Since π is a stationary distribution of (X_n),

$$\sum_{l \in E_{q+1}} \sum_{k \in E_q} \pi(k) P(k, l) = \sum_{l \in E_{q+1}} \pi(l).$$

Both above relations imply that

$$\sum_{k \in E_q} \pi(k) = c, \quad 0 \le q \le d - 1,$$

and thus $c = 1/d$, because the chain has d cyclic classes.

The distribution π is stationary for (X_n), hence for Q; moreover, $Q(k, h) = 0$ if $k \in E_q$ and $h \notin E_q$. Therefore

$$\sum_{k \in E_q} \pi(k) Q(k, h) = \pi(h), \quad h \in E_q,$$

and $(d\pi(k))_{k \in E_q}$ is a stationary distribution for the restriction Q_p of Q to $E_p \times E_p$. This transition matrix Q_p is aperiodic and, thanks to Proposition 3.55, irreducible. Using the ergodic theorem yields (3.12), and the result follows. □

▷ *Example 3.57 (An Ehrenfest Chain with Four States)* Let us consider the Ehrenfest chain with four states—say $E = \{0, 1, 2, 3\}$, and transition matrix

$$P = \begin{pmatrix} 0 & 1 & 0 & 0 \\ 1/3 & 0 & 2/3 & 0 \\ 0 & 2/3 & 0 & 1/3 \\ 0 & 0 & 1 & 0 \end{pmatrix}.$$

This chain is 2-periodic, because

$$P^2 = \begin{pmatrix} 1/3 & 0 & 2/3 & 0 \\ 0 & 7/9 & 0 & 2/9 \\ 2/9 & 0 & 7/9 & 0 \\ 0 & 2/3 & 0 & 1/3 \end{pmatrix}.$$

Its cyclic classes are $E_0 = \{0, 2\}$ and $E_1 = \{1, 3\}$. Solving $\pi P = \pi$, we determine its stationary distribution, that is $\pi = (1/8, 3/8, 3/8, 1/8)$. The transition matrix of the chain (Y_m), defined by $Y_m = X_{2m}$ for $m \geq 0$, is $Q = P^2$, so (Y_m) is an aperiodic chain but is not irreducible. Restricted either to E_0 or to E_1, the chain (Y_n) is ergodic, with respective transition matrices

$$Q_0 = \begin{pmatrix} 1/3 & 2/3 \\ 2/9 & 7/9 \end{pmatrix} \quad \text{and} \quad Q_1 = \begin{pmatrix} 7/9 & 2/9 \\ 2/3 & 1/3 \end{pmatrix},$$

and stationary distributions $(\pi'_0, \pi'_2) = (1/4, 3/4)$ and $(\pi'_1, \pi'_3) = (3/4, 1/4)$. Normalizing the vector $(1/4, 3/4, 3/4, 1/4)$ gives the stationary distribution of (X_n), precisely $(1/8, 3/8, 3/8, 1/8)$.

Finally, by Theorem 3.56, for $r = 0$,

$$P^{2n} \longrightarrow \begin{pmatrix} 1/4 & 0 & 3/4 & 0 \\ 0 & 3/4 & 0 & 1/4 \\ 1/4 & 0 & 3/4 & 0 \\ 0 & 3/4 & 0 & 1/4 \end{pmatrix}, \quad n \to +\infty,$$

and for $r = 1$,

$$P^{2n+1} \longrightarrow \begin{pmatrix} 0 & 3/4 & 0 & 1/4 \\ 1/4 & 0 & 3/4 & 0 \\ 0 & 3/4 & 0 & 1/4 \\ 1/4 & 0 & 3/4 & 0 \end{pmatrix}, \quad n \to +\infty.$$

The graphs of (X_n) and of (Y_m) are shown in Fig. 3.7.

◁

Fig. 3.7 Graphs of the chains of Example 3.57, (X_n) top and (Y_n) bottom

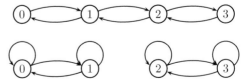

3.5   Finite Markov Chains

Thanks to their specific properties, finite chains are easier to study than the general ones. Note that Markov chains encountered in reliability theory are generally finite, the two-state case being especially pertinent for modeling reliability processes.

3.5.1   Specific Properties

A Markov chain (X_n) whose state space E is finite is said to be finite. Its transition function is represented by a finite matrix P. Due to the properties of the transition probabilities, its entries are nonnegative and the sum of each row is equal to one. Such a matrix is said to be stochastic.

A stochastic matrix having a power without null entries is said to be regular. An aperiodic finite Markov chain can be proven to be irreducible if and only if its transition matrix is regular.

Theorem 3.58 *If its state space E is finite, then the chain has at least one recurrent state.*

Proof Let $j \in E_t$ and $i \in E$. Since $\mathbb{E}_i N_j = \sum_{n \geq 1} P^n(i, j)$ is finite, $P^n(i, j)$ tends to zero when n tends to infinity. If E contained only transient states, then $\sum_{j \in E} P^n(i, j)$ would tend to 0 too; the matrix P is stochastic, so $\sum_{j \in E} P^n(i, j) = \sum_{j \in E} \mathbb{P}_i(X_n = j) = 1$, which yields a contradiction.   □

Therefore any irreducible finite chain is necessarily recurrent. Since its transition matrix is stochastic, $\mathbf{1} = (1, \dots, 1)'$ is one of the eigenvectors of the matrix, associated with the eigenvalue 1. Thus $\mathbf{1}$ is solution of the system (3.10) p. 131 and is absolutely summable; Theorem 3.35 implies that any irreducible finite chain is positive recurrent, so has a unique stationary distribution—by Corollary 3.46.

Proposition 3.59 *For a finite Markov chain, $\mathbb{P}_i(X_n \in E_r)$ tends to 1 when n tends to infinity, for $i \in E$.*

In other words, the chain will almost surely reach the set of recurrent states and will remain there forever.

Proof Since E is finite, if $\mathbb{P}(X_0 = i) = 1$ for a transient i, then there exists $n_i \geq 1$ such that $\mathbb{P}_i(X_{n_i} \in E_r) > 0$. Hence, $p = \mathbb{P}(X_M \in E_r \mid X_0 \in E_t) > 0$, where $M = \sup_{i \in E_t} n_i$.

Therefore, $\mathbb{P}(X_M \in E_t \mid X_0 \in E_t) = 1 - p$, and by the strong Markov property,

$$\mathbb{P}(X_{kM} \in E_t \mid X_0 \in E_t) = (1 - p)^k \longrightarrow 0, \quad k \to +\infty,$$

which yields the result.   □

Fig. 3.8 Graph of the chain
of Example 3.60

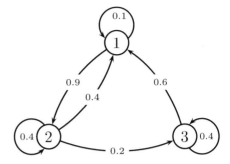

Note that the proof shows that the result holds true for any chain, as soon as E_t is finite.

▷ *Example 3.60* Let (X_n) be a Markov chain with state space $E = \{1, 2, 3\}$ and transition matrix

$$P = \begin{pmatrix} 0.1 & 0.9 & 0 \\ 0.4 & 0.4 & 0.2 \\ 0.6 & 0 & 0.4 \end{pmatrix}.$$

Its graph is given in Fig. 3.8. All the states are communicating, the chain is irreducible and finite, so is recurrent thanks to Theorem 3.58. Its limit distribution satisfies $\pi P = \pi$, that is

$$\begin{cases} 0.1\pi(1) + 0.4\pi(2) + 0.6\pi(3) = \pi(1) \\ 0.9\pi(1) + 0.4\pi(2) = \pi(2) \\ 0.2\pi(2) + 0.4\pi(3) = \pi(3). \end{cases}$$

with $\pi(1) + \pi(2) + \pi(3) = 1$. Thus, $(\pi(1), \pi(2), \pi(3)) = (1/6, 1/2, 1/3)$.   ◁

Proposition 3.61 *A distribution π is stationary for a finite chain (X_n) if and only if π is a limit distribution of (X_n).*

Proof The direct sense holds true for any ergodic discrete chain thanks to the ergodic theorem.

Conversely, let π be a limit distribution of the chain. We have

$$\pi(j) = \lim_{n \to +\infty} P^n(i, j) = \lim_{n \to +\infty} \sum_{k \in E} P^{n-1}(i, k) P(k, j)$$

$$= \sum_{k \in E} \lim_{n \to +\infty} P^{n-1}(i, k) P(k, j) = \sum_{k \in E} \pi(k) P(k, j).$$

Since E is finite, inverting sum and limit is possible.   □

▷ *Example 3.62* Let (X_n) be a finite Markov chain with state space E. Suppose its transition matrix P is doubly stochastic, meaning that both its columns and rows sum to one. Clearly, the uniform distribution on E is a stationary—and limit—distribution of the chain.   ◁

The rate of convergence of a transition matrix to its limit is exponential.

Proposition 3.63 *If (X_n) is a finite aperiodic irreducible Markov chain with transition matrix P and stationary distribution π, then P^n tends with exponential rate to the matrix Π, where $\Pi(i, j) = \pi(j)$, for $(i, j) \in E \times E$.*

Proof Since (X_n) is irreducible and aperiodic, P is regular and some $r \in \mathbb{N}$ exists such that $\varrho = \min_{(i,j) \in E \times E} P^r(i, j) > 0$.

Let (Y_n) be an ergodic Markov chain independent of (X_n), with the same state space, say $E = \{1, \ldots, N\}$ and the same transition matrix P. Let us suppose that π is the initial distribution of (Y_n), so that it is the distribution of Y_n for all n. Set $s = [n/r]$ for $n \geq r$. We have

$$\mathbb{P}_{(k,j)}(X_i \neq Y_i, i \leq n) \leq \mathbb{P}_{(k,j)}(X_i \neq Y_i, \ i = r, 2r, \ldots, sr) \leq$$

$$\leq \mathbb{P}_{(k,j)}(X_r \neq Y_r)\mathbb{P}_{(k,j)}(X_{2r} \neq Y_{2r} \mid X_r \neq Y_r) \cdots \qquad (3.13)$$

$$\cdots \mathbb{P}_{(k,j)}(X_{sr} \neq Y_{sr} \mid X_{tr} \neq Y_{tr}, \ t = 1, \ldots, s - 1).$$

We compute

$$\mathbb{P}(X_r = Y_r \mid X_0 = k, Y_0 = j) = \sum_{i \in E} \mathbb{P}_{(k,j)}[(X_r, Y_r) = (i, i)]$$

$$= \sum_{i=1}^{N} P^r(k, j) P^r(i, i) \geq N \varrho^2 > 0,$$

and hence $\mathbb{P}_{(k,j)}(X_r \neq Y_r) = 1 - \mathbb{P}_{(k,j)}(X_r = Y_r) \leq 1 - N\varrho^2$. Since the same is true for all the other terms of (3.13), we get

$$\mathbb{P}_{(k,j)}(X_i \neq Y_i, i \leq n) \leq (1 - N\varrho^2)^s.$$

Moreover, $\mathbb{P}_{(k,j)}(X_i \neq Y_i, i \leq n) = \mathbb{P}_{(k,j)}(T_{(i,i)} > n)$, where $T_{(i,i)}$ denotes the hitting time of state (i, i) of the two-dimensional chain (X_n, Y_n). Since π is the distribution of Y_n for all n, the result follows.   □

▷ *Example 3.64* Let (X_n) be a Markov chain with state space $E = \{1 \ldots, N\}$ and transition matrix $P = (1 - \theta)I + \theta \Pi$, where $\theta \in [0, 1]$, I is the $N \times N$-identity matrix and $\Pi(i, j) = \pi(j)$ for all i, for a given probability π on E.

Let us prove that π is the stationary distribution of (X_n). It is sufficient to prove that π is the limit distribution of the chain. Since $\Pi^2 = \Pi$, we get

$$P^n = (1 - \theta)^n I + [1 - (1 - \theta)^n]\Pi.$$

Hence P^n tends to Π with geometric rate, equal to $(1 - \theta)^n$.   ◁

If the chain (X_n) has both transient and recurrent states, then its transition matrix P can be written as

$$P = \begin{matrix} & E_t \ E_r \\ \begin{pmatrix} Q & B \\ 0 & A \end{pmatrix} & \begin{matrix} E_t \\ E_r \end{matrix} \end{matrix},$$

where $E_r \subset E$ is the set of recurrent states, and $E_t \subset E$ the set transient states. The potential matrix U of Q,

$$U = I + Q + Q^2 + \cdots + Q^n + \cdots \tag{3.14}$$

with coefficients in $\overline{\mathbb{R}}_+$, is referred to as the fundamental matrix of (X_n).

Theorem 3.65 *Let (X_n) be a reducible finite Markov chain. Let $T = \inf\{n \in \mathbb{N} : X_n \in E_r\}$ be the hitting time of E_r, and let Q be the restriction of P to $E_t \times E_t$. Then:*

1. *Q^n tends to the null matrix when n tends to infinity;*
2. *$I - Q$ is non singular and $U = (I - Q)^{-1}$;*
3. *$\mathbb{E}_i T = e_i (I - Q)^{-1} \mathbb{1}_{|E_t|}$ for any $i \in E_t$, where $e_i = (0, \ldots, 0, 1, 0, \ldots, 0)$ is the i-th line vector of the canonical basis of $\mathbb{R}^{|E_t|}$.*

Proof

1. Set $h_n^i = \sum_{j \in E_t} Q^n(i, j)$ for $n \in \mathbb{N}$ and $i \in E$. Then, for all $i \in E_t$, some $n \in \mathbb{N}^*$ exists such that $h_n^i > 0$. Since

$$h_{n+1}^i = \sum_{j \in E_t} \sum_{\ell \in E_t} Q^n(i, \ell) Q(\ell, j) \le \sum_{\ell \in E_t} Q^n(i, \ell) = h_n^i,$$

the sequence (h_n^i) is decreasing in n. Therefore, some n_i and C_i exist such that $h_n^i < C_i < 1$, for all $n > n_i$. Thus, since E_t is finite, some C independent from i exists such that $h_n^i < C < 1$.

Hence, for a given n,

$$h^i_{mn+n} = \sum_{j \in E_t} \sum_{\ell \in E_t} Q^{mn}(i, \ell) Q^n(\ell, j) \leq C h^i_{mn} \leq C^{m+1}.$$

Moreover, C^{m+1} tends to zero when m tends to infinity. The sequence (h^i_n) is decreasing and has a subsequence $(h^i_{mn})_{m \in \mathbb{N}}$ tending to zero, hence tends to zero too. Since

$$Q^n(i, k) = \sum_{j \in E_t} Q^{n-1}(i, j) Q(j, k) \leq h^i_{n-1},$$

Q^n tends to the null matrix.

2. Since Q^n tends to the null matrix, for n large enough, all the eigenvalues of $I - Q^n$ are different from zero, so this matrix is non singular. We deduce from

$$I - Q^n = (I - Q)(I + Q + \cdots + Q^{n-1})$$

that

$$0 \neq \det(I - Q^n) = \det(I - Q) \det(I + Q + \cdots + Q^{n-1}),$$

and hence

$$I + Q + \cdots + Q^{n-1} = (I - Q)^{-1}(I - Q^n).$$

We get $U = (I - Q)^{-1}$ by letting n go to infinity in the above equality and using Point 1.

3. For all $i \in E_t$, we have

$$\mathbb{E}_i T = \sum_{j \in E} \mathbb{E}_i [T \mathbb{1}_{(X_1 = j)}] = \sum_{j \in E} \mathbb{E}_i(T \mid X_1 = j) P(i, j).$$

We compute for recurrent states

$$\sum_{j \in E_r} \mathbb{E}_i(T \mid X_1 = j) P(i, j) = \sum_{j \in E_r} P(i, j),$$

and for transient states

$$\sum_{j \in E_t} \mathbb{E}_i(T \mid X_1 = j) P(i, j) = \sum_{j \in E_t} \mathbb{E}_j(T + 1) P(i, j)$$

$$= \sum_{j \in E_t} \mathbb{E}_j T P(i, j) + \sum_{j \in E_t} P(i, j).$$

Thus, $\mathbb{E}_i T = 1 + \sum_{j \in E_t} \mathbb{E}_j T P(i, j)$, or, in matrix form, $L = \mathbf{1}_{|E_t|} + QL$, where $L = (\mathbb{E}_1 T, \ldots, \mathbb{E}_{|E_t|} T)'$, that is $(I_{|E_t|} - Q)L = \mathbf{1}_{|E_t|}$. Since the matrix $I_{|E_t|} - Q$ is non singular, the result follows. □

A fundamental matrix can also be defined for any ergodic finite Markov chain (X_n), with transition matrix P and stationary distribution π. Let Π be the matrix with all lines equal to π, that is $\Pi(i, j) = \pi(j)$. The matrix

$$Z = [I - (P - \Pi)]^{-1}$$

is referred to as the fundamental matrix of the ergodic chain (X_n).

Theorem 3.66 *The fundamental matrix of an ergodic finite chain satisfies*

$$Z = I + \sum_{n \geq 0} (P^n - \Pi).$$

Proof Since P^n tends to Π, letting n tend to infinity in $P^{2n} P = P P^{2n} = P^{n+1} P^n = P^{2n+1}$ yields that $\Pi P = P\Pi = \Pi^2 = \Pi$, so that

$$(P - \Pi)^n = \sum_{k=0}^{n} \binom{n}{k} (-1)^{n-k} P^k \Pi^{n-k} = P^n + \sum_{k=0}^{n-1} \binom{n}{k} (-1)^{n-k} \Pi = P^n - \Pi.$$

Therefore $(P - \Pi)^n$ tends to 0, and the result follows. □

Since $P\mathbf{1}_N = \mathbf{1}_N$, the value 1 is always an eigenvalue of P. If the chain is regular, it can be proven to be the unique eigenvalue of the matrix with modulus one.

Proposition 3.67 *The modulus of the eigenvalues of the transition matrix of a Markov chain are less than or equal to 1.*

Proof Set $E = [\![1, N]\!]$ and consider the norm $\|a\| = \max_{1 \leq i \leq N} |a_i|$ on $\mathbb{C}^N$. For all $a \in \mathbb{C}^N$,

$$\|Pa\| = \max_{1 \leq i \leq N} \left| \sum_{j=1}^{N} P(i, j) a_j \right| \leq \left[\max_{1 \leq i \leq N} \sum_{j=1}^{N} P(i, j) \right] \left[\max_{1 \leq i \leq N} |a_i| \right],$$

and $\sum_{j=1}^{N} P(i, j) = 1$ for all i, so that $\|Pa\| \leq \|a\|$.
For $Pa = \lambda a$, we get $|\lambda| \leq 1$. □

Assume that P has N distinct eigenvalues $1 = \lambda_1 > \lambda_2 > \cdots > \lambda_N$. Let $x_1, \ldots, x_N$ denote the right eigenvectors and $y_1, \ldots, y_N$ the left eigenvectors associated with the eigenvalues $\lambda_1, \ldots, \lambda_N$ of P. Set $Q = (x_1, \ldots, x_N)$ and $D = \text{diag}(\lambda_i)$. We have $P = QDQ^{-1}$ and therefore $P^n = QD^nQ^{-1}$. This implies that

$$P^n = \sum_{k=1}^{N} \lambda_k^n A_k = \sum_{k=1}^{N} \frac{\lambda_k^n}{< x_k, y_k >} x_k . y_k', \tag{3.15}$$

where $< x_k, y_k >$ denotes the scalar product of x_k and y_k, and $x_k . y_k'$ their matrix product.

If $|\lambda_k| < 1$, then λ_k^n tends to 0 when n tends to infinity. Since this holds true for $k \geq 2$ if the chain is ergodic, we obtain by the spectral representation (3.15) that P^n tends to A_1—with geometric rate λ_2^n, so $A_1 = \Pi$.

3.5.2 Application to Reliability

The goal is here to investigate the stochastic behavior of Markovian systems with failures—break-down or illness—by observing them along time. The term system is very large and includes technological or economic systems, or living organisms.

▷ *Example 3.68 (Some Typical Systems Studied in Reliability Theory)*

1. the functioning time of a lamp with one or several bulbs;
2. an individual, for whom failure may mean illness or death;
3. careers of individuals, for whom failure may mean unemployment;
4. fatigue models: materials (pieces of chain, wire or cable) subject to stress, up to break. ◁

In order to make the investigation easier, we suppose the system starts operating at the time $t = 0$. Reliability is here investigated for a Markovian system observed at integer times $n \in \mathbb{N}$. The associated techniques are naturally linked to Markov chain theory, while for investigating the case of continuous-time observations ($t \in \mathbb{R}_+$) the theory of continuous-time Markov processes presented below in Chap. 4 will be necessary.

Let T be the random variable equal to the lifetime of the system, taking values in $\mathbb{N}$. The failure rate is defined by

$$h(n) = \mathbb{P}(T = n \mid T \geq n),$$

where, necessarily, $0 \leq h(n) < 1$ for $n \in \mathbb{N}$ and the series with general term $h(n)$ is divergent.

The distribution of T is given by

$$f(n) = \mathbb{P}(T = n) = [1 - h(0)][1 - h(1)] \ldots [1 - h(n-1)]h(n).$$

The reliability of the system is defined by

$$R(n) = \mathbb{P}(T > n) = [1 - h(0)][1 - h(1)] \ldots [1 - h(n)].$$

Therefore,

$$h(n) = \frac{f(n)}{R(n-1)}, \quad n > 0.$$

▷ *Example 3.69 (A Binary Unit)* Suppose a binary component starts operating at time $n = 0$. Let its lifetime be geometrically distributed, with parameter $p \in]0, 1[$. At each failure, the component is replaced by an identical new one. The switching time (time for replacement) is a geometrically distributed random variable too, with parameter $q \in]0, 1[$.

Let X_n be the random variable equal to 0 if the component is functioning at time n and equal to 1 otherwise. The sequence (X_n) is a Markov chain with state space $E = \{0, 1\}$ and transition matrix

$$P = \begin{pmatrix} 1 - p & p \\ q & 1 - q \end{pmatrix}.$$

We compute

$$P^n = \begin{pmatrix} 1 - p & p \\ q & 1 - q \end{pmatrix}^n = \frac{1}{p + q} \begin{pmatrix} q & p \\ q & p \end{pmatrix} + \frac{(1 - p - q)^n}{p + q} \begin{pmatrix} p & -p \\ -q & q \end{pmatrix}.$$

Let μ denote the initial distribution of the chain. The distribution of X_n is given by (3.3) p. 119, that is

$$\mathbb{P}_\mu(X_n = 0) = \mu(0) P^n(0, 0) + \mu(1) P^n(1, 0)$$

$$= \frac{q}{p + q} + \frac{(1 - p - q)^n}{p + q} [p\mu(0) - q\mu(1)],$$

and $\mathbb{P}_\mu(X_n = 1) = 1 - \mathbb{P}_\mu(X_n = 0)$.   ◁

Even when a complex system—with several units and several states—is studied, it is often sufficient to determine whether the system is in an operating state or in a failed one. In this aim, E is divided into two subsets, U and D. The subset U contains the functioning states (up-states), and D contains the failed ones (down-states). The reliability of this binary model is given by

$$R(n) = \mathbb{P}(X_k \in U, k \in \{0, \ldots, n\}), \quad n \geq 0.$$

Suppose the sequence (X_n) of the states visited by the system is a Markov chain with finite state space E, transition matrix P, and initial distribution μ. Then the lifetime of the system is the hitting time of D, that is

$$T = \min\{n \geq 0 : X_n \in D\},$$

where $\min \emptyset = +\infty$. Matrices and vectors have to be divided according to U and D. Without loss of generality, we can set $U = \{1, \ldots, m\}$ and $D = \{m+1, \ldots, N\}$, and thus write $\mu = [\mu_1, \mu_2]$ and

$$P = \begin{pmatrix} P_{11} & P_{12} \\ P_{21} & P_{22} \end{pmatrix}.$$

The reliability of the system is then

$$R(n) = \mu_1 P_{11}^n \mathbf{1}_m. \tag{3.16}$$

We deduce from Theorem 3.65 that

$$\mathbb{E}\,T = \sum_{i \in U} \mathbb{E}\,[T \mathbf{1}_{(X_0=i)}] = \sum_{i \in U} \mathbb{P}(X_0 = i)\mathbb{E}_i T = \mu_1 L,$$

where $L = (\mathbb{E}_1 T, \ldots, \mathbb{E}_m T)' = (I - P_{11})^{-1}\mathbf{1}_m$. Thus,

$$\mathbb{E}\,T = \mu_1 (I - P_{11})^{-1}\mathbf{1}_m. \tag{3.17}$$

The variance of the lifetime T is $\mathrm{Var}_i T = V(i) - (L(i))^2$, where

$$V = [\mathbb{E}_1(T^2), \ldots, \mathbb{E}_m(T^2)]' = (I - P_{11})^{-1}[I + 2P_{11}(I - P_{11})^{-1}]\mathbf{1}_m. \tag{3.18}$$

▷ *Example 3.70 (Continuation of Example 3.60)* Let (X_n) be the Markov chain modeling the behavior of a system with three states, with transition matrix P and initial distribution $\mu = (1, 0, 0)$. Assume that states 1 and 2 are the up-states and state 3 is the down-state, namely $U = \{1, 2\}$ and $D = \{3\}$. Under matrix form, the reliability is

$$R(n) = \mu_1 P_{11}^n \mathbf{1}_2 = (1, 0) \begin{pmatrix} 0.1 & 0.9 \\ 0.4 & 0.4 \end{pmatrix}^n \begin{pmatrix} 1 \\ 1 \end{pmatrix},$$

where $\mu_1 = (\mu(1), \mu(2))$. The mean functioning time is

$$\mathbb{E}\,T = \mu_1(I - P_{11})^{-1}\mathbf{1}_2 = (1, 0) \begin{pmatrix} 0.9 & -0.9 \\ -0.4 & 0.6 \end{pmatrix}^{-1} \begin{pmatrix} 1 \\ 1 \end{pmatrix} = \frac{25}{3}.$$

Note that, due to the spectral representation (3.15) p. 151, the reliability tends to zero with geometric rate $[(5 + 3\sqrt{17})/20]^n$, given by the largest eigenvalue of P_{11}.

$\lhd$

3.6   Branching Processes

Branching processes are a particular type of Markov chains. They model for example the emission of electrons by some substance, queuing problems, radioactive chain reactions, the survival of a genetic characteristic along generations, etc. Historically, they have been developed for investigating the transmission of family names along generations.

We are interested in the offspring of one given individual, called generation 0, that gives birth to τ children. Assume that the random variable τ takes values in $\mathbb{N}$, with offspring distribution $\mathbb{P}(\tau = k) = p_k$, mean μ, variance σ^2 and generating function g.

Definition 3.71 Let X_n be the size of the population at the n-th generation and let τ_j^n denote the number of children of the j-th individual of the n-th generation. In other words, $X_n = \sum_{j=1}^{X_{n-1}} \tau_j^{n-1}$ for $n \geq 1$, with $X_0 = 1$, where empty sums are set to 0.

If $(\tau_j^n)_{(n,j) \in \mathbb{N}^* \times \mathbb{N}^*}$ is i.i.d. with the same distribution as τ, the random sequence (X_n) is called a Galton-Watson process.

This definition is related to the assumption that each individual gives birth to descendants independently of the others. Moreover, we will assume that $p_0 \notin \{0, 1\}$ and $p_1 \notin \{0, 1\}$ so that the process is not trivial. We will also assume that σ^2 is finite. Trajectories of branching processes are shown in Fig. 3.9 for different usual distributions of τ.

Theorem 3.72 *The generating function of a Galton-Watson process (X_n) is $g_{X_n}(t) = g^{\circ n}(t)$, its mean is μ^n and its variance is*

$$\mathbb{V}\mathrm{ar}\, X_n = \begin{cases} \sigma^2 \mu^{n-1} \dfrac{1 - \mu^n}{1 - \mu} & \text{if } \mu \neq 1 \\ n\sigma^2 & \text{if } \mu = 1 \end{cases}.$$

Proof We have $\mathbb{E}\,(t^{X_n} \mid X_{n-1} = h) = g_{\sum_{j=1}^{h} \tau_j^{n-1}}(t) = g(t)^h$, from which we deduce that $g_{X_n}(t) = \mathbb{E}\,[g(t)^{X_{n-1}}] = g_{X_{n-1}}(g(t))$, and, by induction, that $g_{X_n}(t) = g^{\circ n}(t)$.

We know by Proposition 1.33 that $\mathbb{E}\, X_n = g'_{X_n}(1)$. Since we compute $g'_{X_n}(t) = g'_{X_{n-1}}(g(t))g'(t)$, we get that $g'_{X_n}(1) = g'_{X_{n-1}}(1)\mu$, and thus, by induction, $\mathbb{E}\, X_n = \mu^n$.

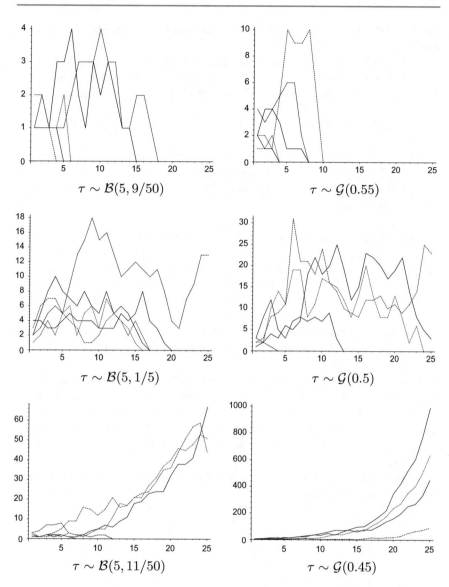

Fig. 3.9 Several trajectories of branching processes

The variance can be computed in the same way, using the second derivative of g_{X_n}.   □

From the expression of the mean, we deduce that the population increases in mean if $\mu > 1$, and this characterizes the supercritical case. It decreases if $\mu < 1$, and this characterizes the subcritical case. When $\mu = 1$, the process is said to be critical.

The probability of extinction of the population before time n is $u_n = \mathbb{P}(X_n = 0) = \mathbb{P}(T \leq n)$, where $T = \inf\{n \in \mathbb{N} : X_n = 0\}$. By definition, $u_n = g_{X_n}(0) = g^{\circ n}(0)$. The probability that the population dies out at the n-th generation is

$$\mathbb{P}(T = n) = \mathbb{P}(T \leq n) - \mathbb{P}(T \leq n - 1) = g^{\circ n}(0) - g^{\circ(n-1)}(0).$$

▷ *Example 3.73 (A Binary Galton-Watson Process)* If $p_0 = q$, $p_1 = p$ with $p + q = 1$, then $g(s) = q + ps$, and hence $g_{X_n}(s) = 1 - p^n + p^n s$. Therefore, the probability of extinction at the n-th generation is $\mathbb{P}(T = n) = p^{n-1} - p^n$, that is to say $T \sim \mathcal{G}(1 - p)$.   ◁

The overall probability of extinction of the population is $\varrho = \mathbb{P}(X_n \longrightarrow 0)$. Since $(X_n = 0) \subset (X_{n+1} = 0)$, we have $(X_n \longrightarrow 0) = \bigcup_{n \geq 0}(X_n = 0)$. Thus ϱ is the limit of the sequence (u_n), and hence is a fixed point of g, precisely the smallest solution of the equation $g(t) = t$. The function g is strictly convex, with $g(1) = 1$ and $g(0) = \mathbb{P}(\tau = 0) > 0$. The slope of the graph of g at 1 is given by $g'(1) = \mu$. If $\mu \leq 1$, then $\varrho = 1$, and, if $\mu > 1$, then $0 < \varrho < 1$, as shown in Fig. 3.10. When $\mu = 1$, the population will die out with probability one; this shows that the mean is not always a good estimator of the behavior of a random variable.

▷ *Example 3.74 (A Supercritical Galton-Watson Process)* The evolution of a biological population is described by a Galton-Watson process, with offspring probabilities $p_0 = 1/4$, $p_1 = 1/4$ et $p_2 = 1/2$. We compute $\mu = 5/4$, a supercritical case. Further, the generating function is given by $g(t) = 1/4 + t/4 + t^2/2$. If $X_0 = 1$, the probability of extinction is the smallest solution of $1/4 - 3t/4 + t^2/2 = 0$, that is $\varrho = 1/2$.

Note that for $X_0 = N$, where $N \in \mathbb{N}^*$ is a fixed integer, the probability of extinction becomes $(1/2)^N$.   ◁

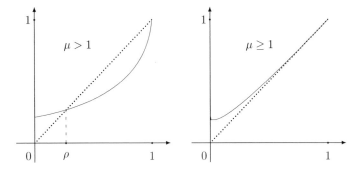

Fig. 3.10 The different critical cases of a Galton-Watson process

The total population issued from one given individual, that is $N = \sum_{n \geq 0} X_n$, is also of interest. If $\mu > 1$, then N is clearly infinite. Otherwise, the total population at the n-th generation, that is $N_n = \sum_{k=0}^{n} X_k$, tends to N, and

$$g_{N_n}(t) = g_{X_0}(t) g_{\sum_{k=0}^{n} X_k}(t) = t g_{\sum_{k=0}^{n} X_k}(t) = t g(g_{N_{n-1}}(t)),$$

hence $g_N(t) = t g(g_N(t))$ by taking limits. This functional equation can be proven to have only one solution when t is fixed. Moreover, $0 < g_N(t) < \varrho$ with $g_N(1) = \varrho$. Therefore, if $\varrho = 1$, then $\sum_{k \geq 0} \mathbb{P}(N = k) = 1$. Thus, $\mathbb{E} N = 1/(1 - \mu)$ if $\mu < 1$.

▷ *Example 3.75 (Branching Process with Geometric Distribution)* Suppose τ has a geometric distribution on $\mathbb{N}$ with parameter $q = 1 - p$, that is $\mathbb{P}(\tau = k) = p q^k$ for $k \geq 0$ and $\mu = p/q$. We have $g(s) = q/(1 - ps)$. This homographic transformation has two fixed points, 1 and q/p, and hence

$$\frac{g(s) - g(q/p)}{g(s) - g(1)} = \frac{1}{\mu} \cdot \frac{s - q/p}{s - 1}.$$

It follows that

$$\frac{g^{\circ n}(s) - g(q/p)}{g^{\circ n}(s) - g(1)} = \frac{1}{\mu^n} \cdot \frac{s - q/p}{s - 1},$$

or

$$g_{X_n}(s) = \begin{cases} q \dfrac{p^n - q^n - (p^{n-1} - q^{n-1}) p s}{p^{n+1} - q^{n+1} - (p^n - q^n) p s} & \text{if } p \neq 1/2, \\[2ex] \dfrac{n - (n-1)s}{n + 1 - ns} & \text{if } p = 1/2. \end{cases}$$

Therefore, the probability of extinction up to the n-th generation is

$$\mathbb{P}(X_n = 0) = g^{\circ n}(0) = \begin{cases} q \dfrac{p^n - q^n}{p^{n+1} - q^{n+1}} & \text{if } p \neq 1/2, \\[2ex] \dfrac{n}{n + 1} & \text{if } p = 1/2, \end{cases}$$

and $g_N(s)$ satisfies $p x^2 - x + q s = 0$, or $g_N(s) = (1 \pm \sqrt{1 - 4pqs})/2p$. ◁

The Galton-Watson process is a Markov chain with state space $\mathbb{N}$, initial state 1 and transition matrix P given by the convolution of its birth distribution; in other words,

$$P(i, j) = \sum_{\sum_{l=1}^{i} k_l = j} p_{k_1} \cdots p_{k_i}, \quad i \neq 0,$$

with $P(0,0) = 1$ and $P(0, j) = 0$ for $j \neq 0$. The extinction problems thus appear as problems of absorption in state 0.

Moreover, the process is a martingale if $\mu = 1$, a submartingale if $\mu > 1$ and a supermartingale if $\mu < 1$.

Proposition 3.76 *The sequence (X_n/μ^n) is a square integrable martingale for its natural filtration and converges a.s. and in mean square.*

Proof We have $\mathbb{E}(X_n \mid X_0, \ldots, X_{n-1}) = \mathbb{E}(X_n \mid X_{n-1})$. We compute

$$\mathbb{E}(X_n \mid X_{n-1} = k) = \sum_{i=1}^{k} \mathbb{E}(\tau_j^{n-1} \mid X_{n-1} = k) = k\mu,$$

and hence $\mathbb{E}(X_n \mid X_0, \ldots, X_{n-1}) = \mu X_{n-1}$, so $(M_n) = (X_n/\mu^n)$ is a martingale. Moreover, $\mathbb{E}(M_n^2) \leq \sigma^2/\mu(\mu - 1)$ for all n, so (M_n) is L^2-bounded. We deduce from the theorem of strong convergence of square integrable martingales that (M_n) converges a.s. and in L^2. $\square$

The Galton-Watson process can be generalized to involve immigration or emigration factors. A spatial—geographical—repartition of the individuals may also be considered, for example in investigating the successive generations of a population of plants.

3.7   Exercises

∇ **Exercise 3.1 (Visit Times)** Let (X_n) be a Markov chain with state space E. Let $T_i^n = \inf\{k > T_i^{n-1} : X_k = i\}$ be the time of the n-th visit to a recurrent state $i \in E$, with $T_i^1 = T_i$—see Definition 3.18.

Prove that $\mathbb{P}_i(T_j^n < +\infty) = (\varrho_{ii})^n$—see Definition 3.22 (setting $Y_r = X_{T_i+r}$ may be of help).

Solution We have

$$\mathbb{P}_i(T_i^2 < +\infty) = \sum_{n\geq 1, m\geq 1} \mathbb{P}_i(T_i = n, T_i^2 = n + m).$$

Thanks to the strong Markov property, (Y_r) is a Markov chain with the same transition function as (X_n). Let T_i^Y denote the time of the first visit of (Y_r) to state i. We compute

$$\mathbb{P}_i(T_i = n, T_i^2 = n + m) \stackrel{(1)}{=} \mathbb{P}_i(T_i = n, T_i^Y = m)$$

$$\stackrel{(2)}{=} \mathbb{P}_i(T_i = n)\mathbb{P}_i(T_i^Y = m)$$

$$\stackrel{(3)}{=} \mathbb{P}_i(T_i = n)\mathbb{P}_i(T_i = m).$$

(1) by the strong Markov property, (2) by Corollary 3.20 and (3) because both chains have the same transition function. Thus,

$$\mathbb{P}_i(T_i^2 < +\infty) = \sum_{n \geq 1, m \geq 1} \mathbb{P}_i(T_i = n)\mathbb{P}_i(T_i = m)$$

$$= \sum_{n \geq 1} \mathbb{P}_i(T_i = n) \sum_{m \geq 1} \mathbb{P}_i(T_i = m) = \mathbb{P}(T_i < +\infty)^2.$$

The result follows by induction. △

▽ **Exercise 3.2 (Chapman-Kolmogorov Equation)** Let $(\Omega, \mathcal{P}(\Omega), \mathbb{P})$ be a probability space, with

$$\Omega = \{(1, 1, 1), (2, 2, 2), (3, 3, 3)\} \cup \{(\omega_1, \omega_2, \omega_3) : \omega_r \in \{1, 2, 3\}, \ \omega_r \neq \omega_s\}$$

and the uniform probability $\mathbb{P}(\{\omega\}) = 1/9$, for all $\omega \in \Omega$.

Let X_1, X_2 and X_3 be defined on $(\Omega, \mathcal{P}(\Omega), \mathbb{P})$ and take values in $E = \{1, 2, 3\}$, with $X_n(\omega_1, \omega_2, \omega_3) = \omega_n$, for $n = 1, 2, 3$. The random variables X_4, X_5 and X_6 are constructed independently of X_1, X_2 and X_3 by setting $X_n(\omega_1, \omega_2, \omega_3) = \omega_{n-3}$, for $n = 4, 5, 6$. The whole sequence $(X_n)_{n \geq 1}$ is then constructed recursively similarly.

1. Study the dependence relations between X_1, X_2 and X_3.
2. Show that (X_n) satisfies the Chapman-Kolmogorov equation.
3. Is (X_n) a Markov chain?

Solution

1. Since $\mathbb{P}(X_m = i) = 1/3$ and $\mathbb{P}(X_m = i, X_n = j) = 1/9$, the events X_1, X_2, X_3 are pairwise independent. They are not mutually independent, because the values of X_1 and X_2 determine uniquely the value of X_3.
2. The Chapman-Kolmogorov equation is clearly satisfied. For instance, for $n = 2$,

$$\frac{1}{3} = P^2(i, j) = \sum_{k \in E} P(i, k)P(k, j) = \frac{1}{9} + \frac{1}{9} + \frac{1}{9}.$$

3. From 1., we get $\mathbb{P}(X_3 = k \mid X_1 = i, X_2 = j) \neq \mathbb{P}(X_3 = k \mid X_2 = j)$, therefore (X_n) is not a Markov chain.

To satisfy the Chapman-Kolmogorov equation is not a sufficient condition for a random sequence to be a Markov chain. △

▽ **Exercise 3.3 (A Markov Chain Defined by a Recursion Relation)** Let (X_n) be a random sequence taking values in E and let (ε_n) be a random sequence taking values in F. Both E and F are assumed to be denumerable. The two sequences are linked by the recursion relation $X_{n+1} = \varphi(X_n, \varepsilon_{n+1})$, for all $n \geq 0$, where $\varphi : E \times F \longrightarrow E$.

1. If, for all $n \geq 0$, ε_{n+1} is assumed to be independent of $X_0, \ldots, X_{n-1}$ and of $\varepsilon_1, \ldots, \varepsilon_n$ given X_n, show that (X_n) is a Markov chain.
2. If, moreover, the conditional distribution of ε_{n+1} given X_n is independent of n, show that the chain (X_n) is homogeneous.

Solution

1. We compute

$$\mathbb{P}(X_{n+1} = j \mid X_0, \ldots, X_n) = \mathbb{P}[\varphi(X_n, \varepsilon_{n+1}) = j \mid X_0, \ldots, X_n]$$
$$= \sum_{k \in F} \mathbb{P}(\varepsilon_{n+1} = k \mid X_0, \ldots, X_n) \mathbb{1}_{(\varphi(X_n, k) = j)}$$
$$= \sum_{k \in F} \mathbb{P}(\varepsilon_{n+1} = k \mid X_n) \mathbb{1}_{(\varphi(X_n, k) = j)}$$
$$= \mathbb{P}[\varphi(X_n, \varepsilon_{n+1}) = j \mid X_n] = \mathbb{P}(X_{n+1} = j \mid X_n).$$

2. We deduce from the third equality above that the chain is homogeneous if $\mathbb{P}(\varepsilon_{n+1} = k \mid X_n)$ is independent of n. $\triangle$

∇ **Exercise 3.4 (A Production System)** Suppose a production system is modelled by a Markov chain (X_n) with state space E and transition matrix P. Let $g : E \longrightarrow \mathbb{R}_+$ be some function. The rate of production $T_n = g(X_n)$ of the system changes according to its state. The demand of product is given by a sequence of positive real numbers (d_n).

1. Determine the mean demand not satisfied in the interval of time $[\![1, N]\!]$.
2. Application to the Markov chain of Example 3.70, with $g = (1, 0.6, 0)$, supposing a constant demand of 0.8 by time unit, with $N = 10$, and $\mu = (1, 0, 0)$.

Solution

1. The demand not satisfied at time n is $Z_n = [g(X_n) - d_n]^-$, with mean value

$$\mathbb{E}\, Z_n = \sum_{i \in E} \sum_{j \in E} \mu(i) P^n(i, j) [g(j) - d_n]^- = \mu P^n G_n,$$

where $G_n = ([g(j) - d_n]^-; j \in E)$ is a column vector. Set $C_N = \sum_{n=1}^N Z_n$. The mean demand not satisfied in $[\![1, N]\!]$ is

$$\mathbb{E}\, C_N = \sum_{n=1}^N \mu P^n G_n.$$

2. We compute $\mathbb{E}\, C_{10} = 0.1$. $\triangle$

▽ **Exercise 3.5 (Harmonic Functions and Markov Chains)** Let (X_n) be an irreducible Markov chain. Show that (X_n) is transient if and only if a bounded not constant function $h : E \longrightarrow \mathbb{R}$ and a state $i_0 \in E$ exist such that $Ph(i) = h(i)$ for all $i \neq i_0$.

Solution If the chain is transient, h can be defined as follows: $h(i_0) = 1$ and $h(i) = \varrho_{ii_0}$ for $i \neq i_0$; see Definition 3.22. Indeed, $h(i) = Ph(i)$ for $i \neq i_0$. The function is clearly bounded and not constant because the chain is transient.

Conversely, if such a function h exists, it can be supposed to be nonnegative. Set $Y_n = X_{T_{i_0} \wedge n}$. The transition function of (Y_n) is P', with $P'(i, j) = P(i, j)$ for all j if $i \neq i_0$ and $P'(i_0, i_0) = 1$. Hence, h is harmonic for P'. Thanks to Proposition 3.16, $\mathbb{E}_i[h(Y_n)] = h(i)$ for all i. If (X_n) was recurrent, we would have $\varrho_{ii_0} = 1$ for all i and (Y_n) would be absorbed in i_0 a.s.—that is $Y_n \overset{\text{a.s.}}{\longrightarrow} i_0$. Using the dominated convergence theorem, $\mathbb{E}_i[h(Y_n)]$ would tend to $h(i_0)$, for all i, and hence h would be constant, a contradiction. $\triangle$

▽ **Exercise 3.6 (Martingales and Markov Chains)** Let (X_n) be a Markov chain with state space $E = [\![0, N]\!]$ and transition matrix P, and let $\mathbf{F}$ denote the natural filtration of (X_n). States 0 and N are supposed to be absorbing, all other states are transient.

1. Let $T = \inf\{n \in \mathbb{N} : X_n = 0 \text{ or } N\}$ be the hitting time of the set $\{0, N\}$. Show that T is a stopping time of $\mathbf{F}$ and that T is a.s. finite.
2. Let $h : E \longrightarrow \mathbb{R}_+$ be an harmonic function for P.
   a. Deduce from the computation of $\mathbb{E}_1[h(X_{T \wedge n})]$ that

   $$\mathbb{E}_i[h(X_T)] = h(i) \quad i \in E \backslash \{0, N\}.$$

   b. Deduce $\mathbb{P}_i(X_T = 0)$ and $\mathbb{P}_i(X_T = N)$ from the computation of $\mathbb{E}_i[h(X_T)]$ in another way.

Solution

1. Set $F = \{0, N\}$. We have $(T = n) = (X_i \notin F, i = 1, \ldots, n - 1) \cap (X_n \in F) \in \mathcal{F}_n$, hence T is a stopping time of $\mathbf{F}$. Moreover, E is finite and all the states but 0 and N are transient, and hence T is a.s. finite.

2. a. The variable $T \wedge n$ is a bounded stopping time for all integers n and, thanks to Proposition 3.16, $(h(X_n))$ is a martingale; using the stopping theorem, we get

$$\mathbb{E}_i[h(X_{T \wedge n})] = \mathbb{E}_i[h(X_0)] = h(i).$$

The sequence $(X_{T \wedge n})$ converges a.s. to X_T, so $(h(X_{T \wedge n}))$ converges to $h(X_T)$. Since the state space E is finite, h is bounded on E. Thanks to the dominated convergence theorem, $(\mathbb{E}_i[h(X_{T \wedge n})])$ converges to $\mathbb{E}_i[h(X_T)]$, and the result follows.

   b. We can also write

$$\mathbb{E}_i[h(X_T)] = h(0)\mathbb{P}_i(X_T = 0) + h(N)\mathbb{P}_i(X_T = N),$$

and we know that $\mathbb{P}_i(X_T = 0) + \mathbb{P}_i(X_T = N) = 1$. Therefore,

$$\mathbb{P}_i(X_T = 0) = \frac{h(i) - h(N)}{h(0) - h(N)},$$

and $\mathbb{P}_i(X_T = N) = 1 - \mathbb{P}_i(X_T = 0)$.   △

▽ **Exercise 3.7 (Communication Classes)** Let (X_n) be a Markov chain with finite state space $E = [\![1, 8]\!]$ and transition matrix

$$P = \begin{pmatrix} 0 & 0 & 0 & * & 0 & 0 & 0 & * \\ 0 & * & * & 0 & * & 0 & 0 & * \\ 0 & 0 & * & 0 & 0 & 0 & 0 & 0 \\ * & 0 & 0 & 0 & 0 & 0 & 0 & 0 \\ 0 & 0 & 0 & 0 & * & 0 & 0 & 0 \\ 0 & * & 0 & 0 & 0 & 0 & 0 & 0 \\ 0 & * & 0 & 0 & 0 & * & * & 0 \\ 0 & 0 & 0 & * & 0 & 0 & 0 & * \end{pmatrix}$$

where the stars stand for the non null coefficients of the matrix (Fig. 3.11).

Determine the communication classes and the nature of the states of (X_n).

Solution The sum of the coefficients of each line of P is one, hence $P(3, 3) = P(5, 5) = 1$. States 3 and 5 are absorbing and communicate with no other state. Since $\mathbb{P}_6(T_6 < +\infty) = 1 - P(6, 2) = 0$, state 6 is transient. We compute

$$\mathbb{P}_7(T_7 = +\infty) = \mathbb{P}_7(X_1 = 6 \text{ or } X_1 = 2)$$
$$= \mathbb{P}_7(X_1 = 6) + \mathbb{P}_7(X_1 = 2)$$
$$= 1 - P(7, 7) > 0,$$

Fig. 3.11 Graph of the chain
of Exercise 3.7

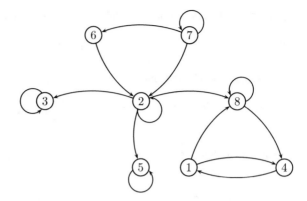

and hence 7 is transient. In the same way, $\mathbb{P}_2(T_2 = +\infty) = \mathbb{P}_2(X_1 \neq 2) = 1 - P(2, 2) > 0$, and hence 2 is transient. States 2, 6, and 7 communicate with no other state.

We compute

$$\mathbb{P}_1(T_1 = +\infty) = \mathbb{P}_1(X_n = 8, n > 0) = \lim_{n \to +\infty} P(1, 8) P(8, 8)^n,$$

with $0 < P(8, 8) < 1$, and hence $\mathbb{P}_1(T_1 < +\infty) = 1$, and 1 is recurrent. Moreover, $P(1, 4) > 0$ and $P(4, 1) > 0$ so that $1 \leftrightarrow 4$ and 4 is also recurrent. Finally, $\mathbb{P}_8(X_2 = 1) = \mathbb{P}_8(X_1 = 4) = P(8, 4) > 0$ and $P(1, 8) > 0$, so $1 \leftrightarrow 8$ and 8 is recurrent too.

Thus, the recurrent states are 1, 3, 4, 5 and 8; the other states are transient. The communication classes are $\{1, 4, 8\}, \{2\}, \{3\}, \{5\}, \{6\}$ and $\{7\}$. △

▽ **Exercise 3.8 (Embedded Chain)** Let (X_n) be a Markov chain with transition matrix P with no absorbing state. Let $S_{m+1} = \min\{n \geq S_m : X_n \neq X_{S_m}\}$, with $S_0 = 0$, be the jump times of (X_n). Set $Y_m = X_{S_m}$ for $m \geq 0$.

1. Show that (Y_m) is a homogeneous Markov chain and give its transition matrix Q.
2. Show that (X_n) is irreducible if and only if (Y_m) is irreducible.
3. If (X_n) has a stationary distribution π, determine a stationary measure for (Y_m).

Solution

1. Clearly, the times (S_n) are stopping times of (X_n). By the strong Markov property, a.s.,

$$\mathbb{P}(Y_{m+1} = j \mid Y_0, Y_1, \ldots, Y_m) = \mathbb{P}(X_{S_{m+1}} = j \mid X_{S_0}, X_{S_1}, \ldots, X_{S_m})$$

$$= \mathbb{P}(X_{S_{m+1}} = j \mid X_{S_m}) = \mathbb{P}(Y_{m+1} = j \mid Y_m),$$

and hence (Y_m) is a Markov chain. Similarly, the chain is homogeneous because

$$\mathbb{P}(Y_{m+1} = j \mid Y_m = i) = \mathbb{P}(X_{S_{m+1}} = j \mid X_{S_m} = i).$$

By definition, if $i \neq j$, then

$$Q(i, j) = \sum_{n \geq 1} \mathbb{P}_i(X_n = j, S_1 = n) = \sum_{n \geq 1} \mathbb{P}_i(X_n = j \mid S_1 = n)\mathbb{P}_i(S_1 = n).$$

On the one hand, $\mathbb{P}_i(S_1 = n) = P(i, i)^{n-1}[1 - P(i, i)]$. On the other hand, for $j \neq i$,

$$\begin{aligned}
\mathbb{P}_i(X_n = j \mid S_1 = n) &= \mathbb{P}_i(X_n = j \mid X_0 = i, \ldots, X_{n-1} = i, X_n \neq i) \\
&= \frac{\mathbb{P}_i(X_0 = i, \ldots, X_{n-1} = i, X_n = j)}{\mathbb{P}_i(X_0 = i, \ldots, X_{n-1} = i, X_n \neq i)} = \frac{P(i, j)}{1 - P(i, i)}.
\end{aligned}$$

Since $\sum_{n \geq 1} P(i, i)^{n-1} = [1 - P(i, i)]^{-1}$, we get

$$Q(i, j) = \begin{cases} P(i, j)/[1 - P(i, i)] & \text{if } j \neq i, \\ 0 & \text{if } j = i. \end{cases}$$

2. Clearly, for $i \neq j$, $P(i, j) > 0$ if and only if $Q(i, j) > 0$, and the result follows.
3. Let $\lambda(i) = \pi(i)[1 - P(i, i)]$ for $i \in E$. For all $j \in E$, we compute

$$\begin{aligned}
\sum_{i \in E} \lambda(i)Q(i, j) &= \sum_{i \neq j} \pi(i)P(i, j) = \left[\sum_{i \in E} \pi(i)P(i, j)\right] - \pi(j)P(j, j) \\
&= \pi(j) - \pi(j)P(j, j) = \lambda(j),
\end{aligned}$$

and hence λ is a stationary measure for Q.   △

▽ **Exercise 3.9 (Birth and Death Chains)** Let (X_n) be a birth and death chain with the notation of Example 3.6. Assume that $P(X_0 = 0) = 1$.

1. For this part, $E = \{-2, -1, 0, 1, 2\}$, thus $q_2 + r_2 = p_{-2} + r_{-2} = 1$.
   a. Assume that $r_0 = q_0 = p_1 = q_2 = 0$, with all other p_i, q_i, and r_i being positive. Classify the states and give the communication classes of the chain (Fig. 3.12).

Fig. 3.12 Graph of the chain of Exercise 3.9, 1. a

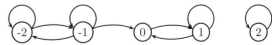

b. Assume that $p_i = 1/2$ for all $x \in E \setminus \{-2, 2\}$ and $r_i = 0$ for all $x \in E$. Show that the chain is irreducible, recurrent, and periodic, and determine its stationary distribution.

2. For this part, $E = \mathbb{Z}$ and $r_i = 0$, $p_i = p$, $q_i = q$, for all $i \in \mathbb{Z}$, with $0 < p < 1$. Set $Y_n = X_n - X_{n-1}$ for $n \in \mathbb{N}^*$, and $Y_0 = X_0$.

   a. Check that (X_n) is a simple random walk.
   b. Show that for all $n \in \mathbb{N}^*$, Y_n takes values in $\{-1, 1\}$, with $\mathbb{P}(Y_n = 1) = p$.
   c. Show that Y_n and Y_{n+1} are independent for all $n \in \mathbb{N}^*$.
   d. Show that Y_n is independent of $(Y_0, \ldots, Y_{n-1})$ for all $n \in \mathbb{N}^*$.

Solution

1. a. The transition matrix of (X_n) can be written

$$
P = \begin{pmatrix}
r_{-2} & p_{-2} & 0 & 0 & 0 \\
q_{-1} & r_{-1} & p_{-1} & 0 & 0 \\
0 & 0 & 0 & p_0 & 0 \\
0 & 0 & q_1 & r_1 & 0 \\
0 & 0 & 0 & 0 & 1
\end{pmatrix}.
$$

We know that $p_i + q_i + r_i = 1$ for all i, so $r_2 = 1$ and 2 is absorbing.

Starting at -1, if the chain reaches state 0, it will never come back to -1, and hence

$$
1 - \varrho_{-1-1} = \mathbb{P}_{-1}(T_{-1} = +\infty) = \mathbb{P}_{-1}(X_1 = 0) \in]0, 1[,
$$

and -1 is transient. Moreover, $P(-1, -2)P(-2, -1) > 0$ so $-2 \leftrightarrow -1$ and -2 is transient, too.

We have $1 - \varrho_{11} = \mathbb{P}_1(T_1 = +\infty) = \mathbb{P}_1(X_n = 0, n \geq 0)$. We compute $\mathbb{P}_1(X_1 = 0, \ldots, X_n = 0) = P(1, 0)P^{n-1}(0, 0)$. The sequence of events $(X_1 = 0, \ldots, X_n = 0)$ is decreasing with limit $(X_n = 0, n > 0)$. Therefore, $1 - \varrho_{11} = \lim_{n \to +\infty} P(1, 0)P^{n-1}(0, 0) = 0$, and hence 1 is recurrent. Moreover, $P(1, 0)P(0, 1) > 0$ hence $1 \leftrightarrow 0$ and 0 is recurrent.

Finally, $0 \leftrightarrow 1$ and $-1 \leftrightarrow -2$, so the partition of E in communication classes is $E = \{-2, -1\} \cup \{0, 1\} \cup \{2\}$. Note that $E_t = \{-2, -1\}$ and that $E_r = \{0, 1\} \cup \{2\}$.

b. The transition matrix of (X_n) is

$$
P = \begin{pmatrix}
0 & 1 & 0 & 0 & 0 \\
1/2 & 0 & 1/2 & 0 & 0 \\
0 & 1/2 & 0 & 1/2 & 0 \\
0 & 0 & 1/2 & 0 & 1/2 \\
0 & 0 & 0 & 1 & 0
\end{pmatrix}.
$$

Fig. 3.13 Graph of the chain
of Exercise 3.9, 1. b

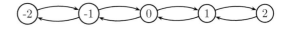

All the states communicate so the chain is irreducible (Fig. 3.13). Since the state space is finite, it is positive recurrent. We compute

$$P^2 = \begin{pmatrix} 1/2 & 0 & 1/2 & 0 & 0 \\ 0 & 3/4 & 0 & 1/4 & 0 \\ 1/4 & 0 & 1/2 & 0 & 1/4 \\ 0 & 1/4 & 0 & 3/4 & 0 \\ 0 & 0 & 1/2 & 0 & 1/2 \end{pmatrix}.$$

Therefore the chain is periodic, with period 2. Its stationary distribution is given by the vector $\pi = (a, b, c, d, e)$, with $a + b + c + d + e = 1$, satisfying $\pi P = \pi$, or

$$b = 2a, \quad 2a + c = 2b, \quad b + d = 2c, \quad c + 2e = 2d, \quad d = 2e.$$

Thus $\pi = (1/8, 1/4, 1/4, 1/4, 1/8)$.

2. a. Since $P(i, j) = 0$ if $|i - j| > 1$, the sequence (X_n) is a simple random walk.
   b. Clearly, Y_n takes values in $\{-1, 1\}$ for all n. We have

$$\mathbb{P}(Y_n = 1) = \mathbb{P}(X_n - X_{n-1} = 1) = \sum_{i \in \mathbb{Z}} \mathbb{P}(X_n = i, X_{n-1} = i - 1).$$

We compute

$$\mathbb{P}(X_n = i \mid X_{n-1} = i - 1)\mathbb{P}(X_{n-1} = i - 1) = P(i - 1, i)\mathbb{P}(X_{n-1} = i - 1),$$

and hence $\mathbb{P}(Y_n = 1) = P(i - 1, i) = p$.
   c. We can write

$$\mathbb{P}(Y_{n+1} = 1, Y_n = 1) = \mathbb{P}(X_{n+1} - X_n = 1, \ X_n - X_{n-1} = 1)$$
$$= \sum_{i \in \mathbb{Z}} \mathbb{P}(X_{n+1} = i + 2, \ X_n = i + 1, \ X_{n-1} = i).$$

We compute $\mathbb{P}(X_n = i + 1 \mid X_{n-1} = i) = P(i, i + 1)$, and

$$\mathbb{P}(X_{n+1} = i + 2 \mid X_n = i + 1, \ X_{n-1} = i)$$
$$= \mathbb{P}(X_{n+1} = i + 2 \mid X_n = i + 1) = P(i + 1, i + 2).$$

The compound probabilities theorem implies that

$$\mathbb{P}(X_{n+1} = i + 2, \ X_n = i + 1, \ X_{n-1} = i) =$$
$$= P(i + 1, i + 2) P(i, i + 1) \mathbb{P}(X_{n-1} = i) = p^2 \mathbb{P}(X_{n-1} = i).$$

Therefore,

$$\mathbb{P}(Y_{n+1} = 1, Y_n = 1) = p^2 \sum_{i \in \mathbb{Z}} \mathbb{P}(X_{n-1} = i) = p^2 = \mathbb{P}(Y_{n+1} = 1) \mathbb{P}(Y_n = 1).$$

d. Let us prove this point by induction. First,

$$\mathbb{P}(Y_1 = y_1, Y_0 = y_0) = \mathbb{P}(X_1 = y_1 + y_0 \mid X_0 = y_0) \mathbb{P}(X_0 = y_0)$$
$$= P(y_0, y_0 + y_1) \mathbb{P}(Y_0 = y_0)$$
$$= \begin{cases} p \mathbb{P}(Y_0 = y_0) & \text{if } y_1 = 1, \\ q \mathbb{P}(Y_0 = y_0) & \text{if } y_1 = -1, \end{cases}$$

and $\mathbb{P}(Y_0 = 1) = p = 1 - \mathbb{P}(Y_0 = -1) = 1 - q$, hence Y_0 and Y_1 are independent. At step n, we have

$$\mathbb{P}(Y_n = y_n, \ldots, Y_0 = y_0) =$$
$$= \mathbb{P}(Y_n = y_n \mid Y_{n-1} = y_{n-1}, \ldots, Y_0 = y_0) \mathbb{P}(Y_{n-1} = y_{n-1}, \ldots, Y_0 = y_0),$$

but $X_i = \sum_{j=0}^{i} Y_j$ and $Y_n = X_n - \sum_{j=0}^{n-1} Y_j$. Thus,

$$\mathbb{P}(Y_n = y_n \mid Y_{n-1} = y_{n-1}, \ldots, Y_0 = y_0) =$$
$$= \mathbb{P}\left(X_n = \sum_{j=0}^{n} y_j \mid X_{n-1} = \sum_{j=0}^{n-1} y_j, \ldots, X_0 = y_0\right)$$
$$= P\left(\sum_{j=0}^{n-1} y_j, \sum_{j=0}^{n} y_j\right),$$

by the Markov property, and

$$P\left(\sum_{j=0}^{n-1} y_j, \sum_{j=0}^{n} y_j\right) = \begin{cases} p & \text{if } y_n = 1, \\ q & \text{if } y_n = -1, \\ 0 & \text{otherwise.} \end{cases}$$

Therefore, $\mathbb{P}(Y_n = y_n \mid Y_{n-1} = y_{n-1}, \ldots, Y_0 = y_0) = \mathbb{P}(Y_n = y_n)$. Since $Y_0, \ldots, Y_{n-1}$ are independent by the induction hypothesis, the result follows. $\triangle$

▽ **Exercise 3.10 (An Alternative Method for Determining Stationary Distributions)** Let (X_n) be a Markov chain with finite state space $E = \{1, \ldots, e\}$ and
transition matrix P.

1. Show that a probability measure π on E is a stationary distribution of (X_n) if and
   only if $\pi(I - P + A) = 1'_e$, where A is the $e \times e$ matrix whose all coefficients
   are equal to 1, and π is a line vector.
2. Deduce from 1. that if (X_n) is irreducible, then $I - P + A$ is invertible.
3. Application: give an alternative method for determining the stationary distribution of an irreducible finite Markov chain.

Solution

1. Since π is a probability, $\pi A = 1'_e$, and hence $\pi(I - P + A) = 1'_e$ is equivalent
   to $\pi(I - P) = 0$, that is to $\pi P = \pi$.
2. It is enough to show that if $(I - P + A)x = 0$ for $x \in \mathbb{R}^e$, then $x = 0$.
   This equation and 1. jointly imply that $0 = \pi(I - P + A)x = 0 + \pi Ax$, hence
   $Ax = 0$. Therefore, $(I - P)x = 0$ or $Px = x$, hence $P^n x = x$, for all $n \geq 1$. So
   $\sum_{n=1}^N P^n x = Nx$, and the ergodic theorem yields $\pi x = x$, or $x_j = \sum_i \pi(i)x_i$
   for all j, and then $x = c1_e$.
   Relations $x = c1_e$ and $Ax = 0$ together imply that $1'_e c1_e = 0$, hence $c = 0$
   and finally $x = 0$.
3. The stationary distribution of the chain is solution of the linear system $(I - P' +
   A)\pi' = 1_e$.   △

▽ **Exercise 3.11 (A Cesaro's Sum)** Let P be a finite stochastic matrix.

1. Set $V_n = \sum_{i=1}^n P^k / n$. Show that the sequence (V_n) has a convergent subsequence. Let Π denote its limit.
2. Show that each line of Π is a stationary distribution of P.
3. Show that (V_n) converges to Π.

Solution

1. Since P is stochastic, V_n is also stochastic, for all n. The coefficients of P, and
   hence the coefficients of V_n belong to $[0, 1]$, which is a compact set. Therefore,
   (V_n) has a convergent subsequence, say (V_{n_j}).
2. Since Π is the limit of (V_{n_j}),

$$\Pi P = \lim_{j \to +\infty} V_{n_j} P = \lim_{j \to +\infty} \frac{1}{n_j} \sum_{k=2}^{n_j+1} P^k = P\Pi,$$

hence $V_{n_j} P - V_{n_j}$ tends to $\Pi P - \Pi$. Also,

$$V_{n_j} P - V_{n_j} = \frac{1}{n_j}(P^{n_j+1} - P) \longrightarrow 0, \quad j \to +\infty,$$

and hence $\Pi P = \Pi = P\Pi$.

3. If another subsequence (V_{n_k}) was converging to another matrix Q, we would get as above $\Pi V_{n_k} = V_{n_k}\Pi = \Pi$ so $\Pi Q = Q\Pi = \Pi$ and, in the same way, $Q\Pi = \Pi Q = Q$, so $\Pi = Q$. Therefore, (V_n) converges to Π.

Thus, for a finite Markov chain, even if P^n has no limit, the weighted sum always converges. $\triangle$

∇ **Exercise 3.12 (Inspection of a Computer)** The inspection of a computer on board is made by the astronauts at times $t_k = k\tau$, for $k \in \mathbb{N}^*$ and a fixed $\tau > 0$. No possibility of repair exists. Inspection can report either of the following results: perfect functioning, minor failures that do not hinder the global functioning of the system, major failures that hinder the development of experiments, or complete failure that makes the spaceship ungovernable.

This is modelled by a Markov chain (Fig. 3.14). The computer is supposed to be in perfect state at the beginning of the flight—at time $t = 0$. During preceding flights, the probabilities of passage from one state to the other between

Fig. 3.14 Graph of the chain of Exercise 3.12

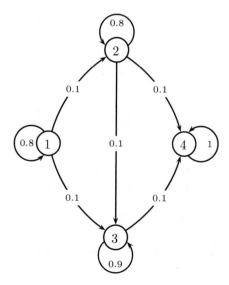

two successive inspections have been estimated by the matrix

$$P = \begin{pmatrix} 0.8 & 0.1 & 0.1 & 0 \\ 0 & 0.8 & 0.1 & 0.1 \\ 0 & 0 & 0.9 & 0.1 \\ 0 & 0 & 0 & 1 \end{pmatrix}.$$

1. Give the state space, the initial distribution and the nature of the states. Is the chain irreducible?
2. Compute the mean number of inspections before complete failure occurs.
3. If $\tau = 50$, and if the flight lasts $1000\,$h, compute the probability that the astronauts will come back on earth.
4. Determine the rate of convergence of the reliability to 0.

Solution

1. We have $E = \{1, 2, 3, 4\}$ and $\mathbb{P}(X_0 = 1) = 1$. State 4 is absorbing and the other states are transient. The chain is not irreducible, but is clearly aperiodic.
2. The state space can be divided into two subsets, that is $E = U \cup D = \{1, 2, 3\} \cup \{4\}$. The lifetime T of the system is the hitting time of state 4. From (3.17) p. 153, we get

$$\mathbb{E}\,T = \begin{pmatrix} 1 & 0 & 0 \end{pmatrix} \begin{pmatrix} 1 - 0.8 & -0.1 & -0.1 \\ 0 & 1 - 0.8 & -0.1 \\ 0 & 0 & 1 - 0.9 \end{pmatrix}^{-1} \begin{pmatrix} 1 \\ 1 \\ 1 \end{pmatrix} = 15.$$

3. The searched probability is the reliability at time $1000/50 = 20$, that is, according to (3.16) p. 153,

$$R(20) = \begin{pmatrix} 1 & 0 & 0 \end{pmatrix} \begin{pmatrix} 0.8 & 0.1 & 0.1 \\ 0 & 0.8 & 0.1 \\ 0 & 0 & 0.9 \end{pmatrix}^{20} \begin{pmatrix} 1 \\ 1 \\ 1 \end{pmatrix} = 0.2316.$$

4. According to (3.15) p. 151, the rate of convergence is given by the largest eigenvalue of the matrix P_{11}, that is $(0.9)^n$.   △

∇ **Exercise 3.13 (Generating Function of the Galton-Watson Process)** With the notation of Sect. 3.6, show without using conditional expectation that the generating function of a Galton-Watson process (X_n) satisfies $g_{X_n} = g_{X_n}^{\circ n}$.

Solution We compute

$$\mathbb{P}(X_n = k) = \sum_{h \geq 0} \mathbb{P}(X_n = k, X_{n-1} = h) = \sum_{h \geq 0} \mathbb{P}\left(\sum_{j=1}^{h} \tau_j^{n-1} = k, X_{n-1} = h\right)$$

$$= \sum_{h \geq 0} \mathbb{P}\left(\sum_{j=1}^{h} \tau_j^{n-1} = k\right) \mathbb{P}(X_{n-1} = h),$$

from which we deduce that

$$g_{X_n}(t) = \sum_{k \geq 0} \sum_{h \geq 0} t^k \mathbb{P}\left(\sum_{j=1}^{h} \tau_j^{n-1} = k\right) \mathbb{P}(X_{n-1} = h).$$

But

$$\sum_{k \geq 0} t^k \mathbb{P}\left(\sum_{j=1}^{h} \tau_j^{n-1} = k\right) = g_{\sum_{j=1}^{h} \tau_j^{n-1}}(t) = g(t)^h,$$

and hence

$$g_{X_n}(t) = \sum_{h \geq 0} g(t)^h \mathbb{P}(X_{n-1} = h) = g_{X_{n-1}}(g(t)),$$

and by induction $g_{X_n}(t) = g^{\circ n}(t).$ △

▽ **Exercise 3.14 (Maximum Likelihood Estimator of a Markov Chain)** Let $(X_n)_{n \in \mathbb{N}}$ be a Markov chain with finite state space $E = [\![1, d]\!]$, transition matrix $P = (P(i, j))_{i,j \in E}$ and initial distribution μ.

Suppose that one trajectory of (X_n) is observed in the time interval $[\![0, m]\!]$, say $(X_0, X_1, \ldots, X_m)$.

1. Give the likelihood of this $m + 1$-sample—that is not i.i.d.
2. Determine the maximum likelihood estimator of the transition matrix P.

Solution

1. The likelihood function of the $m + 1$-sample is

$$L_m(\theta, X_0, \ldots, X_m) = \mu(X_0) \prod_{k=1}^{m} P(X_{k-1}, X_k) = \mu(X_0) \prod_{i,j \in E} P(i, j)^{N_{ij}(m)}.$$

2. The parameter to be estimated is $\theta = (\theta_{ij}) \in \mathbb{R}^{d(d-1)}$, where $\theta_{ij} = P(i, j)$ for $1 \leq i \leq d$ and $1 \leq j \leq d-1$. Let us define

$$N_{ij}(m) = \sum_{k=1}^{m} \mathbb{1}_{\{X_{k-1}=i, X_k=j\}} \quad \text{and} \quad N_i(m) = \sum_{k=0}^{m-1} \mathbb{1}_{\{X_k=i\}}.$$

The log-likelihood function is

$$\ell_m(\theta) = \log L_m(\theta, X_0, \ldots, X_m) = \log \mu(X_0) + \sum_{i,j \in E} N_{ij}(m) \log P(i, j),$$

precisely

$$\ell_m(\theta) = \log \mu(X_0) + \sum_{i=1}^{d} \left(\sum_{j=1}^{d-1} N_{ij}(m) \log P(i, j) + N_{id} \log \left[1 - \sum_{k=1}^{d-1} P(i, k) \right] \right).$$

We obtain from the equations

$$\frac{\partial}{\partial \theta_{ij}} \ell_m(\theta) = \frac{N_{ij}(m)}{\theta_{ij}} - \frac{N_{id}(m)}{\theta_{id}} = 0,$$

first that

$$N_{ij}(m) = \frac{N_{id}(m)}{\theta_{id}} \theta_{ij}, \tag{3.19}$$

and then, by summing over j, that

$$N_i(m) = \sum_{j=1}^{d} N_{ij}(m) = \frac{N_{id}(m)}{\theta_{id}},$$

or $\widehat{\theta}_{id} = N_{id}(m)/N_i(m)$. Hence, we get using (3.19),

$$\widehat{\theta}_{ij} = \frac{N_{ij}(m)}{N_i(m)}, \quad 1 \leq i, j \leq d.$$

We check that

$$\frac{\partial^2}{\partial \theta_{ij}} \ell_m(\widehat{\theta}) = -\frac{N_{ij}(m)}{P(i, j)^2} - \frac{N_{id}(m)}{P(i, d)^2} \leq 0.$$

So, the maximum likelihood estimator of P is

$$\widehat{P}_m(i, j) = \begin{cases} \frac{N_{ij}(m)}{N_i(m)} & \text{if } N_i(m) > 0 \\ 0 & \text{if } N_i(m) = 0, \end{cases}$$

for $1 \leq i, j \leq d$. △

Continuous Time Stochastic Processes

4

A stochastic process represents a system, usually evolving along time, which incorporates an element of randomness, as opposed to a deterministic process.

The independent sequences of random variables, the Markov chains and the martingales have been presented above. In the present chapter, we investigate more general real stochastic processes, indexed by $\mathbb{T} \subset \mathbb{R}$, with $t \in \mathbb{T}$ representing the time, in a wide sense. Depending on the context, when $\mathbb{T} = \mathbb{N}$ or $\mathbb{R}_+$, they are called sequences of random variables—in short random sequences, stochastic processes with continuous or discrete time, signals, times series, etc.

First, we generalize to stochastic processes the notions of distributions, distribution functions, etc., already encountered in the previous chapters for random sequences. Then, we study some typical families of stochastic processes, with a stress on their asymptotic behavior: stationary or ergodic processes, ARMA processes, processes with independent increments such as the Brownian motion, point processes—especially renewal and Poisson processes, jump Markov processes and semi-Markov processes will be especially investigated in the final chapter.

4.1 General Notions

We will first define general stochastic elements, and then extend to stochastic processes notions such as distributions, stopping times, and results such as the law of large numbers, central limit theorem.

Let $(\Omega, \mathcal{F})$ and $(\mathbf{E}, \mathcal{E})$ be two measurable spaces. In probability theory, a measurable function $X : (\Omega, \mathcal{F}) \longrightarrow (\mathbf{E}, \mathcal{E})$, that is such that $X^{-1}(\mathcal{E}) \subset \mathcal{F}$, is called a stochastic element.

The stochastic elements may take values in any measurable space. If $\mathbf{E} = \mathbb{R}$ ($\mathbb{R}^d$, $d > 1$, $\mathbb{R}^{\mathbb{N}} = \mathbb{R} \times \mathbb{R} \times \cdots, \mathbb{R}^{\mathbb{R}}$), then X is a real random variable (random vector, random sequence or stochastic process with discrete time, stochastic process with continuous time).

© Springer Nature Switzerland AG 2018
V. Girardin, N. Limnios, *Applied Probability*,
https://doi.org/10.1007/978-3-319-97412-5_4

It is worth noticing that even in the general theory of stochastic processes, $\mathbb{T}$ is typically a subset of $\mathbb{R}$, as we will assume thereafter. For any fixed $\omega \in \Omega$, the function $t \longrightarrow X_t(\omega)$ is called a trajectory or realization of the process. The value of this function is determined by the result of a random phenomenon at the time t of its observation. For example, one may observe tossing of a coin, the fluctuations of the generations of a population, the day temperature in a given place, etc. If the trajectories of the process are continuous functions (on the left or on the right), the process itself is said to be continuous (on the left or on the right).

In general, if $(\mathbf{E}, \mathcal{E}) = (\mathbb{R}^\mathbb{T}, \mathcal{B}(\mathbb{R}^\mathbb{T}))$, where $\mathbb{T}$ is any convenient set of indices, a stochastic element can be represented as a family of real random variables $\mathbf{X} = (X_t)_{t \in \mathbb{T}}$. It can also be regarded as a function of two variables

$$\mathbf{X} : \Omega \times \mathbb{T} \longrightarrow \mathbb{R}$$
$$(\omega, t) \longrightarrow X_t(\omega)$$

such that X_t is a real random variable defined on $(\Omega, \mathcal{F})$ for any fixed t. If $\mathbb{T}$ is a totally ordered enumerable set (as $\mathbb{N}$ or $\mathbb{R}_+$), then $\mathbf{X}$ is called a time series. If $\mathbb{T} = \mathbb{R}^d$, with $d > 1$ or one of its subsets, the process is a multidimensional process. If $\mathbf{E} = (\mathbb{R}^d)^\mathbb{T}$, with $d > 1$, it is a multivariate or vector process.

The canonical space of a real process $\mathbf{X} = (X_t)_{t \in \mathbb{T}}$ with distribution $\mathbb{P}_\mathbf{X}$ is the triple $(\mathbb{R}^\mathbb{T}, \mathcal{B}(\mathbb{R}^\mathbb{T}), \mathbb{P}_\mathbf{X})$. Generally, we will consider that the real stochastic processes are defined on their canonical spaces.

Note that a function $\theta_s : \mathbb{R}^\mathbb{T} \longrightarrow \mathbb{R}^\mathbb{T}$ is called a translation (or shift) operator on $\mathbb{R}^\mathbb{T}$ if and only if $\theta_s(x_t) = x_{t+s}$ for all $s, t \in \mathbb{T}$ and all $x = (x_t) \in \mathbb{R}^\mathbb{T}$. When $\mathbb{T} = \mathbb{N}$, θ_n is the n-th iterate of the one step translation operator θ_1 (also denoted by θ, see above Definition 3.21).

A stochastic process $\mathbf{X}$ defined on $(\Omega, \mathcal{F}, \mathbb{P})$ is said to be an L^p process if $X_t \in L^p(\Omega, \mathcal{F}, \mathbb{P})$ for all $t \in \mathbb{T}$, for $p \in \mathbb{N}^*$. For $p = 1$ it is also said to be integrable, and for $p = 2$ to be a second order process.

Here are some classical examples of stochastic processes, the sinusoidal signals, the Gaussian processes, and the ARMA processes.

▷ *Example 4.1 (Sinusoidal Signal)*   For the process defined by

$$X_t = A \cos(2\pi \nu t + \varphi), \quad t \in \mathbb{R},$$

φ is called the phase, A the amplitude and ν the frequency. These parameters can be either constant or random. For instance, if A and ν are real constants and φ is a random variable, the signal is a monochromatic wave with random phase.   ◁

▷ *Example 4.2 (Gaussian White Noise)*   Let $(\varepsilon_t)_{t \in \mathbb{Z}}$ be such that $\varepsilon_t \sim \mathcal{N}(0, 1)$ for all $t \in \mathbb{T}$ and $\mathbb{E}(\varepsilon_{t_1} \varepsilon_{t_2}) = 0$ for all $t_1 \neq t_2$. The process (ε_t) is called a Gaussian white noise.   ◁

▷ *Example 4.3 (Gaussian Process)* Let $\mathbf{X} = (X_t)_{t \in \mathbb{T}}$ be such that the vector $(X_{t_1}, \ldots, X_{t_n})$ is a Gaussian vector for all integers n and all $t_1, \ldots, t_n$. The process $\mathbf{X}$ is said to be Gaussian. ◁

Definition 4.4 A sequence of random variables $X = (X_n)_{n \in \mathbb{Z}}$ is called an auto-regressive moving-average process with orders p and q, or ARMA(p, q), if

$$X_n + \sum_{i=1}^{p} a_i X_{n-i} = \varepsilon_n + \sum_{j=1}^{q} b_j \varepsilon_{n-j}, \quad n \in \mathbb{Z},$$

where ε is a Gaussian white noise, p and q are integers and $a_1, \ldots, a_n$ and $b_1, \ldots, b_q$ are all real numbers, with $a_p \neq 0$ and $b_q \neq 0$.

This type of processes is often used for modeling signals. If $q = 0$, the process is said to be an auto-regressive process, or AR(p). This latter models for example economical data depending linearly of the p past values, up to a fluctuation—the noise. If $a = -1$ and $p = 1$, it is a random walk. If $p = 0$, the process is said to be a moving-average process, or MA(q).

The ARMA processes can also be indexed by $\mathbb{N}$, by fixing the value of $X_0, \ldots, X_p$, as shown in Exercise 4.1.

▷ *Example 4.5 (MA(1) Process)* Let ε be a Gaussian white noise with variance σ^2 and let (X_n) be a sequence of random variables such that $X_n = b\varepsilon_{n-1} + \varepsilon_n$, for $n > 0$, with $X_0 = \varepsilon_0$. The vector $(X_0, \ldots, X_n)$ is the linear transform of the Gaussian vector $(\varepsilon_0, \ldots, \varepsilon_n)$, and hence is a Gaussian vector too, that is centered. For $n \geq 1$ and $m \geq 1$,

$$\mathbb{C}\text{ov}\,(X_n, X_m) = b^2 \mathbb{E}\,\varepsilon_{n-1}\varepsilon_{m-1} + b(\mathbb{E}\,\varepsilon_{n-1}\varepsilon_m + \mathbb{E}\,\varepsilon_n\varepsilon_{m-1}) + \mathbb{E}\,\varepsilon_n\varepsilon_m$$

$$= \begin{cases} 0 & \text{if } |n - m| \geq 2, \\ b\sigma^2 & \text{if } |n - m| = 1, \\ (b^2 + 1)\sigma^2 & \text{if } n = m. \end{cases}$$

Finally, we compute $\mathbb{V}\text{ar}\,X_0 = \sigma^2$, $\mathbb{C}\text{ov}\,(X_0, X_1) = b\sigma^2$ and $\mathbb{C}\text{ov}\,(X_0, X_m) = 0$ for all $m > 1$. ◁

Definition 4.6 Let $(\Omega, \mathcal{F})$ be a measurable set.

1. If $\mathcal{F}_t$ is a σ-algebra included in $\mathcal{F}$ for all $t \in \mathbb{T}$, and if $\mathcal{F}_s \subset \mathcal{F}_t$ for all $s < t$, then $\mathbf{F} = (\mathcal{F}_t)_{t \in \mathbb{T}}$ is called a filtration of $(\Omega, \mathcal{F})$. Especially, if $\mathbf{X} = (X_t)_{t \in \mathbb{T}}$ is a stochastic process, the filtration $\mathbf{F} = (\sigma(X_s; s \leq t))_{t \in \mathbb{T}}$ is called the natural filtration of $\mathbf{X}$.

2. A stochastic process **X** is said to be adapted to a filtration $(\mathcal{F}_t)_{t \in \mathbb{T}}$ if for all $t \geq 0$, the random variable X_t is $\mathcal{F}_t$-measurable.

Obviously, every process is adapted to its natural filtration. For investigating stochastic processes, it is necessary to extend the probability space $(\Omega, \mathcal{F}, \mathbb{P})$ by including a filtration $\mathbf{F} = (\mathcal{F}_t)_{t \in \mathbb{T}}$; then $(\Omega, \mathcal{F}, \mathbf{F}, \mathbb{P})$ is called a stochastic basis. Unless otherwise stated, the filtration will be the natural filtration of the studied process.

Theorem-Definition 4.7 *Let* **F** *be a filtration.*
A random variable $T : (\Omega, \mathcal{F}) \to \overline{\mathbb{R}}_+$ *such that* $(T \leq t) \in \mathcal{F}_t$, *for all* $t \in \mathbb{R}_+$, *is called an* **F**-*stopping time. Then, the family of events*

$$\mathcal{F}_T = \{A \in \mathcal{F} : A \cap (T \leq t) \in \mathcal{F}_t, \ t \in \mathbb{R}_+\}$$

is a σ-algebra, called the σ-algebra of events previous to T.

A stopping time adapted to the natural filtration of some stochastic process is said to be adapted to this process. The properties of the stopping times taking values in $\mathbb{R}_+$ derive directly from the properties of stopping times taking values in $\mathbb{N}$ studied in Chap. 2. Note that for any stopping time T, and for the translation operator θ_s for $s \in \mathbb{R}_+$, we have $(T \circ \theta_s = t + s) = (T = t)$.

Definition 4.8 Let $\mathbf{X} = (X_t)_{t \in \mathbb{T}}$ be a stochastic process defined on a probability space $(\Omega, \mathcal{F}, \mathbb{P})$, where $\mathbb{T} \subset \mathbb{R}$.

1. The probability $\mathbb{P}_\mathbf{X} = \mathbb{P} \circ \mathbf{X}^{-1}$ defined on $(\mathbb{R}^\mathbb{T}, \mathcal{B}(\mathbb{R}^\mathbb{T}))$ by

$$\mathbb{P}_\mathbf{X}(B) = \mathbb{P}(\mathbf{X} \in B), \quad B \in \mathcal{B}(\mathbb{R}^\mathbb{T}),$$

   is called the distribution of **X**.
2. The probabilities $\mathbb{P}_{t_1,\dots,t_n}$ defined by

$$\mathbb{P}_{t_1,\dots,t_n}(B_1 \times \dots \times B_n) = \mathbb{P}(X_{t_1} \in B_1, \dots, X_{t_n} \in B_n),$$

   for $t_1 < \dots < t_n$, $t_i \in \mathbb{T}$, are called the finite dimensional distributions of **X**. More generally, the restriction of $\mathbb{P}_\mathbf{X}$ to $\mathbb{R}^\mathbb{S}$ for any $\mathbb{S} \subset \mathbb{T}$ is called the marginal distribution of **X** on $\mathbb{S}$.
3. The functions $F_{t_1,\dots,t_n}$ defined by

$$F_{t_1,\dots,t_n}(x_1, \dots, x_n) = \mathbb{P}(X_{t_1} \leq x_1, \dots, X_{t_n} \leq x_n),$$

   for $t_1 < \dots < t_n$, $t_i \in \mathbb{T}$ and $x_i \in \mathbb{R}$, are called the finite dimensional distribution functions of **X**.

▷ *Example 4.9 (Finite Dimensional Distribution Functions)* Let X be a positive random variable with distribution function F. Let $\mathbf{X}$ be the stochastic process defined by $X_t = X - (t \wedge X)$ for $t \geq 0$. Its one-dimensional distribution functions are given by

$$F_t(x) = P(X_t \leq x) = P(X_t \leq x, X \leq t) + P(X_t \leq x, X > t)$$

$$= \mathbb{P}(0 \leq x, X \leq t) + \mathbb{P}(X - t \leq x, X > t)$$

$$= F(t) + F(t + x) - F(t) = F(t + x), \quad x \geq 0,$$

for $t \geq 0$, and its two-dimensional distribution functions are

$$F_{t_1, t_2}(x_1, x_2) = F((x_1 + t_1) \wedge (x_2 + t_2)) + F((x_1 + t_1) \wedge t_2)$$

$$+ F(t_1 \wedge (x_2 + t_2)) - 2F(t_1 \vee t_2), \quad x_1 \geq 0, x_2 \geq 0,$$

for $t_1 \geq 0$ et $t_2 \geq 0$. ◁

For a given family of distribution functions, does a stochastic process on some probability space exist with these functions as distribution functions? The answer may be positive, for instance under the conditions given by the following theorem, which we state without proof.

Theorem 4.10 (Kolmogorov) *Let $(F_{t_1,\ldots,t_n})$, for $t_1 < \cdots < t_n$ in $\mathbb{T}$, be a family of distribution functions satisfying for any $(x_1, \ldots, x_n) \in \mathbb{R}^n$ the two following coherence conditions:*

1. for any permutation $(i_1, \ldots, i_n)$ of $(1, \ldots, n)$,

$$F_{t_1,\ldots,t_n}(x_1, \ldots, x_n) = F_{t_{i_1},\ldots,t_{i_n}}(x_{i_1}, \ldots, x_{i_n});$$

2. for all $k \in [\![1, n-1]\!]$,

$$F_{t_1,\ldots,t_k,\ldots,t_n}(x_1, \ldots, x_k, +\infty, \ldots, +\infty) = F_{t_1,\ldots,t_k}(x_1, \ldots, x_k).$$

Then, there exists a stochastic process $\mathbf{X} = (X_t)_{t \in \mathbb{T}}$ defined on some probability space $(\Omega, \mathcal{F}, \mathbb{P})$ such that

$$\mathbb{P}(X_{t_1} \leq x_1, \ldots, X_{t_n} \leq x_n) = F_{t_1,\ldots,t_n}(x_1, \ldots, x_n).$$

For stochastic processes, the notion of equivalence takes the following form.

Definition 4.11 Let $\mathbf{X} = (X_t)_{t \in \mathbb{T}}$ and $\mathbf{Y} = (Y_t)_{t \in \mathbb{T}}$ be two stochastic processes both defined on the same probability space $(\Omega, \mathcal{F}, \mathbb{P})$ and taking values in $(\mathbb{R}^d, \mathcal{B}(\mathbb{R}^d))$. They are said to be:

1. weakly stochastically equivalent if for all $(t_1, \ldots, t_n) \in \mathbb{T}^n$, all $(B_1, \ldots, B_n) \in \mathcal{B}(\mathbb{R}^d)^n$ and all $n \in \mathbb{N}^*$,

$$\mathbb{P}(X_{t_1} \in B_1, \ldots, X_{t_n} \in B_n) = \mathbb{P}(Y_{t_1} \in B_1, \ldots, Y_{t_n} \in B_n). \tag{4.1}$$

2. stochastically equivalent if

$$\mathbb{P}(X_t = Y_t) = 1, \quad t \in \mathbb{T}. \tag{4.2}$$

Then, **Y** is called a version of **X**.
3. indistinguishable if

$$\mathbb{P}(X_t = Y_t, \forall t \in \mathbb{T}) = 1. \tag{4.3}$$

The trajectories of two indistinguishable processes are a.s. equal. This latter property is stronger than the two former ones; precisely

$$(4.3) \Longrightarrow (4.2) \Longrightarrow (4.1).$$

The converse implications do not hold, as the following example shows.

▷ *Example 4.12* Let Z be a continuous positive random variable. Let **X** and **Y** be two processes indexed by $\mathbb{R}_+$, defined by $X_t = 0$ for all t and $Y_t = \mathbb{1}_{(Z=t)}$. These two processes are stochastically equivalent, but not indistinguishable. Indeed, $\mathbb{P}(X_t \neq Y_t) = \mathbb{P}(Z = t) = 0$, for all $t \geq 0$, but $\mathbb{P}(X_t = Y_t, \forall t \geq 0) = 0$. ◁

Definition 4.13 A stochastic process $\mathbf{X} = (X_t)_{t \in \mathbb{R}}$ is said to be stochastically continuous at $s \in \mathbb{R}$ if, for all $\varepsilon > 0$,

$$\lim_{t \to s} \mathbb{P}(|X_t - X_s| > \varepsilon) = 0.$$

Note that stochastic continuity does not imply continuity of the trajectories of the process.

Most of the probabilistic notions defined for sequences of random variables indexed by $\mathbb{N}$ in Chap. 1 extend naturally to stochastic processes indexed by $\mathbb{R}_+$.

Definition 4.14 Let **X** be a stochastic process. The quantities

$$\mathbb{E}(X_{t_1}^{n_1} \ldots X_{t_d}^{n_d}), \quad d \in \mathbb{N}^*, \; n_i \in \mathbb{N}^*, \; \sum_{i=1}^d n_i = n,$$

are called the order n moments of **X**, when they are finite.

Especially, $\mathbb{E}\,X_t$ is called the mean value at time t and $\mathbb{V}\mathrm{ar}\,X_t$ the instantaneous variance. The latter is also called power by analogy to the electric tension X_t in a resistance, for which $\mathbb{E}\,(X_t^2)$ is the instantaneous power.

The natural extension of the covariance matrix for finite families is the covariance function.

Definition 4.15 Let $\mathbf{X}$ be a stochastic process. The function $R_\mathbf{X}\colon \mathbb{T}\times\mathbb{T}\longrightarrow \overline{\mathbb{R}}$ defined by

$$R_\mathbf{X}(t_1, t_2) = \mathbb{C}\mathrm{ov}\,(X_1, X_2) = \mathbb{E}\,(X_{t_1} X_{t_2}) - (\mathbb{E}\,X_{t_1})(\mathbb{E}\,X_{t_2}),$$

is called the covariance function of the process.

4.1.1 Properties of Covariance Functions

1. If $R_\mathbf{X}$ takes values in $\mathbb{R}$, then $\mathbf{X}$ is a second order process, and $\mathbb{V}\mathrm{ar}\,X_t = R_\mathbf{X}(t, t)$.
2. A covariance function is a positive semi-definite function. Indeed, for all $(c_1, \ldots, c_n) \in \mathbb{R}^n$, since covariance is bilinear,

$$\sum_{i=1}^{n}\sum_{j=1}^{n} c_i c_j R_\mathbf{X}(t_i, t_j) = \mathbb{C}\mathrm{ov}\,(\sum_{i=1}^{n} c_i X_{t_i}, \sum_{j=1}^{n} c_j X_{t_j})$$

$$= \mathbb{V}\mathrm{ar}\,\left(\sum_{i=1}^{n} c_i X_{t_i}\right) \geq 0.$$

3. For a centered process,

$$R_\mathbf{X}(t_1, t_2)^2 \leq R_\mathbf{X}(t_1, t_1) R_\mathbf{X}(t_2, t_2),$$

by Cauchy-Schwarz inequality.

The moments of a stochastic process are obtained by averaging over Ω. They are called space averages. Another notion of average exists for processes, on the set of indices $\mathbb{T}$; here we take $\mathbb{T} = \mathbb{N}$ for simplification.

Definition 4.16 Let $\mathbf{X} = (X_n)_{n\in\mathbb{N}}$ be a second order random sequence. The random variable

$$\overline{\mathbf{X}} = \lim_{N\to+\infty} \frac{1}{N}\sum_{n=1}^{N} X_n,$$

is called the time mean of $\mathbf{X}$ and the random variable

$$\overline{\mathbf{X}(m)} = \lim_{N \to +\infty} \frac{1}{N} \sum_{n=1}^{N} X_n X_{n+m}$$

is called the (mean) time power of $\mathbf{X}$ on $m \in \mathbb{N}^*$. Limit is taken in the sense of convergence in mean square.

The time mean is linear as is the expectation. If $\mathbf{Y} = a\mathbf{X} + b$, then $\overline{\mathbf{Y}(m)} = a^2 \overline{\mathbf{X}(m)}$ as for the variance. In the next section, we will make precise the links between space and time averages.

The different notions of convergence defined in Chap. 1 for random sequences indexed by $\mathbb{N}$ extend to stochastic processes indexed by $\mathbb{R}_+$ in a natural way. The next extension of the large numbers law has been proven for randomly indexed sequences in Theorem 1.93 in Chap. 1; the next proof is an interesting alternative. It is then completed by a central limit theorem.

Theorem 4.17 *Let (X_n) be an integrable i.i.d. random sequence. If $(N_t)_{t \in \mathbb{R}_+}$ is a stochastic process taking values in $\mathbb{N}^*$, a.s. finite for all t, independent of (X_n) and converging to infinity a.s. when t tends to infinity, then*

$$\frac{1}{N_t}(X_1 + \cdots + X_{N_t}) \xrightarrow{a.s.} \mathbb{E}\,X_1, \quad t \to +\infty.$$

Proof Set $\overline{X}_n = (X_1 + \cdots + X_n)/n$ for $n \geq 1$. By the strong law of large numbers, $\overline{X}_n$ converges to $\mathbb{E}\,X_1$ almost surely, that is on $\Omega \setminus A$, where $A = \{\omega : Y_n(\omega) \not\to \mathbb{E}\,X_1\}$. Note that $(\overline{X}_{N_t(\omega)}(\omega))$ is a sub-sequence of $(\overline{X}_n(\omega))$, and set $B = \{\omega : N_t(\omega) \not\to \infty\}$ and $C = \{\omega : \overline{X}_{N_t(\omega)}(\omega) \not\to \mathbb{E}\,X_1\}$. Then $C \subset A \cap B$, and the proof is completed. $\quad\square$

Theorem 4.18 (Anscombe) *Let $(X_n)_{n \in \mathbb{N}^*}$ be an i.i.d. random sequence with centered distribution with finite variance σ^2. If $(N_t)_{t \in \mathbb{R}_+}$ is a stochastic process taking values in $\mathbb{N}^*$, a.s. finite for all t, independent of (X_n) and converging to infinity a.s. when t tends to infinity, then*

$$\frac{1}{\sigma \sqrt{N_t}}(X_1 + \cdots + X_{N_t}) \xrightarrow{\mathcal{D}} \mathcal{N}(0, 1), \quad t \to +\infty.$$

Proof Set $S_n = X_1 + \cdots + X_n$. We have

$$\mathbb{E}\,(e^{it S_{N_t}/\sigma \sqrt{N_t}}) = \mathbb{E}\left(\sum_{n \geq 1} e^{it S_n/\sigma \sqrt{n}} \mathbb{1}_{(N_t=n)}\right) = \sum_{n \geq 1} \mathbb{P}(N_t = n)\mathbb{E}\,(e^{it S_n/\sigma \sqrt{n}}),$$

so

$$|\mathbb{E}\,(e^{it\,S_{N_t}/\sigma\sqrt{N_t}}) - e^{-\sigma^2 t^2/2}| \le \sum_{n \ge 1} \mathbb{P}(N_t = n)|\mathbb{E}\,(e^{it\,S_n/\sigma\sqrt{n}}) - e^{-\sigma^2 t^2/2}|.$$

On the one hand, due to the central limit theorem, $S_n/\sigma\sqrt{n}$ converges in distribution to a standard Gaussian variable, so, for all $\varepsilon > 0$, we obtain $|\mathbb{E}\,(e^{it\,S_n/\sqrt{n}}) - e^{-\sigma^2 t^2/2}| < \varepsilon$ for $n > n_\varepsilon$. Therefore,

$$\sum_{n > n_\varepsilon} \mathbb{P}(N_t = n)|\mathbb{E}\,(e^{it\,S_n/\sigma\sqrt{n}}) - e^{-t^2/2}| \le \varepsilon \mathbb{P}(N_t > n_\varepsilon) \le \varepsilon.$$

On the other hand,

$$\sum_{n=1}^{n_\varepsilon} \mathbb{P}(N_t = n)|\mathbb{E}\,(e^{it\,S_n/\sigma\sqrt{n}}) - e^{-t^2/2}| \le 2\mathbb{P}(N_t \le n_\varepsilon),$$

and the result follows because N_t converges to infinity and $\mathbb{P}(N_t \le n_\varepsilon)$ tends to zero when t tends to infinity.   □

Results of the same type can be stated for more general functionals, as for example the ergodic theorems.

4.2   Stationarity and Ergodicity

We will study here classical properties of processes taking values in $\mathbb{R}$. First, a stochastic process is stationary if it is invariant by translation of time, that is to say if it has no interne clock.

Definition 4.19 Let $\mathbf{X} = (X_t)_{t \in \mathbb{T}}$ be a stochastic process. It is said to be strictly stationary if

$$(X_{t_1}, \ldots, X_{t_n}) \sim (X_{t_1+s}, \ldots, X_{t_n+s}), \quad (s, t_1, \ldots, t_n) \in \mathbb{T}^{n+1}.$$

▷ *Example 4.20 (Some Stationary Sequences)* An i.i.d. random sequence is stationary. An ergodic Markov chain whose initial distribution is the stationary distribution is stationary.   ◁

Ergodicity, a notion of invariance on the space Ω, has been introduced for Markov chains in Chap. 3. It extends to general random sequences as follows.

Definition 4.21 A random sequence (X_n) is said to be strictly ergodic:

1. at order 1 if for any real number x the random variable

$$\lim_{N \to +\infty} \frac{1}{N} \sum_{n=0}^{N-1} \mathbb{1}_{(X_n \leq x)}$$

   is constant;
2. at order 2 if for all real numbers x_1 and x_2 the random variable

$$\lim_{N \to +\infty} \frac{1}{N} \sum_{n=0}^{N-1} \mathbb{1}_{(X_n \leq x_1, X_{n+m} \leq x_2)}$$

   is constant for all $m \in \mathbb{N}$.

Since $\mathbb{P}(X_n \leq x) = \mathbb{E}[\mathbb{1}_{(X_n \leq x)}]$, a random sequence is ergodic at order 1 (order 2) if the distributions of the random variables X_n (pairs (X_n, X_{n+m})) can be obtained by time average.

▷ *Example 4.22 (Markov Chains)* Any aperiodic positive recurrent Markov chain is ergodic, as seen in Chap. 3.   ◁

▷ *Example 4.23* The stochastic process modeling the sound coming out of a tape recorder, with null ageing but parasite noise, is ergodic. Indeed, the observation of one long-time recording is obviously similar to the observation of several short-time recordings. This process is also obviously stationary.   ◁

The following convergence result is of paramount importance in statistics. We state it without proof.

Theorem 4.24 (Ergodic) *If (X_n) is a strictly stationary and ergodic random sequence and if $g : \mathbb{R} \longrightarrow \mathbb{R}$ is a Borel function, then*

$$\frac{1}{N} \sum_{n=0}^{N-1} g(X_n) \longrightarrow \mathbb{E}\, g(X_0), \quad N \to +\infty.$$

In ergodic theory, the strict stationarity and ergodicity of a stochastic process are expressed using set transformations.

Let $(E, \mathcal{E}, \mu)$ be a measured set. A measurable function $S : E \to E$ is called a transformation of E; we denote by Sx the image of an element $x \in E$ and we set $S^{-1}B = \{x \in E : Sx \in B\}$, for $B \in \mathcal{E}$.

A set B of $\mathcal{E}$ is said to be S-invariant if $\mu(S^{-1}(B)) = \mu(B)$. The collection of S-invariant sets constitutes a σ-field, denoted by $\mathcal{J}$. The function S is said to be

μ-invariant if the elements of $\mathcal{E}$ are all S-invariant. The transformation is said to be strictly ergodic if for all $B \in \mathcal{E}$,

$$S^{-1}(B) = B \implies \mu(B) = 0 \text{ or } 1.$$

The following convergence result, stated without proof, is also called the pointwise ergodic theorem.

Theorem 4.25 (Birkhoff) *Let μ be a finite measure on a measurable set $(E, \mathcal{E})$. If μ is S-invariant for some transformation $S : E \to E$, then, for all μ-integrable functions $f : (E, \mathcal{E}, \mu) \to \mathbb{R}$,*

$$\frac{1}{n} \sum_{i=0}^{n-1} f(S^i x) \longrightarrow \overline{f}(x) \quad \mu - a.e., \quad x \in E,$$

where $\overline{f} : E \to \mathbb{R}$ is a μ-integrable function such that $\overline{f}(x) = \overline{f}(Sx)$ μ-a.e., and

$$\int_E f(t) d\mu(t) = \frac{1}{\mu(E)} \int_E \overline{f}(t) d\mu(t).$$

If, moreover, S is ergodic, then

$$\overline{f}(x) = \int_E f(t) d\mu(t), \quad a.e..$$

Note that if μ is a probability, then by definition, $\overline{f}$ is the conditional expectation of f given $\mathcal{J}$, or $\overline{f} = \mathbb{E}(f \mid \mathcal{J})$. If S is ergodic, then $\overline{f} = \mathbb{E} f$ a.s.

▷ *Example 4.26 (Interpretation in Physics)* Let us observe at discrete times a system evolving in an Euclidean space E.

Let $x_0, x_1, x_2, \ldots$ be the points successively occupied by the system. Let S transform the elements of E through $x_n = Sx_{n-1}$, for $n \geq 1$. Let S^n denote the n-th iterate of S, that is $x_n = S^n x_0$. Thus, the sequence $(x_0, x_1, x_2, \ldots, x_n, \ldots)$ can be written $(x_0, Sx_0, S^2 x_0, \ldots, S^n x_0, \ldots)$ and is called an orbit of the point x_0 (or trajectory).

Let $f : E \to \mathbb{R}$ be a function whose values $f(x)$ express a physical measure (speed, temperature, pressure, etc.) at $x \in E$. Thanks to former experiments, the measure $f(x)$ is known to be subject to error and the empirical mean $[f(x_0) + f(x_1) + \cdots + f(x_{n-1})]/n$ to give a better approximation of the true measured quantity. For n large enough, this mean is close to the limit

$$\lim_{n \to +\infty} \frac{1}{n} \sum_{i=0}^{n-1} f(S^i x),$$

which should be equal to the true physical quantity.   ◁

Any stochastic process $\mathbf{X} = (X_t)_{t \in \mathbb{T}}$ on $(\Omega, \mathcal{F}, \mathbb{P})$, taking values in E, can be considered as defined by

$$X_t(\omega) = X(S_t\omega), \quad \omega \in \Omega, \ t \in \mathbb{T},$$

where $X : \Omega \to E$ is a random variable and $\{S_t : \Omega \to \Omega : t \in \mathbb{T}\}$ is a group of transformations. If $\mathbb{T} = \mathbb{N}$, then S_n is the n-th iterate of some transformation $S : \Omega \to \Omega$. Clearly, the process is strictly stationary if

$$\mathbb{P}((S_t)^{-1}(A)) = \mathbb{P}(A), \quad A \in \mathcal{F}, \ t \in \mathbb{T},$$

and is strictly ergodic at order 1 if

$$\forall A \in \mathcal{F}, \quad (S_t)^{-1}(A) = A, \ t \in \mathbb{T} \implies \mathbb{P}(A) = 0 \text{ or } 1.$$

Any real valued process $\mathbf{X} = (X_t)_{t \in \mathbb{T}}$ can be defined on its canonic space $(\mathbb{R}^{\mathbb{T}}, \mathcal{B}(\mathbb{R}^{\mathbb{T}}), \mathbb{P}_{\mathbf{X}})$ by setting

$$X_t(\omega) = \omega(t), \quad \omega \in \mathbb{R}^{\mathbb{T}}, \ t \in \mathbb{T}.$$

Strict stationarity thus appears as a property of invariance of the probability $\mathbb{P}_{\mathbf{X}}$ with respect to translation operators, that is $\mathbb{P}_{\mathbf{X}} \circ \theta_t^{-1} = \mathbb{P}_{\mathbf{X}}$ for all $t \in \mathbb{T}$, or

$$\mathbb{P}_{\mathbf{X}}(\theta_t^{-1}(B)) = \mathbb{P}(B), \quad B \in \mathcal{B}(\mathbb{R}^{\mathbb{T}}), \ t \in \mathbb{T}.$$

Strict ergodicity can be expressed using θ_t-invariant sets. Precisely, the process is ergodic if

$$\forall B \in \mathcal{B}(\mathbb{R}^{\mathbb{T}}), \quad \theta_t^{-1}(B) = B, \ t \in \mathbb{T} \implies \mathbb{P}_{\mathbf{X}}(B) = 0 \text{ or } 1.$$

For a real stationary and integrable random sequence (X_n), Birkhoff's ergodic theorem yields

$$\frac{1}{n} \sum_{i=0}^{n-1} X_i \xrightarrow{a.s.} \overline{X} = \mathbb{E}(X_0 \mid \mathcal{J}),$$

where $\mathcal{J}$ is the σ-field of θ_1-invariant sets. For an ergodic sequence—such that θ_1-invariant events have probability 0 or 1, the limit is $\overline{X} = \mathbb{E}\,X_0$.

Note that, according to Kolmogorov 0–1 law, any i.i.d. sequence is stationary and ergodic.

Further, the entropy rate of a random sequence has been defined in Chap. 1. The entropy rate of a continuous time stochastic process indexed by $\mathbb{R}_+$ is defined similarly.

Definition 4.27 Let $\mathbf{X} = (X_t)_{t \in \mathbb{R}_+}$ be a stochastic process such that the marginal distribution of $\mathbf{X} = (X_t)_{t \in [0,T]}$ has a density $p_T^{\mathbf{X}}$ with respect either to the Lebesgue or to the counting measure, for all T. The entropy up to time T of $\mathbf{X}$ is the entropy of its marginal distribution, that is

$$\mathbb{H}_T(\mathbf{X}) = -\mathbb{E} \log p_T^{\mathbf{X}}(X).$$

If $\mathbb{H}_T(\mathbf{X})/T$ has a limit when T tends to infinity, this limit is called the entropy rate of the process and is denoted by $\mathbb{H}(\mathbf{X})$.

The next result, stated here for stationary and ergodic sequences is of paramount importance in information theory. Note that its proof is based on Birkhoff's ergodic theorem.

Theorem 4.28 (Ergodic Theorem of Information Theory) *Let* $\mathbf{X} = (X_t)_{t \in \mathbb{T}}$, *with* $\mathbb{T} = \mathbb{R}_+$ *or* $\mathbb{N}$, *be a stationary and ergodic stochastic process. Suppose that the marginal distribution of* $\mathbf{X} = (X_t)_{t \in [0,T]}$ *has a density* $p_T^{\mathbf{X}}$ *with respect to the Lebesgue or counting measure, for all* T. *If the entropy rate* $\mathbb{H}(\mathbf{X})$ *is finite, then* $-\log p_T^{\mathbf{X}}(X)/T$ *converges in mean and a.s. to* $\mathbb{H}(\mathbf{X})$.

Explicit expressions of $\mathbb{H}(\mathbf{X})$ for jump Markov and semi-Markov processes will be developed in the next chapter.

Other notions of stationarity exist, less strict and more convenient for applications; we define here only first and second order stationarity.

Definition 4.29 A second order process $\mathbf{X}$ is said to be weakly stationary:

1. to the order 1 if $\mathbb{E} X_{t_1} = \mathbb{E} X_{t_2} = m_X$ for all $(t_1, t_2) \in \mathbb{T}^2$.
2. to the order 2 if, moreover, $R_{\mathbf{X}}(t_1, t_2) = R_{\mathbf{X}}(t_1 + \tau, t_2 + \tau)$ for all $\tau \in \mathbb{T}$ and $(t_1, t_2) \in \mathbb{T}^2$.

The mean and the variance of X_t are then constant in t, but the X_t do not necessarily have the same distribution.

▷ *Example 4.30 (Sinusoidal Signal (Continuation of Example 4.1))* Let $\mathbf{X}$ be as defined in Example 4.1. Suppose that ν is constant, and that $\varphi \sim \mathcal{U}[-\pi, \pi]$ and A are independent variables. Then $\mathbb{E} X_t = 0$ and $R_{\mathbf{X}}(t_1, t_2) = \mathbb{E}(A^2) \cos[2\pi \nu(t_1 - t_2)]/2$; hence, this sinusoidal signal is second order stationary. Clearly, it is not strictly stationary.   ◁

For a second order process, stationarity induces weak stationarity. If $\mathbf{X}$ is a Gaussian process, the converse holds true. Indeed, then $\mathbb{E}(X_{t_1}, \ldots, X_{t_n}) = M$ is constant and the covariance matrix $\Gamma_{X_{t_1}, \ldots, X_{t_n}}$ is a matrix valued function of the differences $t_j - t_i$. The distribution of $(X_{t_1}, \ldots, X_{t_n})$ depends only on M and Γ.

Further, the covariance function of a weakly stationary process can be written $R_X(t_1, t_2) = r_X(t_2 - t_1)$, where r_X is an even function of only one variable, called the auto-correlation function of the process. The faster X fluctuates, the faster r_X decreases.

If X is a second order process, the Cauchy-Schwarz inequality yields $|r_X(t_2 - t_1)| \leq r_X(0)$. The function r_X is positive definite; if, moreover, it is continuous, then it has a spectral representation—by Bochner's theorem, meaning that it is the Fourier transform of a bounded positive measure. For example, if $\mathbb{T} = \mathbb{R}$ ($\mathbb{Z}$),

$$r_X(t) = \int_\Lambda e^{i\lambda t} \mu(d\lambda)$$

where $\Lambda = \mathbb{R}$ ($[0, 2\pi]$) and μ is called the spectral measure of the process. If μ is absolutely continuous with respect to the Lebesgue measure, its density is called the spectral density of the process. The term "spectral" designs what is connected to frequencies.

Then the integral representation of the process follows—by Karhunen's theorem,

$$X_t = \int_\Lambda e^{i\lambda t} Z(d\lambda),$$

where Z is a second order process with independent increments—see next section, indexed by Λ.

▷ *Example 4.31 (White Noise)*   Let $\varepsilon = (\varepsilon_n)_{n \in \mathbb{Z}}$ be a stationary to the order 2 centered process, with $\mathbb{V}\mathrm{ar}\, \varepsilon_n = 1$ and

$$\mathbb{E}(\varepsilon_n \varepsilon_m) = 0 = \int_0^{2\pi} \frac{1}{2\pi} e^{i\lambda(n-m)} d\lambda, \quad m \neq n.$$

Therefore, the spectral density of ε is constant. All the frequencies are represented with the same power in the signal, hence its name: white noise, by analogy to white light from an incandescent body.   ◁

▷ *Example 4.32 (ARMA Processes)*   An MA(q) process X is obtained from a white noise ε by composition with the function $f(x_1, \ldots, x_n) = \sum_{k=1}^{q} b_k x_{n-k}$, called linear filter. We compute $\mathbb{E}(X_n X_{n+m}) = \sum_{k=1}^{q} b_k b_{m+k}$. The spectral density of X is obtained from that of ε, that is

$$h_X(\lambda) = \frac{1}{2\pi} \left| \sum_{k=1}^{q} b_k e^{-ik\lambda} \right|^2.$$

In the same way, an AR(p) process $\mathbf{Y}$ is obtained by recursive filtering of order p from $\boldsymbol{\varepsilon}$, and

$$h_{\mathbf{Y}}(\lambda) = \frac{1}{2\pi} \left| \sum_{k=1}^{q} a_k e^{-ik\lambda} \right|^{-2}.$$

Finally, the spectral density of an ARMA(p, q) process $\mathbf{Z}$ is

$$h_{\mathbf{Z}}(\lambda) = \frac{1}{2\pi} \frac{|\sum_{k=1}^{q} b_k e^{-ik\lambda}|^2}{|\sum_{k=1}^{p} a_k e^{-ik\lambda}|^2},$$

for $\lambda \in [0, 2\pi]$.   ◁

Just as strict stationarity, strict ergodicity is rarely satisfied in applications, and hence weaker notions of ergodicity are considered.

Definition 4.33 A random sequence (X_n) is said to be:

1. first order (or mean) ergodic if the random variable $\overline{\mathbf{X}}$ is a.s. constant.
2. second order ergodic if, moreover, for all $\tau \in \mathbb{R}$, the random variable $\overline{\mathbf{X}(m)}$ is a.s. constant.

Mean ergodicity is satisfied under the following simple condition.

Proposition 4.34 *A random sequence (X_n) is first order ergodic if and only if its covariance function $R_{\mathbf{X}}$ is summable.*

Proof For a centered sequence $\mathbf{X}$.
We compute

$$\mathbb{E} \left(\frac{1}{N} \sum_{n=0}^{N-1} X_n \right)^2 = \frac{1}{N^2} \sum_{n=1}^{N} \sum_{m=1}^{N} R_{\mathbf{X}}(n, m)$$

If the sequence is first order ergodic, $\frac{1}{N} \sum_{n=1}^{N} X_n$ converges in L^2 to a constant, so

$$\sum_{n=0}^{N-1} \sum_{m=0}^{N-1} R_{\mathbf{X}}(n, m) < +\infty.$$

Conversely, if $R_{\mathbf{X}}$ is summable,

$$\frac{1}{N^2} \sum_{n=0}^{N-1} \sum_{n=0}^{N-1} R_{\mathbf{X}}(n, m) \xrightarrow{L^2} 0, \quad N \to +\infty.$$

and the sequence is first order ergodic.   □

Stationarity and ergodicity are not equivalent, as shown by the following examples.

▷ *Example 4.35* A process $\mathbf{X}$ constant with respect to t, equal to some random variable V, is obviously stationary. Since $\overline{\mathbf{X}} = V$, it is first order ergodic only if V is a.s. constant.

By linearity of integration, the sum of two ergodic processes is ergodic too. The sum of two independent second order stationary processes is second order stationary too. On the contrary, the sum of two dependent second order stationary processes is not stationary in general.

Finally, let $\mathbf{X}$ be a process and let V be a random variable independent of $\mathbf{X}$. If $\mathbf{X}$ is stationary, the processes $(V X_t)$ and $(X_t + V)$ are stationary too; on the contrary, if $\mathbf{X}$ is ergodic, $(V X_t)$ and $(X_t + V)$ are ergodic too only if V is a.s. constant.   ◁

If the process is only weakly stationary and ergodic, the ergodic theorem does not hold in general. Still, the following result holds true. In this case, the mean power is equal to the instantaneous power.

Proposition 4.36 *If (X_n) is stationary and ergodic to the order 2, then $\mathbb{E} X_n = \overline{\mathbf{X}}$ for $n \in \mathbb{N}$ and $\overline{\mathbf{X}(m)} = r_{\mathbf{X}}(m) + \overline{\mathbf{X}}^2$ a.s. for $m \in \mathbb{N}$.*

Proof The random variable $\overline{\mathbf{X}}$ is a.s. constant, and all the X_n have the same distribution, so $\mathbb{E} X_n = \overline{\mathbf{X}}$ by linearity. In the same way, since $\mathbb{E} X_n = \mathbb{E} X_{n+m} = \overline{\mathbf{X}}$ and $r_{\mathbf{X}}(m) = \mathbb{E}(X_n X_{n+m}) - \mathbb{E} X_n \mathbb{E} X_{n+m}$ does not depend on n, we can write

$$r_{\mathbf{X}}(m) = \frac{1}{N} \sum_{n=0}^{N-1} \mathbb{E}(X_n X_{n+m}) - \left(\frac{1}{N} \sum_{n=0}^{N-1} \overline{\mathbf{X}} \right)^2,$$

and $r_{\mathbf{X}}(m) = \overline{\mathbf{X}(m)} - \overline{\mathbf{X}}^2$ a.s.   □

Thus, the moments of the marginal distributions of weakly stationary and ergodic sequences appear to be characterized by time averaging, that is by the knowledge of one—and only one—full trajectory. This explains the importance of ergodicity in the statistical study of marginal distributions.

4.3   Processes with Independent Increments

Numerous types of processes have independent increments, among which we will present the Brownian motion, Poisson processes, compound Poisson processes, etc.

Definition 4.37 Let $\mathbf{X} = (X_t)_{t \in \mathbb{T}}$ be a stochastic process adapted to a filtration $(\mathcal{F}_t)$, where $\mathbb{T} = \mathbb{R}$ or $\mathbb{T} = \mathbb{N}$.

1. The process is said have independent increments if for all $s < t$ in $\mathbb{T}$, the random variable $X_t - X_s$ is independent of the σ-algebra $\mathcal{F}_s$.
2. The process is said to have stationary increments if the distribution of $X_t - X_s$, for $s < t$ in $\mathbb{T}$, depends only on $t - s$.
3. A process with independent and stationary increments is said to be homogeneous (with respect to time).

When $\mathbb{T} = \mathbb{N}$ and $\mathbf{X}$ is a process with independent increments, the random variables $X_{t_0}, X_{t_1} - X_{t_0}, \ldots, X_{t_n} - X_{t_{n-1}}$ are independent for all $n \in \mathbb{N}$ and $t_0 < t_1 < \cdots < t_n$. Since $X_n = \sum_{i=1}^{n}(X_i - X_{i-1})$, knowing a process with independent increments is equivalent to knowing its increments. If $\mathbf{X}$ is a homogeneous process, then necessarily, $X_0 = 0$ a.s.

Let us now present a typical process with independent increments, the Brownian motion. It is an example of Gaussian, ergodic, non stationary process, with independent increments. A trajectory of a Brownian motion is shown in Fig. 4.1.

Definition 4.38 The process $\mathbf{W} = (W_t)_{t \in \mathbb{R}_+}$ taking values in $\mathbb{R}$, with independent increments and such that $W_0 = 0$, and for all $0 \le t_1 < t_2$,

$$W_{t_2} - W_{t_1} \sim \mathcal{N}(0, t_2 - t_1),$$

is called a standard Brownian motion (or Wiener process) with parameters ν and σ^2.

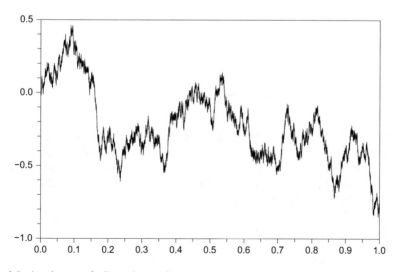

Fig. 4.1 A trajectory of a Brownian motion

The process $\mathbf{W} = (W_t)_{t \in \mathbb{R}_+}$ taking values in $\mathbb{R}$, with independent increments and such that $W_0 = 0$, and for all $0 \le t_1 < t_2$,

$$W_{t_2} - W_{t_1} \sim \mathcal{N}(\nu(t_2 - t_1), \sigma^2(t_2 - t_1)),$$

is called a Brownian motion with drift parameter ν and diffusion parameter σ^2.

Considering the random walk (S_n) of Example 1.77, with $p = 1/2$, yields an elementary construction of a Brownian motion. Indeed, setting

$$X_t = S_{[nT]}, \quad t \in \mathbb{R}_+$$

defines a continuous time process. If $t = nT$, then $\mathbb{E}\, X_t = 0$ and $\mathbb{V}\mathrm{ar}\, X_t = ts^2/T$. Let now t be fixed. If both s and T tend to 0, then the variance of X_t remains fixed and non null if and only if $s \approx \sqrt{T}$.

Let us set $s^2 = \sigma^2 T$ where $\sigma^2 \in \mathbb{R}_+^*$ and define a process $\mathbf{W}$ by

$$W_t(\omega) = \lim_{T \to 0} X_t(\omega), \quad t \in \mathbb{R}_+, \ \omega \in \Omega.$$

Taking the limit yields $\mathbb{E}\, W_t = 0$ and $\mathbb{V}\mathrm{ar}\, W_t = \sigma^2 t$.

Let us show that $W_t \sim \mathcal{N}(0, \sigma^2 t)$ by determining its distribution function at any $w \in \mathbb{R}_+$. Set $r = w/s$ and $T = t/n$. If w and t are fixed and T tends to 0, since $s \approx \sqrt{T}$, we get $r \approx \sqrt{n}$. Since $\mathbb{P}[X_{nT} = (2k-n)s] = \binom{n}{k} p^k (1-p)^{n-k}$, we obtain by applying de Moivre-Laplace theorem

$$\mathbb{P}(S_n \le rs) \approx \int_{-\infty}^{r/\sqrt{n}} \frac{1}{\sqrt{2\pi}} e^{-t^2/2} \, dt,$$

or, since $r/\sqrt{n} = w/\sqrt{\sigma^2 t}$,

$$\mathbb{P}(W_t \le w) \approx \int_{-\infty}^{w/\sqrt{\sigma^2 t}} \frac{1}{\sqrt{2\pi}} e^{-t^2/2} \, dt.$$

Let us show that if $0 \le t_1 < t_2 < t_3$, then $W_{t_2} - W_{t_1}$ and $W_{t_3} - W_{t_2}$ are independent. If $0 < n_1 < n_2 < n_3$, then the number of "heads" obtained between the n_1-th and the n_2-th tossings is independent of the number of "heads" obtained between the n_2-th and n_3-th tossings. Hence $S_{n_2} - S_{n_1}$ and $S_{n_3} - S_{n_2}$ are independent, and taking the limit yields the result.

Finally, let us compute the covariance function of $\mathbf{W}$. If $t_1 < t_2$, then $W_{t_2} - W_{t_1}$ and $W_{t_1} - W_0$ are independent. But $W_0 = 0$, so

$$\mathbb{E}\,[(W_{t_2} - W_{t_1})W_{t_1}] = [\mathbb{E}\,(W_{t_2} - W_{t_1})]\mathbb{E}\, W_{t_1} = 0.$$

Since we also have $\mathbb{E}[(W_{t_2} - W_{t_1})W_{t_1}] = \mathbb{E}(W_{t_2}W_{t_1}) - \mathbb{E}(W_{t_1}^2)$, we obtain $\mathbb{E}(W_{t_2}W_{t_1}) = \mathbb{E}W_{t_1}^2 = \sigma^2 t_1$, or $R_\mathbf{W}(t_1, t_2) = \sigma^2(t_1 \wedge t_2)$.

Note that $X_t/\sqrt{n} = S_{[nt]}/\sqrt{n}$ gives an approximation of W_t that can be simulated as a sum of Bernoulli variables.

4.4 Point Processes on the Line

A point process is a stochastic process consisting of a finite or enumerable family of points set at random in an arbitrary space, for example gravel on a road, stars in a part of the sky, times of occurrences of failures of a given system, positions of one of the basis A, C, G, T of an DNA sequence, etc.

In a mathematical sense, a point can be multiple. Even if the space to which the considered points belong can be any topological space, we will here consider $\mathbb{R}^d$, for $d \geq 1$. After some general notions on point processes we will present the renewal process and the Poisson process on $\mathbb{R}_+$.

4.4.1 Basics on General Point Processes

The point processes are naturally defined through random point measures.

Definition 4.39 Let μ be a measure on $(\mathbb{R}^d, \mathcal{B}(\mathbb{R}^d))$ and let $(x_i)_{i \in I}$ be a sequence of points in $\mathbb{R}^d$, with $I \subset \mathbb{N}$. If

$$\mu = \sum_{i \in I} \delta_{x_i},$$

and if $\mu(K) < +\infty$ for all compact subsets K of $\mathbb{R}^d$, then μ is called a point measure on $\mathbb{R}^d$.

A point measure is a discrete measure. The multiplicity of $x \in \mathbb{R}^d$ is $\mu(\{x\})$. When $\mu(\{x\}) = 0$ or 1 for all $x \in \mathbb{R}^d$, then μ is said to be simple. If $\mu(\{x\}) = 1$ for all $x \in \mathbb{R}^d$, the measure $\mu(A)$ is equal to the number of points belonging to A, for all $A \in \mathcal{B}(\mathbb{R}^d)$.

Definition 4.40 A function $\mu : \Omega \times \mathcal{B}(\mathbb{R}^d) \longrightarrow \overline{\mathbb{R}}$ such that $\mu(\omega, \cdot)$ is a point measure on $\mathbb{R}^d$ for all ω is called a random point measure on $\mathbb{R}^d$.

When $\mu(\omega, \{x\}) = 0$ or 1 for all ω and x, the random measure μ is also said to be simple.

▷ *Example 4.41* Let n points in $\mathbb{R}^d$ be set at random positions $X_1, \ldots, X_n$. A random point measure on $\mathbb{R}^d$ is defined by setting for all $A \in \mathcal{B}(\mathbb{R}^d)$,

$$\mu(A) = \sum_{i=1}^{n} \delta_{X_i}(A),$$

the random number of points belonging to A.   ◁

▷ *Example 4.42* Let $(T_n)_{n \in \mathbb{N}^*}$ be the sequence of times in $\mathbb{R}_+$ of failures of a system. Then setting for all $s < t$ in $\mathbb{R}_+$,

$$\mu([s, t]) = \sum_{i \geq 1} \delta_{T_i}([s, t]),$$

the random number of failures observed in the time interval $[s, t]$, defines a random point measure on $\mathbb{R}_+$.   ◁

Let $M_p(\mathbb{R}^d)$ denote the set of all point measures on $\mathbb{R}^d$.

Definition 4.43 A function $\mathbf{N} : \Omega \times \mathcal{B}(\mathbb{R}^d) \longrightarrow M_p(\mathbb{R}^d)$ such that $N(\cdot, A)$ are random variables for all $A \in \mathcal{B}(\mathbb{R}^d)$ is called a point process.

The variables $N(\cdot, A)$, taking values in $\overline{\mathbb{N}}$, are called the counting random variables of the point process $\mathbf{N}$, and the measure m defined by

$$m(A) = \mathbb{E}[N(\cdot, A)], \quad A \in \mathcal{B}(\mathbb{R}^d),$$

is its mean measure.

Each point process $\mathbf{N}$ is associated with the random point measure μ defined by

$$\mu(\cdot, A) = \int \mathbb{1}_A d\mu = N(\cdot, A).$$

It is said to be simple if m is simple.

Thus, $N(\omega, A)$ counts the number of points of the process belonging to A for the outcome $\omega \in \Omega$. Note that $m(A)$ can be infinite even when $N(\cdot, A)$ is a.s. finite.

If m has a density $\lambda : \mathbb{R}^d \longrightarrow \mathbb{R}_+$, that is $m(dx) = \lambda(x)dx$ or $\mathbb{P}[N(\cdot, dx) = 1] = \lambda(x)dx$, or

$$m(A) = \int_A \lambda(x)dx, \quad A \in \mathcal{B}(\mathbb{R}^d),$$

then the function λ, called the intensity of the process $\mathbf{N}$, is locally summable— meaning that λ is integrable over all bounded rectangles of $\mathbb{R}^d$.

Let us present some properties of integration with respect to a point measure.

Let $f : \mathbb{R}^d \longrightarrow \mathbb{R}$ be a Borel function and μ a point measure on $\mathbb{R}^d$. The integral of f with respect to μ is

$$\mu(f) = \int_{\mathbb{R}^d} f d\mu = \sum_{i \in I} \int_{\mathbb{R}^d} f d\delta_{x_i} = \sum_{i \in I} \delta_{x_i}(f) = \sum_{i \in I} f(x_i).$$

The mean is

$$\mathbb{E} \, N(f) = \int_{\mathbb{R}^d} f dm,$$

and the Laplace functional associated with $\mathbf{N}$ is defined as

$$L_N(f) = \mathbb{E} \left(\exp \left[- \int_{\mathbb{R}^d} f(x) N(\cdot, dx) \right] \right).$$

Especially, if f is positive and if $N = \sum_{n \geq 1} \delta_{X_n}$, then

$$L_N(f) = \mathbb{E} \left(\exp \left[- \sum_{n \geq 1} f(X_n) \right] \right).$$

The Laplace functional characterizes the distribution of a point process. Indeed, the distribution of a point process $\mathbf{N}$ is given by its finite-dimensional distributions, that is by the distributions of all the random vectors $(N(\cdot, A_1), \ldots, N(\cdot, A_n))$ for $n \geq 1$ and $A_1, \ldots, A_n \in \mathcal{B}(\mathbb{R}^d)$. The function

$$L_N \left(\sum_{i=1}^n s_i \mathbb{1}_{A_i} \right) = \mathbb{E} \left(\exp \left[- \sum_{i=1}^n s_i N(\cdot, A_i) \right] \right)$$

is precisely the Laplace transform of the distribution of $(N(\cdot, A_1), \ldots, N(\cdot, A_n))$ that characterizes its distribution.

4.4.2   Renewal Processes

Renewal processes are punctual processes defined on $\mathbb{R}_+$, modeling many experiments in applied probability—in reliability, queues, insurance, risk theory... They are also valuable theoretical tools for investigating more complex processes, such as regenerative, Markov or semi-Markov processes. A renewal process can be regarded as a random walk with positive increments; the times between occurring events are i.i.d. It is not a Markovian process but a semi-Markov process, that is studied by specific methods.

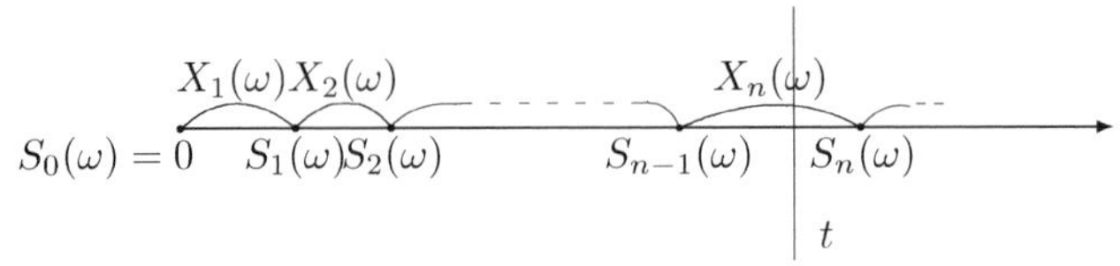

Fig. 4.2 A trajectory of a renewal process, on $(N_t = n)$

Definition 4.44 Let (X_n) be an i.i.d. sequence of positive variables. Set

$$S_0 = 0 \quad \text{and} \quad S_n = X_1 + \cdots + X_n, \quad n \geq 1. \tag{4.4}$$

The random sequence (S_n) is called a renewal process. The random times S_n are called renewal times. The associated counting process is defined by $\mathbf{N} = (N_t)_{t \in \mathbb{R}_+}$, where

$$N_t = \sum_{n \geq 0} \mathbb{1}_{[0,t]}(S_n) = \inf\{n \geq 0 : S_n \leq t\}, \quad t \in \mathbb{R}_+. \tag{4.5}$$

Generally, (X_n) is the sequence of inter-arrival times of some sequence of events, and N_t counts the number of events in the time interval $[0, t]$. Note that $(N_t = n) = (S_{n-1} \leq t < S_n)$ and that $N_0 = 1$. A trajectory of a renewal process is shown in Fig. 4.2.

When each of the variables X_n has an exponential distribution with parameter λ, the counting process $\mathbf{N}$ is called a Poisson process with parameter λ, in which case it is usual to set $N_0 = 0$; the Poisson processes will be especially investigated in the next section.

A counting process can also be regarded as a simple point process. Indeed,

$$N(\cdot, A) = \sum_{n \geq 0} \delta_{S_n}(A), \quad A \in \mathcal{B}(\mathbb{R}_+), \tag{4.6}$$

defines a point process, and we obtain for $A = [0, t]$,

$$N_t = N(\cdot, [0, t]), \quad t \geq 0.$$

▷ *Example 4.45 (A Renewal Process in Reliability)* A new component begins operating at time $S_0 = 0$. Let X_1 denote its lifetime. When it fails, it is automatically and instantly replaced by a new identical component. When the latter fails after a time X_2, it is renewed, and so on.

If (X_n) is supposed to be i.i.d., then (4.4) defines a renewal process (S_n) whose distribution is that of the sum of the life durations of the components. The counting process $(N_t)_{t \in \mathbb{R}_+}$ gives the number of components used in $[0, t]$, of which the last component still works at time t.   ◁

The expectation of the counting process at time t (or expected number of renewals), is

$$m(t) = \mathbb{E} N_t = \sum_{n \geq 0} \mathbb{P}(S_n \leq t),$$

and m is called the renewal function. If the variables X_n are not degenerated, this function is well-defined. Therefore, we will suppose thereafter that $F(0) < 1$, where F is the distribution function of X_n. The distribution function of S_n is the n-th Lebesgue-Stieltjes convolution of F, that is

$$F^{*(n)}(t) = \int_{\mathbb{R}_+} F^{*(n-1)}(t-x) dF(x),$$

with $F^{*(0)}(t) = \mathbb{1}_{\mathbb{R}_+}(t)$ and $F^{*(1)}(t) = F(t)$, so that

$$m(t) = \sum_{n \geq 0} F^{*(n)}(t), \quad t \in \mathbb{R}_+. \tag{4.7}$$

The mean measure m of the point process (4.6) is given by

$$m(A) = \mathbb{E} N(\cdot, A), \quad A \in \mathcal{B}(\mathbb{R}_+),$$

and we have $m(t) = m([0, t])$ pour $t \in \mathbb{R}_+$. Note that we use the same notation for both the renewal function and the mean measure.

When F is absolutely continuous with respect to the Lebesgue measure, the derivative of the renewal function $\lambda(t) = m'(t)$ is called the renewal density (or renewal rate) of the process.

Proposition 4.46 *The renewal function is increasing and finite.*

Proof For all $s > 0$, we have $N_{t+s} \geq N_t$, so m is increasing.

Assume that $F(t) < 1$ for $t > 0$. Then $F^{*(n)}(t) \leq [F(t)]^n$ for all n, and hence

$$m(t) \leq 1 + F(t) + [F(t)]^2 + \cdots \leq \frac{1}{1 - F(t)},$$

so $m(t)$ is finite. The general case is omitted. $\qquad \Box$

Relation (4.7) implies straightforwardly that $m(t)$ is a solution of

$$m(t) = 1 + \int_0^t m(t-x) dF(x), \quad t \in \mathbb{R}_+.$$

This equation is a particular case of the scalar renewal equation

$$h = g + F * h, \tag{4.8}$$

where h and g are functions bounded on the finite intervals of $\mathbb{R}_+$. The solution of this equation is determined by use of the renewal function m, as follows.

Proposition 4.47 *If* $g \colon \mathbb{R}_+ \longrightarrow \mathbb{R}$ *is bounded on the finite intervals of* $\mathbb{R}$, *then* (4.8) *has a unique solution* $h \colon \mathbb{R}_+ \longrightarrow \mathbb{R}$ *that is bounded on the finite intervals of* $\mathbb{R}$, *given by*

$$h(t) = m * g(t) = \int_0^t g(t - x) dm(x),$$

where m *is defined by (4.7), and extended to* $\mathbb{R}_-$ *by 0.*

Proof We deduce from (4.7) that

$$F * (m * g) = F * g + F^{*(2)} * g + \cdots = m * g - g.$$

Thus, $m * g$ is a solution of (4.8).

Suppose that $k \colon \mathbb{R}_+ \longrightarrow \mathbb{R}$ is another solution of (4.8) bounded on the finite intervals of $\mathbb{R}$. Then $h - k = F * (h - k)$. Since $F^{*(n)}$ tends to zero when n tends to infinity and $k - h$ is bounded, it follows that $h - k = F^{*(n)} * (h - k)$ and hence $k = h$. $\qquad\square$

Many extensions of renewal processes exist.

If S_0 is a nonnegative variable not identically zero, independent of (X_n) and with distribution function F_0 different from the renewal distribution function F, then the process (S_n) is said to be delayed or modified. When

$$F_0(x) = \frac{1}{\mu} \int_0^x [1 - F(u)] du, \quad x > 0,$$

the delayed renewal process is said to be stationary.

Definition 4.48 Let (Y_n) and (Z_n) be two i.i.d. nonnegative independent random sequences, with respective distribution functions G and H. The associated process defined by

$$S_0 = 0 \quad \text{and} \quad S_n = S_{n-1} + Y_n + Z_n, \quad n \geq 1,$$

is called an alternated renewal process.

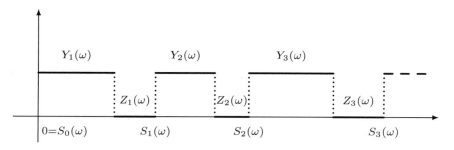

Fig. 4.3 A trajectory of an alternated renewal process

Such an alternated renewal process (S_n) is shown to be a renewal process in the sense of Definition 4.44 by setting $X_n = Y_n + Z_n$ and $F = G * H$. A trajectory of an alternated renewal process is shown in Fig. 4.3.

Still another extension is the stopped renewal process, also called transient renewal process.

Definition 4.49 Let (X_n) be an i.i.d. random sequence taking values in $\overline{\mathbb{R}}_+$, with defective distribution function F. The associated renewal process defined by (4.4) is called a stopped renewal process.

The life duration of this process is $T = \sum_{n=1}^{N} X_n$, where N is the number of events at the stopping time of the process, defined by

$$(N = n) = (X_1 < +\infty, \ldots, X_n < +\infty, X_{n+1} = +\infty).$$

▷ *Example 4.50* If N has a geometric distribution on $\mathbb{N}^*$ with parameter q, then the distribution function of T is

$$F_T(t) = \sum_{n \geq 1} F^{*(n)}(t)(1 - q)^{n-1}q.$$

Indeed, $\mathbb{P}(N = n) = (1 - q)^{n-1}q$ for $n \geq 1$, and the distribution function of T follows by Point 2. of Proposition 1.73.   ◁

▷ *Example 4.51 (Risk Process in Insurance)* Let $u > 0$ be the initial capital of an insurance company. Let (S_n) be the sequence of times at which accidents occur, and let $\mathbf{N} = (N_t)$ denote the associated counting process. Let (Y_n) be the sequence of compensations paid at each accident. The capital of the company at time t is $U_t = u + ct - \sum_{n=1}^{N_t} Y_n$, where c is the rate of subscriptions in $[0, t]$. A trajectory of such a process is shown in Fig. 4.4. The time until ruin is T.

The linked problems are the ruin in a given time interval, that is $\mathbb{P}(U_t \leq 0)$ with limit $\mathbb{P}(\lim_{t \to +\infty} U_t \leq 0)$, and the mean viability of the company, that is

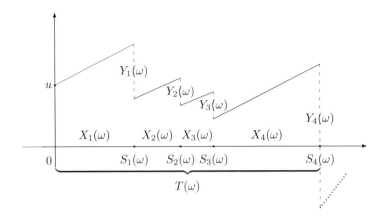

Fig. 4.4 A trajectory of a risk process

$\mathbb{E}\, U_t \geq 0$. Different approaches exist for solving these issues, involving either renewal theory or martingale theory. Note that a particular case of risk process, the Cramér-Lundberg process, will be presented in the next section.   ◁

Renewal process can also be considered in the vector case. Another extension will also be studied below, the Markov renewal process.

4.4.3 Poisson Processes

The renewal process whose counting process is a Poisson process is the most used in modeling real experiments. We keep the notation of the preceding section.

Definition 4.52 If $X_n \sim \mathcal{E}(\lambda)$ for $n \in \mathbb{N}^*$, the counting process $\mathbf{N} = (N_t)_{t \in \mathbb{R}_+}$ defined by (4.5) p. 195 for $t \in \mathbb{R}_+$, with $N_0 = 0$, is called a homogeneous Poisson process with intensity (or parameter) λ.

Thanks to the absence of memory of the exponential distribution, the probability that an event occurs for the first time after time $s + t$ given that it did not occur before time t is equal to the probability that it occurs after time s. More generally, the following result holds true.

Theorem 4.53 *A Poisson process $\mathbf{N}$ with intensity λ is homogeneous—with independent and stationary increments—and satisfies $N_t \sim \mathcal{P}(\lambda t)$ for $t \in \mathbb{R}_+$.*

Proof Let us show first that $N_t \sim \mathcal{P}(\lambda t)$ for all $t > 0$. We have

$$\mathbb{P}(N_t = k) = \mathbb{P}(S_k \leq t) - \mathbb{P}(S_{k+1} \leq t).$$

According to Example 1.66, $S_n \sim \gamma(n, \lambda)$, so

$$\mathbb{P}(S_n \leq x) = \int_0^x \frac{1}{(n-1)!} \lambda^n e^{-\lambda t} t^{n-1} \, dt = 1 - e^{-\lambda x} \sum_{k=1}^{n-1} \frac{(\lambda x)^k}{k!},$$

and hence $\mathbb{P}(N_t = k) = e^{-\lambda t}(\lambda t)^k / k!$.

Let us now show that the increments of the process are independent. We begin by determining the joint distribution of N_t and N_s, for $0 \leq s \leq t$.

$$\mathbb{P}(N_s = k, N_t = n) =$$

$$= \mathbb{P}(S_k \leq s < S_{k+1}, S_n \leq t < S_{n+1})$$

$$= \int_{E_{n+1}} \mathbb{1}_{[x_k, x_{k+1}[}(s) \lambda^{n+1} e^{-\lambda x_{n+1}} \mathbb{1}_{]x_n, x_{n+1}[}(t) dx_1 \ldots dx_{n+1}$$

where $E_n = \{0 < x_1 < \cdots < x_n\}$. Thus, by Fubini's theorem,

$$\mathbb{P}(N_s = k, N_t = n) =$$

$$= \int_{E_n} \mathbb{1}_{[x_k, x_{k+1}[}(s) \lambda^n \mathbb{1}_{[x_n, +\infty[}(t) dx_1 \ldots dx_n \int_t^{+\infty} \lambda e^{-\lambda x_{n+1}} dx_{n+1}$$

$$= \lambda^n e^{-\lambda t} \int_{E_n} \mathbb{1}_{]x_k, x_{k+1}[}(s) \mathbb{1}_{]x_n, +\infty[}(t) dx_1 \ldots dx_n$$

$$= \lambda^n e^{-\lambda t} \int_E dx_1 \ldots dx_k \int_F dx_{k+1} \ldots dx_n = \lambda^n e^{-\lambda t} \frac{s^k}{k!} \frac{(t-s)^{n-k}}{(n-k)!},$$

where $E = \{0 < x_1 < \cdots < x_k < s\}$ and $F = \{s < x_{k+1} < \cdots < x_n < t\}$. Therefore,

$$\mathbb{P}(N_s = k, N_t - N_s = l) = e^{-\lambda s} \frac{(\lambda s)^k}{k!} \cdot e^{-\lambda(t-s)} \frac{[\lambda(t-s)]^l}{l!},$$

and the result follows.   □

Note that Theorem 4.53 can also be taken as an alternative definition of the Poisson process, under the following form.

Definition 4.54 A stochastic process $\mathbf{N} = (N_t)_{t \in \mathbb{R}_+}$ is a homogeneous Poisson process if it is a process with independent and stationary increments such that $N_t - N_s \sim \mathcal{P}(\lambda(t-s))$ for all $t \geq s > 0$.

Indeed, the associated renewal process can be defined by $S_0 = 0$ and then S_n recursively though the relation $(N_t = n) = (S_n \leq t < S_{n+1})$. Setting $X_n = S_n - S_{n-1}$, we get $\mathbb{P}(X_1 > t) = \mathbb{P}(N_t = 0) = e^{-\lambda t}$, and hence $X_1 \sim \mathcal{E}(\lambda)$.

The Poisson process can also be defined more qualitatively, as follows.

Definition 4.55 A stochastic process $\mathbf{N} = (N_t)_{t \in \mathbb{R}_+}$ is a Poisson process if it is a process with independent increments such that $t \longrightarrow N_t(\omega)$ is for almost all ω an increasing step function with jumps of size 1.

Let us now define the compound Poisson process.

Definition 4.56 Let $\mathbf{N}$ be a (homogeneous) Poisson process. Let (Y_n) be an i.i.d. random sequence with finite mean and variance, and independent of $\mathbf{N}$. The stochastic process ξ defined on $\mathbb{R}_+$ by

$$\xi_t = \sum_{n=1}^{N_t} Y_n, \quad t \geq 0, \tag{4.9}$$

with $\xi_t = 0$ if $N_t = 0$, is called a (homogeneous) compound Poisson process.

The compound Poisson process has independent increments.

▷ *Example 4.57 (Cramér-Lundberg Process)* With the notation of Example 4.51, we suppose here that (Y_n) is i.i.d. with distribution function G with mean μ and that $\mathbf{N}$ is a homogeneous Poisson process with intensity λ independent of (Y_n). Thus, the process ξ defined by (4.9) is a homogeneous compound Poisson process, and (U_t) is called a Cramér-Lundberg process.

We compute $\mathbb{E} U_t = u + ct - \mathbb{E} N_t \mathbb{E} Y_1 = u + ct - \lambda \mu t$. This gives a condition of viability of the company, namely $c - \lambda \mu > 0$. The probability of ruin before time t is $r(t) = \mathbb{P}(U_t \leq 0) = \mathbb{P}(\xi_t \geq u + ct)$. Since the distribution function of ξ_t is

$$\mathbb{P}(\xi_t \leq x) = \sum_{n \geq 0} \mathbb{P}\left(\sum_{i=1}^{n} Y_i \leq x, N_t = n \right) = \sum_{n \geq 0} \mathbb{P}\left(\sum_{i=1}^{n} Y_i \leq x \right) \mathbb{P}(N_t = n)$$

$$= \sum_{n \geq 0} e^{-\lambda t} \frac{(\lambda t)^n}{n!} G^{*(n)}(x),$$

we get

$$r(t) = \sum_{n \geq 0} e^{-\lambda t} \frac{(\lambda t)^n}{n!} G^{*(n)}(u + ct).$$

The probability of ruin of the company—in other words the probability that the life duration T of the process is finite—is

$$\mathbb{P}(\lim_{t \to +\infty} U_t \leq 0) = \mathbb{P}(T < +\infty) = \lim_{t \to +\infty} r(t),$$

but this quantity remains difficult to compute under this form.   ◁

4.4.4 Asymptotic Results for Renewal Processes

All throughout this section, (S_n) will be a renewal process such that $0 < \mu = \mathbb{E} X_1 < +\infty$, with counting process $(N_t)_{t \in \mathbb{R}_+}$. Recall that $m(t) = \mathbb{E} N_t$ defines the renewal function of the process, and that we suppose $F(0) < 1$, where F is the distribution function of X_n.

Proposition 4.58 *The following statements hold true:*

1. S_n *tends a.s. to infinity when n tends to infinity.*
2. S_n *and* N_t *are a.s. finite for all* $n \in \mathbb{N}$ *and* $t \in \mathbb{R}_+^*$.
3. N_t *tends a.s. to infinity and* $m(t)$ *tends to infinity when t tends to infinity.*

Proof

1. The law of large numbers implies that S_n/n tends a.s. to μ, so S_n tends a.s. to infinity.
2. Since $(S_n \leq t) = (N_t \geq n + 1)$, it derives from Point 1. that N_t is a.s. finite.
   Moreover, $(S_n < +\infty) = \cap_{i=1}^n (X_i < +\infty)$ for $n \in \mathbb{N}^*$, and X_n is $\mathbb{P}$-a.s. finite, so S_n is a.s. finite for all $n \in \mathbb{N}^*$.
3. We compute

$$\mathbb{P}(\lim_{t \to +\infty} N_t < +\infty) = \mathbb{P}[\cup_{n \geq 1}(X_n = +\infty)] \leq \sum_{n \geq 1} \mathbb{P}(X_n = +\infty) = 0.$$

Therefore, N_t tends a.s. to infinity, from which it follows that $m(t)$ tends to infinity.   □

Proposition 4.59 *The following convergence holds true,*

$$\frac{1}{t} N_t \xrightarrow{a.s.} \frac{1}{\mu}, \quad t \to +\infty.$$

The induced convergence of $m(t)/t$ to $1/\mu$ is known as the elementary renewal theorem.

Proof Thanks to the law of large numbers, S_n/n tends a.s. to μ. Moreover, N_t is a.s. finite and tends a.s. to infinity when t tends to infinity. Thanks to Theorem 4.17, S_{N_t}/N_t tends a.s. to μ, and the inequality

$$\frac{S_{N_t}}{N_t} \leq \frac{t}{N_t} < \frac{S_{N_t+1}}{N_t+1} \frac{N_t+1}{N_t}$$

yields the result.   □

▷ *Example 4.60 (Cramér-Lundberg Process (Continuation of Example 4.57))* We have

$$\frac{1}{t}U_t = \frac{u}{t} + c - \frac{N_t}{t}\frac{1}{N_t}\sum_{n=1}^{N_t} X_n.$$

According to Proposition 4.59, N_t/t tends to $1/\lambda$. Hence, thanks to Theorem 4.17, U_t/t tends a.s. to $c - \mu/\lambda$ when t tends to infinity.   ◁

The next result is an extension of the central limit theorem to renewal processes.

Theorem 4.61 *If* $0 < \sigma^2 = \mathbb{V}\mathrm{ar}\, X_1 < +\infty$, *then*

$$\frac{N_t - t/\mu}{\sqrt{t\sigma^2/\mu^3}} \xrightarrow{\mathcal{L}} \mathcal{N}(0, 1), \quad t \to +\infty.$$

Proof Set

$$Z_t = \sqrt{t}\sqrt{\frac{\mu^3}{\sigma^2}}\left(\frac{N_t}{t} - \frac{1}{\mu}\right).$$

We compute $\mathbb{P}(Z_t \leq x) = \mathbb{P}(N_t \leq t/\mu + x\sqrt{t\sigma^2/\mu^3})$. If n_t denotes the integer part of $t/\mu + x\sqrt{t\sigma^2/\mu^3}$, then

$$\mathbb{P}(Z_t \leq x) = \mathbb{P}(S_{n_t} \geq t) = \mathbb{P}\left(\frac{S_{n_t} - n_t\mu}{\sigma\sqrt{n_t}} \geq \frac{t - n_t\mu}{\sigma\sqrt{n_t}}\right).$$

By the central limit theorem, $(S_{n_t} - n_t\mu)/\sigma\sqrt{n_t}$ tends in distribution to a standard Gaussian variable. Moreover, $n_t \approx x\sqrt{t\sigma^2/\mu^3} + t/\mu$ when t tends to infinity. Hence $t - n_t\mu \approx -x\sqrt{t\sigma^2/\mu}$ and $\sigma\sqrt{n_t} \approx \sigma\sqrt{t/\mu}$, so $(t - n_t\mu)/\sigma\sqrt{n_t} \approx -x$, and the conclusion follows.   □

We state without proof the following two renewal theorems. A distribution function F is said to be arithmetic with period δ if the distribution is concentrated on $\{x_0 + n\delta : n \in \mathbb{N}\}$.

Theorem 4.62 (Blackwell's Renewal) *If F is a non arithmetic distribution function on $\mathbb{R}_+$, then, for all $h > 0$,*

$$m(t) - m(t - h) \longrightarrow \frac{h}{\mu}, \quad t \to +\infty. \tag{4.10}$$

If F is arithmetic with period δ, the above result remains valid provided that h is a multiple of δ.

▷ *Example 4.63 (Poisson Process)* In this case, $F(t) = (1 - e^{-\lambda t})\mathbb{1}_{\mathbb{R}_+}(t)$, whose Laplace transform is $\widetilde{F}(s) = \lambda/(\lambda+s)$. Moreover, $N_0 = 0$, so the Laplace transform of m is

$$\widetilde{m}(t) = \sum_{n \geq 1} (\widetilde{F}(s))^n = \frac{1}{1 - \widetilde{F}(s)} - 1 = \frac{\lambda}{s}.$$

Inverting Laplace transform yields $m(t) = \lambda t$. Thus, the relation $m(t) - m(t - h) = \lambda h$ holds for all $t > h$. ◁

In order to state the key renewal theorem, let us introduce the direct Riemann integrable functions.

Definition 4.64 Let g be a function defined on $\overline{\mathbb{R}}_+$. Let $\overline{m}_n(a)$ denote the maximum and $\underline{m}_n(a)$ the minimum of g on $[(n - 1)a, na]$, for all $a > 0$. Then g is said to be direct Riemann integrable if $\sum_{n \geq 1} \underline{m}_n(a)$ and $\sum_{n \geq 1} \overline{m}_n(a)$ are finite for all $a > 0$ and if $\lim_{a \to 0} \sum_{n=1}^{\infty} \underline{m}_n(a) = \lim_{a \to 0} \sum_{n=1}^{\infty} \overline{m}_n(a)$.

▷ *Example 4.65* Any nonnegative, decreasing function integrable over $\mathbb{R}_+$ is direct Riemann integrable. Indeed, we get

$$\sum_{n \geq 1} (\overline{m}_n(a) - \underline{m}_n(a)) \leq g(0).$$

Any nonnegative function integrable over $\mathbb{R}_+$ and with a compact support is also direct Riemann integrable. ◁

Theorem 4.66 (Key Renewal) *If F is a non arithmetic distribution function on $\mathbb{R}_+$ and if $g : \mathbb{R}_+ \longrightarrow \mathbb{R}_+$ is direct Riemann integrable, then*

$$\int_0^t g(t - x)dm(x) \longrightarrow \frac{1}{\mu} \int_0^{+\infty} g(x)dx, \quad t \to +\infty. \tag{4.11}$$

If F is arithmetic with period δ and if $\sum_{k \geq 0} g(x + k\delta) < +\infty$, then

$$m * g(x + n\delta) \longrightarrow \frac{\delta}{\mu} \sum_{k \geq 0} g(x + k\delta), \quad n \to +\infty.$$

Note that Blackwell's renewal theorem and the key renewal theorem are equivalent in the sense that (4.10) and (4.11) are equivalent.

For stopping renewal processes, the key renewal theorem takes the following form.

Proposition 4.67 *Let F be a defective distribution function on $\mathbb{R}_+$. If $g : \mathbb{R}_+ \longrightarrow \mathbb{R}_+$ is direct Riemann integrable and such that $g(+\infty) = \lim_{t \to +\infty} g(t)$ exists, then the solution of the renewal equation (4.8) p. 198 satisfies*

$$h(t) = m * g(t) \longrightarrow \frac{g(+\infty)}{q}, \quad t \to +\infty,$$

where $q = 1 - F(+\infty)$.

Proof According to Proposition 4.47, $h(t) = m * g(t)$. According to relation (4.7) p. 197, the limit of $m(t)$ when t tends to infinity is

$$1 + F(+\infty) + F(+\infty)^2 + \cdots = \frac{1}{1 - F(+\infty)} = \frac{1}{q},$$

and the result follows.   □

▷ *Example 4.68 (Cramér-Lundberg Process (Continuation or Example 4.57))* Let us determine the probability of ruin of the Cramér-Lundberg process. Set $\zeta(u) = \mathbb{P}(T = +\infty) = 1 - \mathbb{P}(T < +\infty)$, where T is the time of ruin of the process. We compute

$$\begin{aligned}
\zeta(u) &= \int_0^{+\infty} \int_0^{u+cs} \mathbb{P}(S_1 \in ds, Y_1 \in dy, T \circ \theta_s = +\infty) \\
&= \int_0^{+\infty} \int_0^{u+cs} \mathbb{P}(S_1 \in ds)\mathbb{P}(Y_1 \in dy, T \circ \theta_s = +\infty \mid S_1 = s) \\
&= \int_0^{+\infty} \mathbb{P}(S_1 \in ds) \int_0^{u+cs} \mathbb{P}(T \circ \theta_s = +\infty \mid S_1 = s, Y_1 = y)\mathbb{P}(Y_1 \in dy) \\
&= \int_0^{+\infty} \lambda e^{-\lambda s} ds \int_0^{u+cs} \zeta(u + cs - y)dG(y).
\end{aligned}$$

By the change of variable $v = u + cs$, we get

$$\zeta(u) = \lambda_0 \int_u^{+\infty} e^{-\lambda_0(v-u)} dv \int_0^v \zeta(v - y) dG(y) = \lambda_0 e^{\lambda_0 u} g(u),$$

where

$$g(u) = \int_u^{+\infty} \int_0^v \lambda_0 e^{-\lambda_0(v-u)} dv \zeta(v - y) dG(y),$$

and $\lambda_0 = \lambda/c$. Differentiating this relation gives

$$\zeta'(u) = \lambda_0 e^{\lambda_0 u} [\lambda_0 g(u) + g'(u)] = \lambda_0 \zeta(u) - \lambda_0 G * \zeta(u) = \lambda_0 [1 - G] * \zeta(u).$$

Integrating the above differential equation on $[0, u]$ yields

$$\zeta(u) = \zeta(0) + \lambda_0 \int_0^u \zeta(u - y)[1 - G(y)] dy, \quad u \geq 0. \tag{4.12}$$

This equation is a renewal equation with defective distribution function with density $L(y) = \lambda_0[1 - G(y)]$, where $L(+\infty) = \lambda_0 \mu < 1$. The case $L(+\infty) = 1$ is excluded because then $\zeta(u) = 0$ for all $u \in \mathbb{R}_+$. Thanks to the key renewal theorem for stopped renewal processes,

$$\zeta(+\infty) = \frac{\zeta(0)}{1 - L(+\infty)},$$

or, finally, since $\zeta(+\infty) = 1$,

$$\zeta(0) = 1 - \frac{\lambda \mu}{c}.$$

This allows the computation of $\zeta(u)$ for all $u \in \mathbb{R}_+$ through (4.12).

We have considered only nonnegative Y_1. The result remains valid for any variable Y_1: it is sufficient to take $-\infty$ and $u + cs$ instead of 0 and $u + cs$ as bounds of the second integral in the above computation of $\zeta(u)$. ◁

4.5 Exercises

▽ **Exercise 4.1 (The AR(1) Process on $\mathbb{N}$)** Let $(\varepsilon_n)_{n \in \mathbb{N}}$ be a white noise with variance 1. Let $a \in]-1, 1[$. An AR(1) process on $\mathbb{N}$ is defined by setting $X_n = aX_{n-1} + \varepsilon_n$ for $n > 0$ and $X_0 = \varepsilon_0$.

1. a. Write X_n as a function of $\varepsilon_0, \ldots, \varepsilon_n$ and determine its distribution.
   b. Determine the characteristic function of X_n.
   c. Give the distribution of $(X_0, \ldots, X_n)$.

2. Show that:
   a. (X_n) converges in distribution and give the limit.
   b. $(X_0 + \cdots + X_n)/n$ converges in probability to 0;
   c. $(X_0 + \cdots + X_n)/\sqrt{n}$ converges in distribution; give the limit.

Solution

1. a. For $n \geq 1$, we have $X_n = \sum_{p=1}^{n} a^{n-p}\varepsilon_p$, and hence $\mathbb{E}\, X_n = 0$ and

$$\mathbb{V}\mathrm{ar}\, X_n = a^{2n} + \cdots + a^2 + 1 = \frac{1 - a^{2(n+1)}}{1 - a^2}.$$

   b. Since $(\varepsilon_1, \ldots, \varepsilon_n)$ is a Gaussian vector, X_n is a Gaussian variable, so the characteristic function of X_n is

$$\phi_{X_n}(t) = \exp\left(-\frac{1 - a^{2(n+1)}}{1 - a^2}t^2\right).$$

   c. The vector $(X_0, \ldots, X_n)$ is a linear transform of the standard Gaussian vector $(\varepsilon_1, \ldots, \varepsilon_n)$, so is a Gaussian vector too, with mean $(0, \ldots, 0)$ and covariance matrix given by

$$\mathbb{C}\mathrm{ov}\,(X_j, X_{j+k}) = \mathbb{E}\,(X_j X_{j+k}) = \mathbb{E}\left[\left(\sum_{p=0}^{j} a^{j-p}\varepsilon_p\right)\left(\sum_{q=0}^{j+k} a^{j+k-q}\varepsilon_q\right)\right]$$

$$= a^k \sum_{l=0}^{j} a^{2l} = \frac{1 - a^{2(j+1)}}{1 - a^2}a^k.$$

2. a. Clearly, $\phi_{X_n}(t)$ tends to $\exp[-t^2/(1-a^2)]$, so X_n converges in distribution to a random variable with distribution $\mathcal{N}(0, 1/(1-a^2))$.
   b. We compute

$$\frac{1}{n}(X_0 + \cdots + X_n) = \frac{X_0}{n} + \frac{1}{n}\sum_{i=1}^{n}(aX_{i-1} + \varepsilon_i) = \frac{a}{n}\sum_{i=0}^{n-1}X_i + \frac{1}{n}\sum_{i=0}^{n}\varepsilon_i$$

$$= \frac{a}{n}(X_0 + \cdots + X_n) - \frac{a}{n}X_n + \frac{1}{n}\sum_{i=0}^{n}\varepsilon_i.$$

   We know that $X_n \sim \mathcal{N}(0, (1 - a^{2(n+1)})/(1 - a^2))$, so X_n/n converges to 0 in probability. Moreover, by the strong law of large numbers, $\sum_{i=0}^{n}\varepsilon_i/n$ converges a.s. to 0. Therefore, $(1 - a)(X_0 + \cdots + X_n)/n$ converges to 0 in probability, and hence $(X_0 + \cdots + X_n)/n$ too.

c. In the same way, by the central limit theorem, $\sum_{i=0}^{n} \varepsilon_i / \sqrt{n}$ converges in distribution to $\mathcal{N}(0, 1)$, so $(1-a)(X_0 + \cdots + X_n)/\sqrt{n}$ too, from which it follows that $(X_0 + \cdots + X_n)/\sqrt{n}$ converges in distribution to $\mathcal{N}(0, 1/(1-a)^2)$. $\triangle$

$\triangledown$ **Exercise 4.2 (Generalization of AR and MA Processes)** Let (γ_n) be a sequence of i.i.d. standard random variables. Let $\theta \in\,]-1, 1[$.

1. Set $V_n = \gamma_1 + \theta \gamma_2 + \cdots + \theta^{n-1} \gamma_n$, for $n \in \mathbb{N}^*$.
   a. Show that V_n converges in square mean—use the Cauchy criterion.
   b. Set $V = \sum_{i \geq 1} \theta^{i-1} \gamma_i$. Show that V_n tends a.s. to V.
2. Let X_0 be a random variable independent of (γ_n). Set $X_n = \theta X_{n-1} + \gamma_n$ for $n \geq 1$.
   a. Show that V_n and $X_n - \theta^n X_0$ have the same distribution for $n \geq 1$.
   b. Let ρ denote the distribution of V. Compute the mean and the variance of ρ. Show that X_n tends in distribution to ρ.
   c. Assume that $X_0 \sim \rho$. Show that $X_n \sim \rho$ for all n.

Solution

1. a. We have

$$\mathbb{E}\,(V_m - V_{n+m})^2 = \mathbb{E}\,(\theta^m \gamma_{m+1} + \cdots + \theta^{n+m-1} \gamma_{n+m})^2$$

$$= \sum_{i=0}^{n-1} \theta^{2(m+i)} \mathbb{E}\,(\gamma_{m+i}^2) + 2 \sum_{i<j} \theta^{m+i} \theta^{m+j} \mathbb{E}\,(\gamma_{m+i} \gamma_{m+j}),$$

so that $\mathbb{E}\,(V_m - V_{n+m})^2 = \sum_{i=0}^{n-1} \theta^{2(m+i)}$, which converges to 0.

   b. According to Proposition 1.80, it is sufficient to show that $\mathbb{P}(\overline{\lim}\, |V_n - V| > \varepsilon) = 0$, or, using Borel-Cantelli lemma, that $\sum_{n \geq 0} \mathbb{P}(|V_n - V| > \varepsilon)$ is finite for all $\varepsilon > 0$.

   Chebyshev's inequality gives $\mathbb{P}(|V_n - V| > \varepsilon) \leq \mathbb{E}\,[(V_n - V)^2]/\varepsilon^2$. Moreover,

$$\mathbb{E}\,[(V_n - V)^2] = \mathbb{E}\,[(\theta^{n+1} \gamma_{n+2} + \theta^{n+2} \gamma_{n+3} + \cdots)^2] = \frac{\theta^{2n}}{1 - \theta^2},$$

   so $\mathbb{P}(|V_n - V| > \varepsilon) \leq \theta^{2n}/(1 - \theta^2)\varepsilon^2$, and the sum of the series is finite.
2. a. We can write $X_n - \theta^n X_0 = \theta^{n-1} \gamma_1 + \cdots + \theta \gamma_{n-1} + \gamma_n$, from which the result follows, because all the γ_i have the same distribution.

b. The sequence (V_n) converges to V in square mean—so also in mean, hence $\mathbb{E}\,V = \lim_{n\to+\infty}\mathbb{E}\,V_n = 0$ and

$$\mathbb{V}\text{ar}\,V = \mathbb{E}\,(V^2) = \lim_{n\to+\infty}\mathbb{E}\,(V_n^2) = \lim_{n\to+\infty}\sum_{i=0}^{n}\theta^{2i}\mathbb{E}\,(\gamma_{i+1}^2)$$

$$= \lim_{n\to+\infty}\frac{1-\theta^{2n}}{1-\theta^2} = \frac{1}{1-\theta^2}.$$

Since $X_n - \theta^n X_0 \sim V_n$, we can write $X_n = \theta^n X_0 + U_n$, where U_n is a variable with the same distribution as V_n. Therefore, (U_n) converges in distribution to ρ too, and since $(\theta^n X_0)$ converges a.s. to 0, (X_n) converges in distribution to ρ.

c. We have $X_0 \sim V$ so $\theta X \sim \sum_{n\geq0}\theta^{n+1}\gamma_{n+1}$ or $\theta X \sim \sum_{n\geq1}\theta^n\gamma_n$. Since $\gamma_1 \sim \gamma_0$, it follows that $X_1 \sim \sum_{n\geq0}\theta^n\gamma_n$; since $\gamma_n \sim \gamma_{n+1}$ for all n, we obtain $X_1 \sim \sum_{n\geq0}\theta^n\gamma_{n+1}$, that is $X_1 \sim V \sim \rho$. The result follows by induction.

Note that if $\rho = \mathcal{N}(0, 1)$ and $\gamma_n \sim \mathcal{N}(0, 1)$, then X_n is an AR(1) process and (V_n) is an MA$(n-1)$ process on $\mathbb{N}$. $\triangle$

∇ **Exercise 4.3 (Sinusoidal Signals and Stationarity)** Let $\mathbf{X}$ be the stochastic process defined in Example 4.1. The variables v, A and ϕ are not supposed to be constant, unless specifically stated. The variable φ takes values in $[0, 2\pi[$.

1. Assume that $A = 1$ and $\varphi \sim \mathcal{U}(0, 2\pi)$.
   a. Show that if v is a.s. constant, then $\mathbf{X}$ is strictly stationary to the order 1.
   b. Show that if v is continuous with density f and is independent of φ, then $\mathbf{X}$ is weakly stationary to the order 2; determine its spectral density h.
2. Suppose that v and φ are constant.
   a. Give a necessary and sufficient condition on A for $\mathbf{X}$ to be weakly stationary to the order 1.
   b. Can $\mathbf{X}$ be weakly stationary to the order 2?
   c. Let $\mathbf{S} = \mathbf{X} + \mathbf{Y}$. Give a necessary and sufficient condition on A and B for S to be weakly stationary to the order 1, and then to the order 2.
3. Suppose that v is a.s. constant, that A is nonnegative and that A and φ are independent.
   a. Give a necessary and sufficient condition on φ for $\mathbf{X}$ to be weakly stationary to the order 1 and then 2.
   b. Give a necessary and sufficient condition on φ for $\mathbf{X}$ to be strictly stationary to the order 1.
   c. Let $\mathbf{Z}$ be the stochastic process defined by

$$Z_t = A\cos(vt + \varphi) + B\sin(vt + \varphi),$$

where $\varphi \sim \mathcal{U}(0, 2\pi)$ is independent of A and B. Show that $\mathbf{Z}$ is weakly stationary to the order 2.

Solution

1. a. Setting $\psi = \nu\tau + \varphi$, we obtain $X_{t+\tau} = \cos(\nu t + \psi)$. Since $\varphi \sim \mathcal{U}[0, 2\pi]$ and $\nu\tau$ is constant for a fixed τ, we have $\psi \sim \mathcal{U}[\nu\tau, \nu\tau + 2\pi]$, and hence $X_t \sim X_{t+\tau}$.

   b. We have $\mathbb{E}\, X_t = 0$ and $2\mathbb{E}\,(X_t X_{t+\tau}) = \mathbb{E}\,(\cos[\nu(2t+\tau)+2\varphi]) + \mathbb{E}\,[\cos(\nu\tau)]$. We compute

   $$\mathbb{E}\,(\cos[\nu(2t + \tau) + 2\varphi]) =$$
   $$= \mathbb{E}\,(\cos[\nu(2t + \tau)])\mathbb{E}\,[\cos(2\varphi)] - \mathbb{E}\,(\sin[\nu(2t + \tau)])\mathbb{E}\,[\sin(2\varphi)] = 0,$$

   and $\mathbb{E}\,[\cos(2\varphi)] = \mathbb{E}\,[\sin(2\varphi)] = 0$, so $\mathbf{X}$ is weakly stationary to the order 2, with $r(\tau) = \int_{\mathbb{R}} \cos(\lambda\tau) f(\lambda) d\lambda$. Since r is even and real,

   $$r(\tau) = \int_{\mathbb{R}} e^{i\lambda\tau} h(\lambda) d\lambda = \int_{\mathbb{R}_+} 2\cos(\lambda\tau) h(\lambda) d\lambda,$$

   and hence $h(\lambda) = [f(-\lambda) + f(\lambda)]/4$.

2. a. We have $\mathbb{E}\, X_t = [\cos(\nu t)]\mathbb{E}\, A$, so $\mathbf{X}$ is weakly stationary to the order 1 if and only if A is centered.

   b. We compute $2\mathbb{E}\,(X_t X_{t+\tau}) = [\cos(2\nu t + \nu\tau) + \cos(\nu t)]\mathbb{E}\,(A^2)$. This is a function of τ only if $\mathbb{E}\,(A^2) = 0$, that is to say if A is null. Then the signal itself is null.

   c. Therefore, $\mathbb{E}\, S_t = [\cos(\nu t)]\mathbb{E}\, A + [\sin(\nu t)]\mathbb{E}\, B$, which is constant in t if and only if A and B are centered. Moreover,

   $$\mathbb{E}\,(S_t S_{t+\tau}) = (\mathbb{V}\mathrm{ar}\, A + \mathbb{V}\mathrm{ar}\, B)\cos(\nu t) + (\mathbb{V}\mathrm{ar}\, A - \mathbb{V}\mathrm{ar}\, B)\cos(2\nu t + \nu\tau)$$
   $$+ \mathbb{C}\mathrm{ov}\,(A, B)\sin(2\nu t + \nu\tau),$$

   so S is weakly stationary to the order 2 if A and B are uncorrelated and have the same variance.

3. a. We have $\mathbb{E}\, X_t = [\cos(\nu t)\mathbb{E}\,(\cos\varphi) - \sin(\nu t)\mathbb{E}\,(\sin\varphi)]\mathbb{E}\, A$, which is constant in t if $\mathbb{E}\,(\cos\varphi) = \mathbb{E}\,(\sin\varphi) = 0$. Similarly,

   $$2\mathbb{E}\,(X_t X_{t+\tau}) =$$
   $$= [\cos(\nu\tau) + \cos(2\nu t + \nu\tau)\mathbb{E}\,(\cos 2\varphi) - \sin(2\nu t + \nu\tau)\mathbb{E}\,(\sin 2\varphi)]\mathbb{E}\,(A^2)$$

   so $\mathbf{X}$ is weakly stationary to the order 2 if $\mathbb{E}\,(\cos\varphi) = \mathbb{E}\,(\sin\varphi) = 0$ and $\mathbb{E}\,(\cos 2\varphi) = \mathbb{E}\,(\sin 2\varphi) = 0$.

b. The variables $X_{t+\tau} = A \cos(\nu t + \varphi + \nu \tau)$ and $X_t = A \cos(\nu t + \varphi)$ have the same distribution if $(A, \varphi + \nu \tau) \sim (A, \varphi)$ for all τ. This condition is fulfilled if and only if $\varphi \sim \mathcal{U}(0, 2\pi)$ and is independent of A.

c. Since $\mathbb{E}[\cos(\nu t + \varphi)] = \mathbb{E}[\sin(\nu t + \varphi)] = 0$, the process is centered. Set $X_t = A \cos(\nu t + \varphi)$ and $Y_t = B \sin(\nu t + \varphi)$. We have

$$\mathbb{E}(Z_t Z_{t+\tau}) = \mathbb{E}(X_t X_{t+\tau}) + \mathbb{E}(Y_t Y_{t+\tau}) + \mathbb{E}(X_t Y_{t+\tau}) + \mathbb{E}(Y_t X_{t+\tau}).$$

We compute $2\mathbb{E}(X_t X_{t+\tau}) = \cos(\nu \tau)\mathbb{E}(A^2)$ and $2\mathbb{E}(Y_t Y_{t+\tau}) = \cos(\nu \tau)\mathbb{E}(B^2)$, and also $2\mathbb{E}(X_t Y_{t+\tau}) = \sin(\nu \tau)\mathbb{E}(AB) = -2\mathbb{E}(Y_t X_{t+\tau})$, so **Z** is indeed weakly stationary to the order 2.   △

∇ **Exercise 4.4 (Alternated Renewal Process and Availability)** Consider a component starting operating at time $S_0 = 0$. When it fails, it is renewed. When the second fails, it is renewed and so on. Suppose the sequence of time durations (X_n) of the successive components is i.i.d. and that of the replacing times (Y_n) is also i.i.d. and is independent of (X_n).

1. Show that (S_n), defined by $S_0 = 0$ and $S_n = S_{n-1} + X_n + Y_n$ for $n \geq 1$, is a renewal process.
2. a. Write the event $E_t =$"the system is in good shape at time t" as a function of (X_n) and (S_n).
   b. Infer from a. the instantaneous availability $A(t) = \mathbb{P}(E_t)$.
   c. If $\mathbb{E} X_1 + \mathbb{E} Y_1 < +\infty$, compute the limit availability $A = \lim_{t \to +\infty} A(t)$.

Solution

1. The sequence $(T_n) = (X_n + Y_n)$ is i.i.d., and, according to Definition 4.44, (S_n) is indeed a renewal process.
2. a. We can write

$$E_t = (X_1 > t) \bigcup \left[\bigcup_{n \geq 1} (S_n \leq t) \cap (X_{n+1} > t - S_n)) \right].$$

   b. Let F and G denote the respective distribution functions of X_1 and Y_1. The distribution function of T_n is

$$H(t) = F * G(t) = \int_0^t F(t - x) dG(x), \quad t \geq 0.$$

   Therefore, setting $R(t) = 1 - F(t)$,

$$A(t) = \mathbb{P}(E_t) = \mathbb{P}(X_1 > t) + \sum_{n \geq 1} \mathbb{P}(S_n \leq t, X_{n+1} > t - S_n).$$

We compute

$$\mathbb{P}(S_n \le t, X_{n+1} > t - S_n) = \mathbb{E}\left[\mathbb{1}_{(S_n \le t)}\mathbb{P}(X_{n+1} > t - S_n \mid S_n)\right]$$
$$= \int_0^t R(t - y) dH^{*(n)}(y).$$

Thus

$$A(t) = \int_0^t R(t - y) dM(y) = R * m(t),$$

where $M(t) = \sum_{n \ge 1} H^{*(n)}(t)$ and $m(t) = \mathbb{1}_{\mathbb{R}_+}(t) + M(t)$. Finally, the key renewal theorem yields

$$A = \lim_{t \to +\infty} A(t) = \frac{\mathbb{E}\,X}{\mathbb{E}\,X + \mathbb{E}\,Y},$$

Note that the same problem will be modelled by a semi-Markov process in Exercise 5.5. △

∇ **Exercise 4.5 (Poisson Process)** The notation is that of Sect. 4.4.3.

1. Set $U_n = S_n/n$, for $n \in \mathbb{N}^*$. Show that (U_n) and $(n^{1/2}(\lambda U_n - 1))$ converge in distribution and give the limits.
2. Set $V_n = N_n/n$, for $n \in \mathbb{N}^*$. Determine the characteristic function of V_n. Show that (V_n) and $((n/\lambda)^{1/2}(V_n - \lambda))$ converge in distribution; give the limits.

Solution

1. Thanks to the law of large numbers, U_n converges a.s. to $1/\lambda$, and, thanks to the central limit theorem, $n^{1/2}(\lambda U_n - 1)$ converges in distribution to the standard normal distribution.
2. We compute

$$\mathbb{E}\,(e^{iuN_t}) = \sum_{k \ge 0} e^{iuk}\mathbb{P}(N_t = k) = e^{-\lambda t}\sum_{k \ge 0}\frac{(\lambda t e^{iu})^k}{k!} = e^{-\lambda t(1 - e^{iu})}.$$

Hence $\phi_{V_n}(u) = e^{-\lambda n(1 - e^{iu})}$ converges to $e^{iu\lambda}$, and V_n converges in distribution to λ. Moreover, $Z_n = \sqrt{n}(V_n - \lambda)/\sqrt{\lambda} = N_n/\sqrt{n\lambda} - \sqrt{n\lambda}$, so $\phi_{Z_n}(u) = \exp[-\lambda n(1 - e^{iu/\sqrt{n\lambda}}) - iu\sqrt{n\lambda}]$, which converges to $e^{-u^2/2}$, and Z_n converges in distribution to the standard normal distribution. △

∇ **Exercise 4.6 (Superposition and Decomposition of Poisson Processes)**

1. Let $\mathbf{N}$ and $\widetilde{\mathbf{N}}$ be two independent Poisson processes with respective intensities $\lambda > 0$ and $\mu > 0$. Show that the process $\mathbf{K}$, defined by $K_t = N_t + \widetilde{N}_t$, is a Poisson process, called superposition. Give its intensity.
2. Let $\mathbf{N}$ be a Poisson process with intensity $\lambda > 0$, whose arrivals are of two different types, A and B, with respective probabilities p and $1 - p$ independent of the arrivals times. Show that the counting process $\mathbf{M} = (M_t)_{t \in \mathbb{R}_+}$ of the arrivals of type A is a Poisson process. Give its intensity.

Solution

1. We can write $K_t - K_s = (N_t - N_s) + (\widetilde{N}_t - \widetilde{N}_s)$, so $\mathbf{K}$ is indeed a process with independent increments. Moreover,

$$\mathbb{P}(K_t = n) = \sum_{m=0}^{n} \mathbb{P}(N_t = m)\mathbb{P}(\widetilde{N}_t = n - m)$$

$$= \sum_{m=0}^{n} e^{-\lambda t} \frac{(\lambda t)^m}{m!} e^{-\mu t} \frac{(\mu t)^{n-m}}{(n-m)!} = e^{-(\lambda + \mu)t} \frac{[(\lambda + \mu)t]^n}{n!}.$$

   One can show similarly that $K_{t+s} - K_t \sim \mathcal{P}((\lambda + \mu)s)$. Proposition 4.54 yields that $\mathbf{K}$ is a Poisson process, with intensity $\lambda + \mu$.
2. By definition, the process $\mathbf{N}$ is the counting process of the renewal process associated with the sequence of arrivals times of A and B. These times are i.i.d., hence in particular the arrivals times of A are i.i.d., and the associated counting process is $\mathbf{M}$. Moreover,

$$\mathbb{P}(M_t = k) = \sum_{n \geq k} \mathbb{P}(M_t = k \mid N_t = n)\mathbb{P}(N_t = n)$$

$$= \sum_{n \geq k} \binom{n}{k} p^k (1-p)^{n-k} e^{-\lambda t} \frac{(\lambda t)^n}{n!} = e^{-p\lambda t} \frac{(p\lambda t)^k}{k!}.$$

One can show similarly that $M_{t+s} - M_t \sim \mathcal{P}(p\lambda s)$ so that $\mathbf{M}$ is a Poisson process with intensity $p\lambda$.   △

Markov and Semi-Markov Processes

5

This chapter is devoted to jump Markov processes and finite semi-Markov processes. In both cases, the index is considered as the calender time, continuously counted over the positive real line. Markov processes are continuous-time processes that share the Markov property with the discrete-time Markov chains. Their future evolution conditional to the past depends only on the last occupied state. Their extension to the so-called semi-Markov processes naturally arises in many types of applications. The future evolution of a semi-Markov process given the past depends on the occupied state too, but also on the time elapsed since the last transition.

Detailed notions on homogeneous jump Markov processes with discrete (countable) state spaces will be presented. Some basic notions on semi-Markov processes with finite state spaces will follow, illustrated through typical examples.

5.1 Jump Markov Processes

We will investigate in this section mainly the jump Markov process. Since a jump Markov process is a Markov process which is constant between two successive jumps, we will first present some basic notions on Markov processes with continuous time.

Thereafter, $\mathbf{X} = (X_t)_{t \in \mathbb{R}_+}$ will denote a process defined on a stochastic basis $(\Omega, \mathcal{F}, (\mathcal{F}_t)_{t \in \mathbb{R}_+}, \mathbb{P})$, taking values in a finite or enumerable set E. The filtration will be supposed to be the natural filtration of the process, even if, of course, $\mathbf{X}$ can satisfy the Markov property below with respect to a larger filtration.

5.1.1 Markov Processes

A stochastic process is called a Markov process if its future values given the past and the present depend only on the present. Especially, processes with independent increments—Brownian motion, Poisson processes—are Markov processes.

© Springer Nature Switzerland AG 2018
V. Girardin, N. Limnios, *Applied Probability*,
https://doi.org/10.1007/978-3-319-97412-5_5

Definition 5.1 A stochastic process $\mathbf{X} = (X_t)_{t \in \mathbb{R}_+}$, with state space E, is called a Markov process—with respect to its natural filtration $(\mathcal{F}_t)_{t \in \mathbb{R}_+}$ if it satisfies the Markov property

$$\mathbb{P}(X_{s+t} = j \mid \mathcal{F}_s) = \mathbb{P}(X_{s+t} = j \mid X_s) \quad \text{a.s.,}$$

for all non-negative real numbers t and s and all $j \in E$.

The Markov property can also be written

$$\mathbb{P}(X_{s+t} = j \mid X_{s_1} = i_1, \ldots, X_{s_n} = i_n, X_s = i) = \mathbb{P}(X_{s+t} = j \mid X_s = i),$$

for all $n \in \mathbb{N}$, all $0 \leq s_1 < \cdots < s_n < s, 0 \leq t$, and all states $i_1, \ldots, i_n, i, j$ in E.

If moreover the above conditional probability does not depend on s, then

$$\mathbb{P}(X_{s+t} = j \mid X_s = i) = P_t(i, j), \quad i \in E, j \in E \text{ and } s \geq 0, t \geq 0,$$

and the process $\mathbf{X}$ is said to be homogeneous with respect to time.

We will study here only homogeneous processes. The trajectories $t \longrightarrow X_t(\omega)$ will be assumed to be continuous on the right for the discrete topology, with a.s. finite limits on the left; such a process is said to be cadlag. Similarly to Markov chains, the distribution of X_0 is called the initial distribution of the process, and we will set $\mathbb{P}_i(\cdot) = \mathbb{P}(\cdot \mid X_0 = i)$ and $\mathbb{E}_i(\cdot) = \mathbb{E}(\cdot \mid X_0 = i)$.

$\triangleright$ *Example 5.2 (Process with Independent Increments)* Any continuous-time process $\mathbf{X} = (X_t)_{t \in \mathbb{R}_+}$ with independent increments and taking values in $\mathbb{Z}$ is a Markov process. Indeed, for all nonnegative real numbers s and t,

$$\mathbb{P}(X_{s+t} = j \mid \mathcal{F}_s) = \sum_{i \in \mathbb{Z}} \mathbb{P}(X_{s+t} = j, X_s = i \mid \mathcal{F}_s)$$

$$\overset{(1)}{=} \sum_{i \in \mathbb{Z}} \mathbb{1}_{(X_s = i)} \mathbb{P}(X_{s+t} - X_s = j - i \mid X_s)$$

$$= \sum_{i \in \mathbb{Z}} \mathbb{P}(X_{s+t} = j, X_s = i \mid X_s) = \mathbb{P}(X_{s+t} = j \mid X_s).$$

Note that we can also write

$$\mathbb{P}(X_{s+t} = j \mid \mathcal{F}_s) = \sum_{i \in \mathbb{Z}} \mathbb{E}\left[\mathbb{1}_{(X_s = i)} \mathbb{1}_{(X_{s+t} - X_s = j - i)} \mid \mathcal{F}_s\right]$$

$$\overset{(1)}{=} \sum_{i \in \mathbb{Z}} \mathbb{1}_{(X_s = i)} \mathbb{E}\left[\mathbb{1}_{(X_{s+t} - X_s = j - i)}\right]. \tag{5.1}$$

(1) because X_s is $\mathcal{F}_s$-measurable and $X_{s+t} - X_s$ is independent of $\mathcal{F}_s$.

The random variable defined in (5.1) is measurable for the σ-algebra generated by X_s, so is equal to $\mathbb{P}(X_{s+t} = j \mid X_s)$, and the process satisfies the Markov property. ◁

Proposition 5.3 *The past and the future of a Markov process are independent given the present, that is, for all $A \in \mathcal{F}_t = \sigma(X_s, s \leq t)$ and $B \in \sigma(X_s, s \geq t)$,*

$$\mathbb{P}(A \cap B \mid X_t) = \mathbb{P}(A \mid X_t)\mathbb{P}(B \mid X_t) \quad a.s..$$

Proof We compute

$$\mathbb{P}(A \cap B \mid X_t) = \mathbb{E}\left[\mathbb{E}\left(\mathbb{1}_A \mathbb{1}_B \mid \mathcal{F}_t\right) \mid X_t\right] = \mathbb{E}\left[\mathbb{1}_A \mathbb{E}\left(\mathbb{1}_B \mid \mathcal{F}_t\right) \mid X_t\right]$$
$$= \mathbb{E}\left[\mathbb{1}_A \mathbb{E}\left(\mathbb{1}_B \mid X_t\right) \mid X_t\right] = \mathbb{E}\left[\mathbb{1}_A \mid X_t\right]\mathbb{E}\left(\mathbb{1}_B \mid X_t\right)$$
$$= \mathbb{P}(A \mid X_t)\mathbb{P}(B \mid X_t),$$

where all equalities hold a.s. □

Definition 5.4 A Markov process $\mathbf{X}$ is said to satisfy the strong Markov property if for any stopping time T adapted to $\mathbf{X}$ and all $s \geq 0$, on $(T < +\infty)$,

$$\mathbb{P}(X_{T+s} = j \mid \mathcal{F}_T) = \mathbb{P}_{X_T}(X_s = j) \quad a.s.,$$

where $\mathcal{F}_T = \{A \in \mathcal{F} : \forall n \in \overline{\mathbb{N}},\ A \cap (T = n) \in \mathcal{F}_n\}$ is the σ-algebra of events previous to T. Then, the Markov process is said to be strong.

Theorem 5.5 *Let $\mathbf{X}$ be a strong Markov process, with state space E. If $f : E^{\mathbb{R}_+} \longrightarrow \mathbb{R}_+^d$ is a Borel function and T is a finite stopping time for $\mathbf{X}$, then*

$$\mathbb{E}_i(f \circ \mathbf{X} \circ \theta_T \mid \mathcal{F}_T) = \mathbb{E}_{X_T}(f \circ \mathbf{X}), \quad a.s.,\ i \in E.$$

where θ_s is the shift operator such that $X_t \circ \theta_s = X_{t+s}$.

Proof For a function $f \circ \mathbf{X} = f(X_{t_1}, \ldots, X_{t_n})$ with $0 \leq t_1 < \cdots < t_n$.
We compute

$$\mathbb{E}_i[f(X_{t_1+T}, \ldots, X_{t_n+T}) \mid \mathcal{F}_T] =$$
$$= \sum_{(i_1,\ldots,i_n)\in e^n} \mathbb{P}_i(X_{t_1+T} = i_1, \ldots, X_{t_n+T} = i_n \mid \mathcal{F}_T)f(i_1,\ldots,i_n).$$

Thanks to the strong Markov property,

$$\mathbb{P}_i(X_{t_1+T} = i_1, \ldots, X_{t_n+T} = i_n \mid \mathcal{F}_T) =$$
$$= \mathbb{P}_i(X_{t_n+T} = i_n \mid X_{t_{n-1}+T} = i_{n-1}) \ldots \mathbb{P}_i(X_{t_1+T} = i_n \mid X_T)$$
$$= \mathbb{P}_i(X_{t_n} = i_n \mid X_{t_{n-1}} = i_{n-1}) \ldots \mathbb{P}_{X_T}(X_{t_1} = i_n)$$
$$= \mathbb{P}_{X_T}(X_{t_1} = i_1, \ldots, X_{t_n} = i_n),$$

Therefore,

$$\mathbb{E}_i[f(X_{t_1+T}, \ldots, X_{t_n+T}) \mid \mathcal{F}_T] = \mathbb{E}_{X_T}[f(X_{t_1}, \ldots, X_{t_n})],$$

and the result follows.   □

5.1.2 Transition Functions

Definition 5.6 A Markov process **X** is said to be a (pure) jump Markov process if, whatever be its initial state, it evolves through isolated jumps from state to state, and its trajectories are a.s. constant between the jumps.

To be exact, a jump time is a time where a change of state occurs. If (T_n) is the sequence of successive jump times of a jump Markov process **X**, for all $n \in \mathbb{N}$, we have $T_n < T_{n+1}$ if $T_n < +\infty$ and $T_n = T_{n+1}$ if $T_n = +\infty$, and

$$X_t = X_{T_n}, \quad T_n \leq t < T_{n+1}, \quad T_n < +\infty, \quad n \geq 0.$$

A typical trajectory of such a process is shown in Fig. 5.1.
Clearly, any jump Markov process is a strong Markov process.

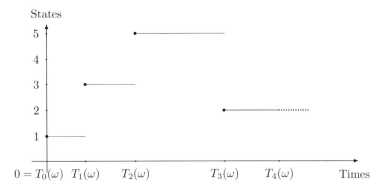

Fig. 5.1 A trajectory of a jump Markov process

Definition 5.7 A jump Markov process is said to be regular, or non explosive, if for any initial state, the number of jumps in any finite time interval is a.s. finite.

Let $\zeta = \sup_{n \geq 1} T_n$ be the time duration of the process, a random variable taking values in $\overline{\mathbb{R}}_+$. If $\mathbb{P}(\zeta = +\infty) = 1$, the process is regular; otherwise, that is if $\mathbb{P}(\zeta < +\infty) > 0$, it is explosive.

For regularity criteria, see Theorem 5.31 and Proposition 5.33 below. The next example explains the phenomenon of explosion.

▷ *Example 5.8* Let (X_n) be a sequence of independent random variables with exponential distributions with parameters (λ_n). Setting $S = \sum_{n \geq 1} X_n$, we have

$$\mathbb{P}(S = +\infty) > 0 \quad \text{if and only if} \quad \sum_{n \geq 1} \frac{1}{\lambda_n} = +\infty.$$

Indeed, if $\mathbb{P}(S = +\infty) > 0$, then $\sum_{n \geq 1} 1/\lambda_n = \mathbb{E}\,S = +\infty$, and

$$\mathbb{E}\,(e^{-S}) = \frac{1}{\prod_{n \geq 1}(1 + 1/\lambda_n)} \leq \frac{1}{\sum_{n \geq 1} 1/\lambda_n}$$

yields the converse. ◁

Definition 5.9 Let $\mathbf{X}$ be a homogeneous jump Markov process. The family of functions defined on $\mathbb{R}_+$ by $t \longrightarrow P_t(i, j) = \mathbb{P}(X_{t+h} = j \mid X_h = i)$, for i and j in E, are called the transition functions of the process on E.

We will denote by P_t the (possibly infinite) matrix $(P_t(i, j))_{(i,j) \in E \times E}$, and consider only processes such that $P_0(i, i) = 1$.

Properties of Transition Functions

1. $0 \leq P_t(i, j) \leq 1$, for all i and j in E and $t \geq 0$, because $P_t(i, \cdot)$ is a probability.
2. $\sum_{j \in E} P_t(i, j) = 1$, for all $i \in E$ and $t \geq 0$, because E is the set of all values taken by $\mathbf{X}$.
3. **(Chapman-Kolmogorov equation)**

$$\sum_{k \in E} P_t(i, k) P_s(k, j) = P_{t+s}(i, j), \quad i, j \in E, \ s \geq 0, \ t \geq 0.$$

4. If $\lim_{t \to 0^+} P_t(i, j) = \delta_{ij}$ for all $i \in E$, the process is said to be stochastically continuous. If this property is satisfied uniformly in i, the transition function (or semi-group) is said to be uniform.

Thanks to Chapman-Kolmogorov equation, the family $\{P_t : t \geq 0\}$ equipped with the composition $P_t P_h = P_{t+h}$ is a semi-group. Indeed, the operation is commutative and associative, and I (the identity $|E| \times |E|$-matrix) is a neutral element; on the contrary, in general, a given element has no inverse in the family, which therefore is not a group.

▷ *Example 5.10 (Birth Process)*   Let **X** be a stochastic process with state space $E = \mathbb{N}$ and transition function such that, when $h \to 0^+$,

$$P_h(i, j) = \begin{cases} \lambda_i h + o(h) & \text{if } j = i + 1, \\ 1 - \lambda_i h + o(h) & \text{if } j = i, \\ 0 & \text{otherwise,} \end{cases}$$

This is indeed a transition function satisfying the four above properties.

If, for instance, $\lambda_i = \lambda / i$, then P_t is uniform; on the contrary, it is not uniform if $\lambda_i = i\lambda$. If $\lambda_i = \lambda$ for all i, the process is a Poisson process.   ◁

The following proposition is a straightforward consequence of both the compound probabilities formula and Markov property.

Proposition 5.11 *For all $n \in \mathbb{N}$, all nonnegative reals $0 \le t_0 < t_1 < \cdots < t_n$, and all finite sequence of states $i_0, i_1, \ldots, i_n$, we have*

$$\mathbb{P}(X_{t_1} = i_1, \ldots, X_{t_n} = i_n \mid X_{t_0} = i_0) = P_{t_1 - t_0}(i_0, i_1) \ldots P_{t_n - t_{n-1}}(i_{n-1}, i_n),$$

and

$$\mathbb{P}(X_0 = i_0, X_{t_1} = i_1, \ldots, X_{t_n} = i_n) = \alpha(i_0) P_{t_1}(i_0, i_1) \ldots P_{t_n - t_{n-1}}(i_{n-1}, i_n),$$

where α denotes the initial distribution of the process.

Thus, the finite-dimensional distributions of a jump Markov process are characterized by its initial distribution and its transition function.

We state without proof the next result, necessary for proving the following one.

Theorem 5.12 (Lévy) *For all given states i and j, $P_t(i, j)$ is either identically null, or never null on $\mathbb{R}_+$.*

Proposition 5.13 *Let P_t be the transition function of a jump Markov process.*

1. *If some $t > 0$ exists such that $P_t(i, i) = 1$, then $P_s(i, i) = 1$ for all $s \in \mathbb{R}_+$.*
2. *$| P_{t+\varepsilon}(i, j) - P_t(i, j) | \le 1 - P_{|\varepsilon|}(i, i)$ for all $t \ge 0$ and $(i, j) \in E \times E$, and hence $P_t(i, j)$ is uniformly continuous with respect to t.*

Proof

1. Let $t > 0$ be such that $P_t(i, i) = 1$. Then, for any $s < t$,

$$0 = 1 - P_t(i, i) = \sum_{j \neq i} P_t(i, j) \geq P_{t-s}(i, i) P_s(i, j) \geq 0.$$

But, thanks to Lévy's theorem, $P_{t-s}(i, i) > 0$, so $P_s(i, j) = 0$ for all $j \neq i$, and hence $P_s(i, i) = 1$.

For $s > t$, it is sufficient to choose n such that $s/n < t$ for getting $P_s(i, i) \geq [P_{s/n}(i, i)]^n \geq 1$.

2. Let $\varepsilon > 0$. We deduce from

$$P_{t+\varepsilon}(i, j) - P_t(i, j) = \sum_{k \neq i} P_\varepsilon(i, k) P_t(k, j) - P_t(i, j)[1 - P_\varepsilon(i, i)]$$

that

$$-[1 - P_\varepsilon(i, i)] \leq -P_t(i, j)[1 - P_\varepsilon(i, i)] \leq P_{t+\varepsilon}(i, j) - P_t(i, j)$$
$$\leq \sum_{k \neq i} P_\varepsilon(i, k) P_t(k, j) \leq \sum_{k \neq i} P_\varepsilon(i, k) \leq 1 - P_\varepsilon(i, i),$$

and hence

$$\mid P_{t+\varepsilon}(i, j) - P_t(i, j) \mid \leq 1 - P_\varepsilon(i, i). \tag{5.2}$$

Replacing t by $t - \varepsilon$ in the above inequality for $0 < \varepsilon < t$, we get

$$\mid P_{t-\varepsilon}(i, j) - P_t(i, j) \mid = \mid P_t(i, j) - P_{t-\varepsilon}(i, j) \mid \leq 1 - P_\varepsilon(i, i). \tag{5.3}$$

The result follows from (5.2) and (5.3). $\qquad\square$

5.1.3 Infinitesimal Generators and Kolmogorov's Equations

The transition function of a Markov process is identified from its generator through the Kolmogorov's equations.

Definition 5.14 The (infinitesimal) generator $A = (a_{ij})_{(i,j) \in E \times E}$ of a Markov process $\mathbf{X}$ is given by the derivative on the right of the transition function P_t at time $t = 0$, or

$$a_{ij} = \lim_{t \to 0^+} \frac{P_t(i, j) - I(i, j)}{t},$$

where I is the identity matrix.

These quantities are always well-defined, but when $i = j$ they may be equal to $-\infty$. The generator of a Markov process is such that $a_{ij} \geq 0$ for all $i \neq j$ and $\sum_{j \in E} a_{ij} = 0$ for all i. We set $a_i = -a_{ii} \geq 0$.

Definition 5.15 A state i is said to be stable if $0 < a_i < +\infty$, instantaneous if $a_i = +\infty$, absorbing if $a_i = 0$, and conservative if $\sum_{j \in E} a_{ij} = 0$.

A generator—or the associated process—of which all states are stable (instantaneous, conservative) is said to be stable (instantaneous, conservative).

At each passage in a stable state, the process will spend a.s. a non null and finite time. On the contrary, it will a.s. jump instantaneously from an instantaneous state. Finally, reaching an absorbing state, the process will remain there forever.

We state without proof the next result.

Theorem 5.16 *Let* **X** *be a jump Markov process.*

1. *The trajectories of* **X** *are a.s. continuous on the right if and only if no instantaneous states exist.*
2. *If E is finite, then no instantaneous states exist.*

▷ *Example 5.17 (Birth Process on* $\mathbb{N}$*)* The generator of this process is

$$a_{ij} = \begin{cases} \lambda_i & \text{if } j = i + 1, \\ -\lambda_i & \text{if } j = i, \\ 0 & \text{otherwise,} \end{cases}$$

for all integers i and j. Hence it is a conservative process.   ◁

Theorem 5.18 (Kolmogorov's Equations) *If* **X** *is regular, then the transition functions* $t \longrightarrow P_t(i, j)$ *are continuously differentiable on* $\mathbb{R}_+^*$ *for all states i and j, and satisfy the equations*

$$P_t'(i, j) = \sum_{k \in E} a_{ik} P_t(k, j) \quad \text{and} \quad P_t'(i, j) = \sum_{k \in E} P_t(i, k) a_{kj}.$$

In matrix form, the above equations become

$$\frac{d}{dt} P_t = A P_t \quad \text{and} \quad \frac{d}{dt} P_t = P_t A, \tag{5.4}$$

and are respectively called backward and forward Kolmogorov's equations.

Proof We prove only the second part of the theorem for a finite E.

We deduce by differentiating with respect to s the Chapman-Kolmogorov equation that

$$P'_{s+t}(i, j) = \sum_{k \in E} P'_s(i, k) P_t(k, j),$$

or, when $s \to 0^+$,

$$P'_t(i, j) = \sum_{k \in E} a_{ik} P_t(k, j).$$

The second equation is obtained symmetrically.   □

The definition of uniformisable processes is necessary for stating the following result.

Definition 5.19 A jump Markov process is said to be uniformisable if

$$\sup_{i \in E} a_i < +\infty.$$

The next result is a straightforward consequence of the theory of linear differential equations.

Theorem 5.20 *When the process is uniformisable, the common solution of Kolmogorov's equations (5.4) is*

$$P_t = e^{tA} = I + \sum_{k \geq 1} \frac{t^k}{k!} A^k. \tag{5.5}$$

Numerous methods for computing numerically the above solution of Kolmogorov's equations exist: direct computation of the series (5.5) truncated at a certain value of k, uniformisation—see Example 5.51 below, Laplace transform, determination of the eigen-values and eigen-vectors, ...

Fig. 5.2 Graph of the
conservative process of
Example 5.21

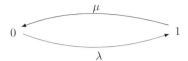

▷ *Example 5.21* Consider a conservative process with two states, say $E = \{0, 1\}$, with generator

$$A = \begin{pmatrix} -\lambda & \lambda \\ \mu & -\mu \end{pmatrix},$$

with $\lambda > 0$ and $\mu > 0$ (Fig. 5.2). The two eigen-values of the generator A are $s_1 = 0$ and $s_2 = -\lambda - \mu$, and $(1, 1)'$ and $(\lambda, -\mu)'$ are two associated eigen-vectors. Therefore,

$$e^{tA} = QDQ^{-1},$$

where

$$Q = \begin{pmatrix} 1 & \lambda \\ 1 & -\mu \end{pmatrix}, \quad Q^{-1} = \frac{1}{\lambda + \mu} \begin{pmatrix} \mu & \lambda \\ 1 & -1 \end{pmatrix} \quad \text{and} \quad D = \begin{pmatrix} 1 & 0 \\ 0 & e^{-(\lambda+\mu)t} \end{pmatrix}.$$

So, finally,

$$P_t = e^{tA} = \frac{1}{\lambda + \mu} \begin{pmatrix} \mu & \lambda \\ \mu & \lambda \end{pmatrix} + \frac{e^{-(\lambda+\mu)t}}{\lambda + \mu} \begin{pmatrix} \lambda & -\lambda \\ -\mu & \mu \end{pmatrix},$$

that is a closed-form expression.   ◁

5.1.4   Embedded Chains and Classification of States

Let us begin by showing that the exit time of a given state of a Markov process has an exponential distribution. The nature of the parameter and its connection with the generator of the process will be specified later in Corollary 5.27. Note the exit time of i is also the sojourn time in i (before first exit), or the hitting time of $E \setminus \{i\}$.

Proposition 5.22 *If the state i is not an absorbing state, then the exit time of i has an exponential distribution with respect to $\mathbb{P}_i$, with parameter λ_i depending on i.*

Proof Let T_1 denote this first jump time. If θ_s denotes the shift operator, then $T_1 \circ \theta_s$ is the time of the first jump after time s, and we have

$$\mathbb{P}_i(T_1 > s + t) = \mathbb{P}_i(T_1 > s, T_1 \circ \theta_s > t) = \mathbb{E}_i[\mathbb{P}_i(T_1 > s, T_1 \circ \theta_s > t \mid \mathcal{F}_s)]$$

$$= \mathbb{E}_i[\mathbb{1}_{(T_1 > s)}\mathbb{P}_i(T_1 \circ \theta_s > t \mid \mathcal{F}_s)]$$

$$\overset{(1)}{=} \mathbb{E}_i[\mathbb{1}_{(T_1 > s)}\mathbb{P}_i(T_1 > t)] = \mathbb{P}_i(T_1 > s)\mathbb{P}_i(T_1 > t).$$

(1) by the Markov property.

Set $R(t) = \mathbb{P}_i(T_1 > t)$. The above equation can be written $R(s + t) = R(s)R(t)$, for all nonnegative reals s and t. This is Cauchy functional equation on $\mathbb{R}_+$, whose solution is well-known to be an exponential function, precisely

$$R(u) = \mathbb{P}_i(T_1 > u) = e^{-\lambda_i u}\mathbb{1}_{\mathbb{R}_+}(u),$$

with $\lambda_i \geq 0$; in other words, $T_1 \sim \mathcal{E}(\lambda_i)$. □

Note that the expected sojourn time in an instantaneous state is null.

Proposition 5.23 *The random variables T_1 and X_{T_1} are independent with respect to $\mathbb{P}_i$ for all non absorbing $i \in E$.*

Proof Let B denote a subset of $E \setminus \{i\}$. We have

$$\mathbb{P}_i(X_{T_1} \in B, T_1 > s) = \mathbb{E}_i[\mathbb{P}_i(X_{T_1} \circ \theta_s \in B, T_1 > s \mid \mathcal{F}_s)]$$

$$= \mathbb{E}_i[\mathbb{1}_{(T_1 > s)}\mathbb{P}_i(X_{T_1} \circ \theta_s \in B \mid \mathcal{F}_s)]$$

$$= \mathbb{P}_i(T_1 > s)\mathbb{P}_i(X_{T_1} \in B),$$

if $s > 0$. □

Now, let us consider the sequence of random variables (J_n) defined by

$$J_n = X_{T_n}, \quad n \text{ such that } T_n < +\infty.$$

This is the sequence of the successive states visited by the process **X**. Clearly, it is defined up to the explosion of the process—if explosion occurs; in this regard, it is said to be minimal.

Theorem 5.24 *For all $n \in \mathbb{N}$, all $i, j \in E$ and all $t \geq 0$, we have:*

1. $\mathbb{P}(J_{n+1} = j, T_{n+1} - T_n \leq t \mid \mathcal{F}_{T_n}) = \mathbb{P}(J_{n+1} = j, T_{n+1} - T_n \leq t \mid J_n)$;
2. $\mathbb{P}(J_{n+1} = j, T_{n+1} - T_n \leq t \mid J_n = i) = \mathbb{P}_i(X_{T_1} = j)(1 - e^{-\lambda_i t})$.
3. *Moreover, the sequence (J_n) is a Markov chain.*

Proof

1. We compute

$$\mathbb{P}(J_{n+1} = j, T_{n+1} - T_n \leq t \mid \mathcal{F}_{T_n}) = \mathbb{P}(X_{T_{n+1}} = j, T_{n+1} - T_n \leq t \mid \mathcal{F}_{T_n})$$

$$\overset{(1)}{=} \mathbb{P}(X_{T_{n+1}} = j, T_{n+1} - T_n \leq t \mid X_{T_n}).$$

(1) by the strong Markov property.

2. Since $\mathbb{P}(J_{n+1} = j, T_{n+1} - T_n \leq t \mid J_n = i) = \mathbb{P}(X_{T_{n+1}} = j, T_{n+1} - T_n \leq t \mid X_{T_n} = i)$, we have

$$\mathbb{P}(J_{n+1} = j, T_{n+1} - T_n \leq t \mid J_n = i) \overset{(1)}{=} \mathbb{P}_i(X_{T_1} = j, T_1 \leq t)$$

$$\overset{(2)}{=} \mathbb{P}_i(X_{T_1} = j)\mathbb{P}_i(T_1 \leq t)$$

$$\overset{(3)}{=} \mathbb{P}_i(X_{T_1} = j)(1 - e^{-\lambda_i t}).$$

(1) by homogeneity, (2) by independence of X_{T_1} and T_1 and (3) by Proposition 5.22.

3. Letting t go to infinity in 2. yields the result. □

The chain (J_n) is called the embedded Markov chain of the process, with transition function P defined by $P(i, j) = \mathbb{P}_i(X_{T_1} = j)$; see Corollary 5.27 below for a closed-form expression. We check that $P(i, j) \geq 0$, $P(i, i) = 0$ and $\sum_{j \in E} P(i, j) = 1$.

The sojourn times in the different states are mutually independent given the successive states visited by the process. Applying iteratively Theorem 5.24 and Proposition 5.22 yields the following closed-form expression.

Corollary 5.25 *For all $n \in \mathbb{N}^*$, all $i_0, \ldots, i_{n-1} \in E$, all $k = 1, \ldots, n$, and all $t_k \geq 0$, we have*

$$\mathbb{P}(T_1 - T_0 \leq t_1, \ldots, T_n - T_{n-1} \leq t_n \mid J_k = i_k, k \geq 0) =$$

$$= \mathbb{P}(T_1 - T_0 \leq t_1 \mid J_0 = i_0) \ldots \mathbb{P}(T_n - T_{n-1} \leq t_n \mid J_{n-1} = i_{n-1})$$

$$= \prod_{k=0}^{n-1}(1 - e^{-\lambda_{i_k} t_{k+1}}).$$

Theorem 5.26 (Kolmogorov's Integral Equation) *For any non absorbing state i,*

$$P_t(i, j) = I(i, j)e^{-\lambda_i t} + \sum_{k \in E} \int_0^t \lambda_i e^{-\lambda_i s} P(i, k) P_{t-s}(k, j) ds, \quad t \geq 0, \ j \in E.$$

For an absorbing state i, the above theorem amounts to $P_t(i, j) = I(i, j)$.

Proof We have

$$P_t(i, j) = \mathbb{P}_i(X_t = j, T_1 > t) + \mathbb{P}_i(X_t = j, T_1 \leq t).$$

We compute $\mathbb{P}_i(X_t = j, T_1 > t) = I(i, j)e^{-\lambda_i t}$ and

$$
\begin{aligned}
\mathbb{P}_i(X_t = j, T_1 \leq t) &= \mathbb{E}_i[\mathbb{P}_i(X_t = j, T_1 \leq t \mid \mathcal{F}_{T_1})] \\
&= \mathbb{E}_i[\mathbb{1}_{(T_1 \leq t)} \mathbb{P}_{X_{T_1}}(X_{t-T_1} = j)] \\
&= \int_0^{+\infty} \mathbb{1}_{]0,t]}(s) \sum_{k \in E} \mathbb{P}_k(X_{t-s} = j) \mathbb{P}_i(T_1 \in ds, X_{T_1} = k) \\
&= \int_0^t \sum_{k \in E} P_{t-s}(k, j) P(i, k) \lambda_i e^{-\lambda_i s} ds.
\end{aligned}
$$
$\square$

We can now link the distribution of the first jump time to the distribution of the hitting time of the complementary set of i, in other words compute the transition matrix P of the embedded chain in terms of the generator of the jump Markov process.

Corollary 5.27 *For any state i, we have $\lambda_i = a_i$. If i is non absorbing, then*

$$P(i, j) = \mathbb{P}_i(X_{T_1} = j) = \begin{cases} a_{ij}/a_i & \text{if } j \neq i \\ 0 & \text{if } j = i. \end{cases} \tag{5.6}$$

Proof According to Proposition 5.13, the transition function is continuous with respect to t. Differentiating the Kolmogorov's integral equation yields

$$P_t'(i, j) = -\lambda_i e^{-\lambda_i t} I(i, j) + P(i, j) \lambda_i e^{-\lambda_i t}. \tag{5.7}$$

Thus, for $t \to 0^+$ and $i = j$, we get $a_{ii} = -\lambda_i$, and, for $t \to 0^+$ and $i \neq j$, we get $a_{ij} = P(i, j) a_i$.
$\square$

Equation (5.7) implies that $a_{ij} = -a_i I(i, j) + a_i P(i, j)$, or under matrix form,

$$A = \text{diag}(a_i)(P - I). \tag{5.8}$$

The above corollary yields the stochastic simulation of a trajectory of a jump Markov process in a given interval of time $[0, T]$. Indeed, the method presented in Sect. 3.1.1 applies to the embedded chain, and simulation of the sojourn times amounts to simulation of the exponential distribution as follows.

1. Let x_0 be the realization of a random variable $J_0 \sim \alpha$. $n := 0$. $T_0(\omega) := 0$.
2. $n := n + 1$. Let $W_n(\omega)$ be the realization of an $\mathcal{E}(a_{x_{n-1}})$-distributed random variable. $T_n(\omega) := T_{n-1}(\omega) + W_n(\omega)$.
3. If $T_n(\omega) \geq T$, then end.
4. Let $J_n(\omega)$ be the realization of a random variable whose distribution is given by (5.6); set $x_n := J_n(\omega)$.
5. Continue at Step 2.

$\triangleright$ *Example 5.28 (Birth Process (Continuation of Example 5.10))* The transition matrix of the embedded chain of a birth process is given by

$$P(i, j) = \begin{cases} 1 & \text{if } j = i + 1, \\ 0 & \text{otherwise.} \end{cases}$$

The sojourn times have exponential distributions with parameters λ_i. $\triangleleft$

The following example is an application of Markov processes to reliability. Another example is presented in Exercise 5.4.

$\triangleright$ *Example 5.29 (Cold Stand-by System)* A cold stand-by system generally contains one component (or sub-system) functioning and one or several components (or sub-systems) in stand-by, all identical. The stand-by component is tried when the functioning component fails, and then begins working successfully or not according to a given probability.

Consider a stand-by system with two components. When the functioning component fails, the stand-by component is connected through a switching device. This latter commutes successfully with probability $p \in]0, 1]$. The failure rate of the functioning component is λ. The failure rate of the stand-by component is null. This system can be modelled by a Markov process with three states:

state 1: one component works, the second is in stand-by;
state 2: one component is failed, the second works;
state 3: either both components or one component and the commutator are failed.

The repairing rate between state 3 and state 2, and between state 2 and state 1 is μ; the direct transition from state 3 to state 1 is impossible; see Fig. 5.3.

The generator of the process is

$$A = \begin{pmatrix} -\lambda & p\lambda & (1-p)\lambda \\ \mu & -(\lambda + \mu) & \lambda \\ 0 & \mu & -\mu \end{pmatrix},$$

Fig. 5.3 Graph of the 3-state Markov process of Example 5.29

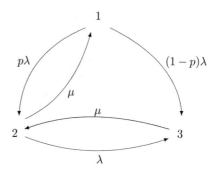

and

$$P = \begin{pmatrix} 0 & p & (1-p) \\ \mu/(\lambda + \mu) & 0 & \lambda/(\lambda + \mu) \\ 0 & 1 & 0 \end{pmatrix}$$

is the transition matrix of its embedded chain. ◁

Let us now state some criterion linked to regularity.

Proposition 5.30 *For a uniformisable jump Markov process, the probability of more than one jump occurring in a time interval $[0, h]$ is $o(h)$ when $h \to 0^+$, uniformly in $i \in E$.*

Proof We have

$$\mathbb{P}_i(T_2 \leq h) = \mathbb{P}_i[T_1 + (T_2 - T_1) \leq h] \leq \mathbb{P}_i(T_1 \leq h, T_2 - T_1 \leq h)$$

and

$$\begin{aligned}
\mathbb{P}_i(T_1 \leq h, T_2 - T_1 \leq h) &= \mathbb{E}_i[\mathbb{P}_i(T_1 \leq h, T_2 - T_1 \leq h \mid \mathcal{F}_{T_1})] \\
&= \mathbb{E}_i[\mathbb{1}_{(T_1 \leq h)} \mathbb{P}_i(T_2 - T_1 \leq h \mid \mathcal{F}_{T_1})] \\
&\overset{(1)}{=} \mathbb{E}_i(\mathbb{1}_{(T_1 \leq h)}[1 - \exp(-a_{X_{T_1}} h)]) \\
&\leq \mathbb{E}_i[\mathbb{1}_{(T_1 \leq h)}(1 - e^{-ah})] \leq (1 - e^{-ah})^2 = o(h),
\end{aligned}$$

where $a_{X_{T_1}} = a_j$ on the event $(X_{T_1} = j)$ and $a = \sup_{i \in E} a_i < +\infty$.
(1) by Proposition 5.22. □

The following two criterion for a jump Markov process to be regular are stated without proofs.

Theorem 5.31 *A jump Markov process is regular if and only if one of the following conditions is fulfilled:*

1. *(**Reuter's condition of explosion**) the only bounded nonnegative solution of the equation $Ay = y$ is the null solution.*
2. $\sum_{n \geq 1} 1/a_{J_n} = +\infty$ *a.s., where $a_{J_n} = a_j$ on the event $(J_n = j)$.*

▷ *Example 5.32 (Birth Process (Continuation of Example 5.10))* If $\sum_{i \in E} 1/a_i$ is infinite, the birth process is regular; if the sum is finite, it is explosive.   ◁

Proposition 5.33 *The process is regular if one of the two following conditions is fulfilled:*

1. *The process is uniformisable.*
2. *Its embedded chain is recurrent.*

Thus, a finite (such that $|E| < +\infty$) process is regular.

As for Markov chains, the states of a jump Markov process are classified according to their nature, and a communication relation can be defined. Recall that T_1 denotes the first jump time of the process.

Definition 5.34 Let $i \in E$. If $\mathbb{P}_i(\sup\{t \geq 0 : X_t = i\} = +\infty) = 1$, the state i is said to be recurrent. Otherwise, that is if $\mathbb{P}_i(\sup\{t \geq 0 : X_t = i\} < +\infty) = 1$, it is said to be transient.

If i is recurrent, then either $\mu_i = \mathbb{E}_i(\inf\{t \geq T_1 : X_t = i\}) < +\infty$ and i is said to be positive recurrent, or $\mu_i = +\infty$ and i is said to be null recurrent. The quantity μ_i is called the mean recurrence time to i.

If all states are positive recurrent, the process is said to be positive recurrent too. Such a process is regular.

Theorem 5.35 *A state is recurrent (transient) for the jump Markov process if and only if it is recurrent (transient) for its embedded Markov chain.*

Proof The absorbing case is clear. If i is recurrent and not absorbing, then $\sup\{t \in \mathbb{R}_+^* : X_t = i\} = +\infty$, a.s.

If $N = \sup\{n \in \mathbb{N}^* : X_{T_n} = i\}$ was a.s. finite, then T_{N+1} would be finite too, hence a contradiction.

Similar arguments yield the converse and the transient case.   □

If i and j are two states in E, then i is said to lead to j if $P_t(i, j) > 0$ for some $t > 0$. If i leads to j and j leads to i, the states i and j are said to be communicating. The communication relation is an equivalence relation on E. If all states are communicating, the process is said to be irreducible.

Considering the embedded Markov chain, the following result is clear.

Corollary 5.36 *Recurrence and transience are class properties.*

5.1.5 Stationary Distribution and Asymptotic Behavior

The stationary measures are linked to the asymptotic behavior of the jump Markov processes, exactly as for Markov chains. Note that measures and distributions are represented for finite state spaces by line vectors.

Definition 5.37 Let **X** be a jump Markov process, with generator A and transition function P_t. A measure π on $(E, \mathcal{P}(E))$ is said to be stationary or invariant (for A or **X**) if $\pi P_t = \pi$ for all real numbers $t \geq 0$. If, moreover π is a probability measure, it is called a stationary distribution of the process.

For uniformisable processes, stationary measures are solutions of a homogeneous linear system.

Proposition 5.38 *If the process is uniformisable, a measure π is stationary if and only if $\pi A = 0$.*

Proof Thanks to (5.5) p. 223, $\pi = \pi P_t$ is equivalent to

$$\pi = \pi\left(I + \sum_{k \geq 1} \frac{t^t}{k!} A^k\right) = \pi + \pi \sum_{k \geq 1} \frac{t^t}{k!} A^k, \quad t \geq 0.$$

This is satisfied if and only if

$$\sum_{k \geq 1} \frac{t^k}{k!} \pi A^k = 0, \quad t \geq 0,$$

or $\pi A^k = 0$ for all k, from which the result follows.   □

▷ *Example 5.39 (Cold Stand-by System (Continuation of Example 5.29))* The stationary distribution of the Markov process modeling the cold stand-by system is solution of $\pi A = 0$. In other words,

$$\pi(1) = \frac{2\mu^2}{d}, \quad \pi(2) = \frac{2\lambda\mu}{d}, \quad \pi(3) = \frac{\lambda(\lambda + \mu - p\mu)}{d},$$

where $d = 3\lambda\mu - p\mu\lambda + 2\mu^2 + \lambda^2$.   ◁

Definition 5.40 A measure (or distribution) λ on E is said to be reversible (for A or $\mathbf{X}$) if

$$\lambda(i)a_{ij} = \lambda(j)a_{ji}, \quad i \in E, \ j \in E. \tag{5.9}$$

Proposition 5.41 *All reversible distributions are stationary.*

Proof If λ is reversible then summing both sides of (5.9) on $i \in E$, we get $\sum_{i \in E} \lambda(i)a_{ij} = 0$, and hence, according to Proposition 5.38, λ is stationary. $\qquad\square$

The stationary distributions of the process $\mathbf{X}$ and its embedded chain (J_n) are not equal, but they are closely linked.

Proposition 5.42 *If π is the stationary distribution of the process $\mathbf{X}$ and v the stationary distribution of the embedded chain (J_n), then*

$$\pi(i)a_i = v(i) \sum_{k \in E} a_k \pi(k), \quad i \in E,$$

Proof We deduce from both $\pi A = 0$ and (5.8) p. 227 that $\pi DP = \pi D$, where $D = \text{diag}(a_i)$. Therefore, πD is an invariant measure of P, and hence $v = \pi D / \sum_{i \in E} a_i \pi(i)$. $\qquad\square$

▷ *Example 5.43 (Continuation of Example 5.21)* This two-state process has a reversible distribution satisfying the equations $\pi(0)\lambda = \pi(1)\mu$ and $\pi(1) + \pi(2) = 1$, so

$$\pi(0) = \frac{\mu}{\lambda + \mu} \quad \text{and } \pi(1) = \frac{\lambda}{\lambda + \mu}.$$

This distribution is stationary for the process. $\qquad\qquad\qquad\qquad\qquad\triangleleft$

Definition 5.44 An irreducible jump Markov process whose all states are positive recurrent is said to be ergodic.

Note that the embedded Markov chain of an ergodic jump Markov process is not ergodic itself in general, as shown in the next example.

▷ *Example 5.45 (Continuation of Example 5.21)* An irreducible two-state jump Markov process is ergodic, but its embedded chain is never ergodic because it is 2-periodic. $\qquad\qquad\qquad\qquad\qquad\triangleleft$

The entropy rate of an ergodic jump Markov process has an explicit expression given by the following result that we state without proof.

Proposition 5.46 *Let* $\mathbf{X}$ *be a jump Markov process with generator* (a_{ij}) *and stationary distribution* π. *Its entropy rate is*

$$\mathbb{H}(\mathbf{X}) = -\sum_{i \in E} \pi(i) \sum_{j \neq i} a_{ij} \log a_{ij} + \sum_{i \in E} \pi(i) \sum_{j \neq i} a_{ij},$$

if this quantity is finite.

The Lagrange multipliers method yields that the jump Markov process with finite state space E having the maximum entropy rate is that with a uniform generator.

The next result leads to characterize the asymptotic behavior of an ergodic jump Markov process in the following two theorems.

Lemma 5.47 *If* X *is an ergodic Markov process with stationary distribution* π, *then the mean recurrence time of any state* $i \in E$ *is given by*

$$\mu_i = \frac{1}{a_i \pi(i)}.$$

Proof The embedded chain (J_n) is an irreducible and recurrent Markov chain. By Proposition 5.42, its stationary distribution ν is given by $\nu(j) = \pi(j)a_j$ for all $j \in E$.

Suppose J_n starts from state j. The expectation of the first jump time—or mean sojourn time in j—is $1/a_j$. Further, the expectation of the number of visits of (J_n) to state i before return to j, is $\nu(j)/\nu(i) = \pi(j)a_j/\pi(i)a_i$; see Theorem 3.44 and Proposition 3.45.

Therefore

$$\mu_i = \sum_{j \in E} \frac{\pi(j)a_j}{\pi(i)a_i} \frac{1}{a_j} = \frac{1}{\pi(i)a_i}$$

for all states $i \in E$.   □

Theorem 5.48 *Let* $\mathbf{X}$ *be an ergodic jump Markov process. For all states i and j, we have*

$$P_t(i, j) \longrightarrow \frac{1}{a_j \mu_j} = \pi(j), \quad t \to +\infty,$$

where μ_j *is the mean recurrence time of state* j.

Proof Thanks to Chapman-Kolmogorov equation, $P_{nh}(i, j) = (P_h)^n(i, j)$ for any fixed $h > 0$ and $n \in \mathbb{N}$. Thanks to the ergodic Theorem 3.50, we know that $(P_h)^n(i, j)$ converges to $\pi(j)$ when n tends to infinity.

For any $\varepsilon > 0$, some integer N exists such that

$$|P_{nh}(i, j) - \pi(j)| \leq \varepsilon/2 \qquad \text{for all} \quad n \geq N.$$

From Lévy's theorem, for any $t \geq 0$, some $h > 0$ exists such that

$$|P_{t+h}(i, j) - P_t(i, j)| \leq 1 - P_h(i, i) \leq \varepsilon/2.$$

Thus, for $nh \leq t < (n + 1)h$ and $n > N$, we get $|P_t(i, j) - P_{nh}(i, j)| \leq \varepsilon/2$.
Finally,

$$|P_t(i, j) - \pi(j)| \leq |P_t(i, j) - P_{nh}(i, j)| + |P_{nh}(i, j) - \pi(j)| \leq \varepsilon,$$

and the result follows.   □

Theorem 5.49 (Ergodic) *If* **X** *is an ergodic jump Markov process, then, for all states i and j,*

$$\frac{1}{t} \int_0^t \mathbb{1}_{(X_u=i)} du \longrightarrow \pi(i), \quad t \to +\infty, \ \mathbb{P}_j - \text{a.s.}$$

Proof Let (W_n) denote the i.i.d. sequence of successive sojourn times and let $N_i(t)$ be the number of visits in the time interval $]0, t]$ to a given state i. Then

$$\frac{W_1 + \cdots + W_{N_i(t)-1}}{t} \leq \frac{1}{t} \int_0^t \mathbb{1}_{(X_u=i)} du \leq \frac{W_1 + \cdots + W_{N_i(t)}}{t}$$

or

$$\frac{W_1 + \cdots + W_{N_i(t)-1}}{N_i(t) - 1} \frac{N_i(t) - 1}{t} \leq \frac{1}{t} \int_0^t \mathbb{1}_{(X_u=i)} du \tag{5.10}$$

and

$$\frac{1}{t} \int_0^t \mathbb{1}_{(X_u=i)} du \leq \frac{W_1 + \cdots + W_{N_i(t)}}{N_i(t)} \frac{N_i(t)}{t}. \tag{5.11}$$

Thanks to Theorem 4.17,

$$\frac{W_1 + \cdots + W_{N_i(t)}}{N_i(t)} \xrightarrow{\text{a.s.}} \mathbb{E}\, W_1 = \frac{1}{a_i}.$$

Thanks to Proposition 4.59, $N_i(t)/t$ converges a.s. to $1/\mu_i$. The result follows from inequalities (5.10) and (5.11) for t tending to infinity.   □

Therefore, for ergodic jump Markov processes, if g is a real function defined on E, the time mean is equal to the space mean of the function, that is

$$\frac{1}{t}\int_0^t g(X_u)du = \sum_{i\in E} g(i)\frac{1}{t}\int_0^t \mathbb{1}_{(X_u=i)}du \longrightarrow \sum_{i\in E} g(i)\pi(i), \quad t\to+\infty, \ \mathbb{P}_j-\text{a.s.,}$$

provided that $\sum_{i\in E}|g(i)|\pi(i) < +\infty$.

Thanks to the dominated convergence theorem, the following result is clear.

Corollary 5.50 *If* X *is an ergodic jump Markov process, then, for all states i and j,*

$$\frac{1}{t}\int_0^t P_u(j,i)du \longrightarrow \pi(i), \quad t\to+\infty.$$

▷ *Example 5.51 (Uniformisation method)* Using the stationary distribution provides a numerical method for solving Kolmogorov's equations for an ergodic process.

Let X be an ergodic uniformisable process, with stationary distribution π, with $a = \sup_{i\in E} a_i < +\infty$. The matrix $Q = I + a^{-1}A$ is stochastic. We compute

$$P_t = e^{tA} = e^{ta(Q-I)} = e^{-at}e^{atQ} = e^{-at}\sum_{n\geq 0}\frac{(at)^n}{n!}Q^n.$$

Let Π be the $E\times E$-matrix defined by $\Pi(i,j) = \pi(j)$, for all i and j. We can write

$$P_t = \Pi + e^{-at}\sum_{n\geq 0}\frac{(at)^n}{n!}(Q^n - \Pi).$$

The system $\pi A = 0$ is equivalent to $\pi a(Q-I) = 0$, or to $\pi Q = \pi$. Thus, Q has the same invariant distribution π as P_t. Therefore, $Q^n - \Pi$ converges to zero when n tends to infinity.

If α is a distribution on E, one can show that

$$\sup_{t\geq 0}\|\alpha P_t - \alpha P_t(k)\| \longrightarrow 0, \quad k\to+\infty,$$

where

$$P_t(k) = \Pi + e^{-at}\sum_{n=0}^k\frac{(at)^n}{n!}(Q^n - \Pi).$$

Note that the truncating level k can be chosen such that the error is bounded for some t by an ε, and then it will be bounded for all $t \geq 0$. ◁

5.2   Semi-Markov Processes

This section is dedicated to the investigation of semi-Markov processes, mainly with finite state spaces. The semi-Markov processes constitute a natural generalization of the Markov and renewal processes. Their future evolution depends on both the occupied state and the time elapsed since the last transition. This time, called local time, is measured by a watch that comes back to zero at each transition. Of course, if the watch is considered as an integral part of the system—in other words if $E \times \mathbb{R}_+$ becomes the state space (where E is the state space of the semi-Markov process), then the process becomes a Markov process.

5.2.1   Markov Renewal Processes

In order to define semi-Markov processes, it is easier first to define Markov renewal processes.

Definition 5.52 Let $(J_n, T_n)_{n \in \mathbb{N}}$ be a process defined on $(\Omega, \mathcal{F}, \mathbb{P})$ such that (J_n) is a random sequence taking values in a discrete set E and (T_n) is an increasing random sequence taking values in $\mathbb{R}_+$, with $T_0 = 0$. Set $\mathcal{F}_n = \sigma(J_k, T_k; k \leq n)$. The process (J_n, T_n) is called a Markov renewal process with discrete state space E if

$$\mathbb{P}(J_{n+1} = j, T_{n+1} - T_n \leq t \mid \mathcal{F}_n) = \mathbb{P}(J_{n+1} = j, T_{n+1} - T_n \leq t \mid J_n) \quad \text{a.s.,}$$

for all $n \in \mathbb{N}$, all $j \in E$ and all $t \in \mathbb{R}_+$.

If the above conditional probability does not depend on n, the process is said to be homogeneous, and we set

$$Q_{ij}(t) = \mathbb{P}(J_{n+1} = j, T_{n+1} - T_n \leq t \mid J_n = i) \tag{5.12}$$

for all $n \in \mathbb{N}$, $i \in E$, $j \in E$ and $t \in \mathbb{R}_+$. The family $Q = \{Q_{ij}(t); i, j \in E, t \in \mathbb{R}_+\}$ is called a semi-Markov kernel on E and the square matrix $Q(t) = (Q_{ij}(t))_{(i,j) \in E \times E}$ is a semi-Markov matrix.

We will study here only homogeneous Markov renewal processes with finite state spaces, say $E = \{1, \ldots, e\}$.

The process (J_n, T_n) is a two-dimensional Markov chain, with state space $E \times \mathbb{R}_+$; its transition function is the semi-Markov kernel Q. Letting t go to infinity in (5.12) shows that (J_n) is a Markov chain with state space E and transition matrix $P = (P(i, j))$ where

$$P(i, j) = \lim_{t \to +\infty} Q_{ij}(t), \quad i, j \in E.$$

▷ *Example 5.53* The following processes are Markov renewal processes:

1. For a renewal process, only one state is visited—say $E = \{1\}$, and $Q(t) = Q_{11}(t) = F(t)$, which is a scalar function.
2. For an alternated renewal process, $E = \{1, 2\}$, $P(1, 2) = 1$, $P(2, 1) = 1$ and

$$Q(t) = \begin{pmatrix} 0 & F(t) \\ G(t) & 0 \end{pmatrix};$$

see Exercise 4.4 below for further details on this process.   ◁

Definition 5.54 The stochastic process $\mathbf{Z} = (Z_t)_{t \in \mathbb{R}_+}$, defined by

$$Z_t = J_n \quad \text{if} \quad T_n \leq t < T_{n+1}$$

is called the semi-Markov process associated with the Markov renewal process (J_n, T_n).

The process $\mathbf{Z}$ is continuous on the right. Clearly $J_n = Z_{T_n}$, meaning that (J_n) is the sequence of the successive states visited by $\mathbf{Z}$. As for Markov processes, (J_n) is called the embedded chain of the process.

The initial distribution α of $\mathbf{Z}$ (that is the distribution of Z_0) is also the initial distribution of (J_n). Knowledge of both α and the kernel Q characterizes the distribution of $\mathbf{Z}$.

Properties of Semi-Markov Kernels

1. For all $i, j \in E$, the function $t \longrightarrow Q_{ij}(t)$ is a defective distribution function on $\mathbb{R}_+$. On the contrary, $H_i(t) = \sum_{j \in E} Q_{ij}(t)$ for $t \geq 0$ defines the distribution function of the total time spent by $\mathbf{Z}$ in $[0, t]$ at i, called sojourn time. We will write $H(t) = \text{diag}(H_i(t))_{i \in E}$, and denote by m_i the mean sojourn time in state i, that is $m_i = \mathbb{E}_i(T_1) = \int_0^{+\infty}(1 - H_i(t))\, dt$.
2. $Q_{ij}(t) = P(i, j)F_{ij}(t)$, where $F_{ij}(t) = \mathbb{P}(T_{n+1} - T_n \leq t \mid J_n = i, J_{n+1} = j)$ is the distribution function of the time spent by $\mathbf{Z}$ in state i conditional on transition to state j.
3. $\mathbb{P}(J_n = j, T_n \leq t \mid J_0 = i) = Q_{ij}^{*(n)}(t)$ for $n > 0$, where $Q_{ij}^{*(n)}$ is the n-th Lebesgue-Stieltjes convolution of Q_{ij}, that is

$$Q_{ij}^{*(n)}(t) = \sum_{k \in E} \int_0^t Q_{ik}(ds) Q_{kj}^{*(n-1)}(t - s), \quad n \geq 2,$$

with $Q^{*(1)} = 0$ and $Q_{ij}^{*(0)} = \delta_{ij}$.

▷ *Example 5.55 (Kernel of a jump Markov process)* For a jump Markov process—
with generator (a_{ij}), the sequences (J_n) and (T_n) are $\mathbb{P}_i$-independent. We obtain
from Theorem 5.24 that

$$Q_{ij}(t) = \mathbb{P}_i(J_1 = j)(1 - e^{-a_i t}) = \frac{a_{ij}}{a_i}(1 - e^{-a_i t}), \quad i \neq j \in E,$$

and $Q_{ii}(t) = 0$ for $t \in \mathbb{R}_+$, where $a_i = -a_{ii} = \sum_{j \neq i} a_{ij}$.   ◁

Let $N_j(t)$ be the number of visits of **Z** to state j in the time interval $]0, t]$,
and set $N(t) = (N_1(t), \ldots, N_e(t))$. Note that the Markov renewal process may
alternatively be defined from the process $\mathbf{N} = (N(t))_{t \in \mathbb{R}_+}$.

Definition 5.56 The function

$$t \longrightarrow \psi_{ij}(t) = \mathbb{E}_i N_j(t) = \sum_{n \geq 0} Q_{ij}^{*(n)}(t)$$

is called a Markov renewal function; we will write $\psi(t) = (\psi_{ij}(t))_{(i,j) \in E \times E}$.

We compute

$$\psi_{ij}(t) = I(i, j)\mathbb{1}_{\mathbb{R}_+}(t) + \sum_{k \in E} Q_{ik} * \psi_{kj}(t),$$

or, under matrix form, $\psi(t) = \mathrm{diag}(\mathbb{1}_{\mathbb{R}_+}(t)) + Q * \psi(t)$. This equation is a particular
case of the Markov renewal equation

$$L(t) = G(t) + Q * L(t), \quad t \in \mathbb{R}_+,$$

where G (given) and L (unknown) are matrix functions null on $\mathbb{R}_-$ and bounded on
the finite intervals of $\mathbb{R}_+$. When it exists, the solution takes the form

$$L(t) = \psi * G(t), \quad t \in \mathbb{R}_+.$$

We assume here that none of the functions H_i, for $i \in E$, is degenerated (that is
$H_i(t) \neq \mathbb{1}_{(t \geq 0)}$).

Definition 5.57 The transition function of **Z** is defined by

$$P_t(i, j) = \mathbb{P}(Z_t = j \mid Z_0 = i), \quad i, j \in E, t \in \mathbb{R}_+,$$

and we will write in matrix form $P(t) = (P_t(i, j))_{(i,j) \in E \times E}$.

Proposition 5.58 *The transition function $P(t)$ of $\mathbf{Z}$ is solution of the Markov renewal equation*

$$P(t) = \text{diag}(1 - H_i(t)) + Q * P(t). \tag{5.13}$$

Proof We have

$$P_t(i, j) = \mathbb{P}(Z_t = j \mid Z_0 = i) = \mathbb{P}_i(Z_t = j, T_1 > t) + \mathbb{P}_i(Z_t = j, T_1 \leq t).$$

Since $\mathbb{P}_i(Z_t = j, T_1 > t) = [1 - H_i(t)]I(i, j)$ and

$$\begin{aligned}
\mathbb{P}_i(Z_t = j, T_1 \leq t) &= \mathbb{E}_i\left[\mathbb{P}_i(Z_t = j, T_1 \leq t \mid \mathcal{F}_{T_1})\right] \\
&= \mathbb{E}_i\left[\mathbb{1}_{(T_1 \leq t)}\mathbb{P}_{Z_{T_1}}(Z_{t-T_1} = j)\right] \\
&= \sum_{k \in E}\int_0^t Q_{ik}(ds)P_{t-s}(k, j),
\end{aligned}$$

(5.13) follows, under matrix form. □

The unique solution of (5.13) is given by

$$P(t) = [I - Q(t)]^{*(-1)} * [I - H(t)], \tag{5.14}$$

where $[I - Q(t)]^{*(-1)} = \psi(t) = \sum_{n \geq 0} Q^{*(n)}(t)$.

5.2.2 Classification of States and Asymptotic Behavior

Let $(T_n^j)_{n \in \mathbb{N}}$ be the sequence of successive times of visit of the semi-Markov process $\mathbf{Z}$ to state $j \in E$. It is a renewal process, possibly modified. Thus, T_0^j is the time of the first visit to j and $G_{ij}(t) = \mathbb{P}(T_0^j \leq t \mid Z_0 = i)$ is the distribution function of the time of the first transition from state i to state j. If $i = j$, then $T_0^j = 0$, and hence $G_{jj}(t) = \mathbb{P}(T_1^j \leq t \mid Z_0 = j)$ is the distribution function of the time between two successive visits to j. We have

$$\psi_{jj}(t) = \sum_{n \geq 0} G_{jj}^{*(n)}(t),$$

and for $i \neq j$,

$$\psi_{ij}(t) = \sum_{n \geq 0} G_{ij} * G_{jj}^{*(n)}(t) = G_{ij} * \psi_{jj}(t).$$

The expectation of the hitting time of j starting from i at time 0 is

$$\mu_{ij} = \int_0^{+\infty} (1 - G_{ij}(t))\, dt, \quad i, j \in E. \tag{5.15}$$

When $i = j$, it is the mean return time to i.

As for Markov processes, two states i and j are said to be communicating if $G_{ij}(+\infty)G_{ji}(+\infty) > 0$ or if $i = j$. The communication relation is an equivalence relation on E. If all states are communicating, the process is said to be irreducible.

Definition 5.59 If $G_{ii}(+\infty) = 1$ or $\psi_{ii}(+\infty) = +\infty$, the state i is said to be recurrent. Otherwise, it is said to be transient.

If i is recurrent, then either its mean return time μ_{ii} is finite and i is said to be positive recurrent, or it is infinite and i is said to be null recurrent.

A state i is said to be periodic with period $h > 0$ if G_{ii} is arithmetic with period h. Then, $\psi_{ii}(t)$ is constant on the intervals of the form $[nh, nh + h[$, where h is the largest number sharing this property. Otherwise, i is said to be aperiodic. If all states are aperiodic, the process is said to be aperiodic. Note that this notion of periodicity for semi-Markov processes is different from the notion of periodicity seen in Chap. 3 for Markov chains.

The following result is a straightforward consequence of Blackwell's renewal theorem applied to the renewal process (T_n^i).

Theorem 5.60 (Markov Renewal) *If the state i is aperiodic, then for any $c > 0$,*

$$\psi_{ii}(t) - \psi_{ii}(t - c) \longrightarrow \frac{c}{\mu_{ii}}, \quad t \to +\infty. \tag{5.16}$$

When i is periodic with period h, the result remains valid if c is a multiple of h.

We state the next result without proof.

Theorem 5.61 (Key Markov Renewal) *If (J_n) is irreducible and aperiodic, if v is an invariant measure for P and if $m_i < +\infty$ for all $i \in E$, then for all direct Riemann integrable real functions g_i defined on $\mathbb{R}_+$ for $i \in E$,*

$$\int_0^t g_i(t - y)\psi_{ji}(dy) \longrightarrow \frac{v(i)}{< v, m >} \int_0^{+\infty} g_i(y)dy, \quad t \to +\infty.$$

Thus, thanks to the key Markov renewal theorem and (5.14) p. 239, we get

$$\pi(j) = \lim_{t \to +\infty} P_t(i, j) = \frac{v(j)m_j}{< v, m >}, \quad i, j \in E,$$

which defines the limit distribution π of $\mathbf{Z}$.

The entropy rate of a semi-Markov process has an explicit form under suitable conditions. It is given by the next proposition stated without proof.

Proposition 5.62 *Let $\mathbf{Z}$ be a semi-Markov process such that*

$$Q_{ij}(t) = \int_0^t q_{ij}(x)dx, \quad with \quad \int_{\mathbb{R}_+} q_{ij}(t)|\log q_{ij}(t)|dt < +\infty, \quad i, j \in E.$$

If $m_i < +\infty$ for all $i \in E$, then

$$\mathbb{H}(\mathbf{X}) = \frac{-1}{<\nu, m>} \sum_{i,j \in E} \nu(i) \int_0^{+\infty} q_{ij}(x) \log q_{ij}(x)dx.$$

▷ *Example 5.63 (Analysis of Seismic Risk)* We consider here two simplified models. The intensity of an earthquake is classified according to a discrete ladder of states in $E = \{1, \ldots, N\}$. The process $\mathbf{Z} = (Z(t))_{t \in \mathbb{R}_+}$ is defined by $Z(t) = i$ for $1 \leq i < N$ if the intensity of the last earthquake before time t was $f \in [i, i+1[$ and finally $Z(t) = N$ if $f \geq N$.

For a time predictable model, it is assumed that the stronger is an earthquake, the longer is the time before the next occurs. The stress accumulated on a given rift has a minimum bound. When a certain level of stress is reached, an earthquake occurs. Then the stress decreases to the minimum level; see Fig. 5.4. The semi-Markov kernel of $\mathbf{Z}$ is then

$$Q_{ij}(t) = \nu(j)F_j(t), \quad t \in \mathbb{R}_+.$$

For a slip predictable model, it is assumed that the longer is the time elapsed since the last earthquake, the stronger is the next one. The stress has a maximum bound. When this level is reached, an earthquake occurs. Then the stress decreases of a certain quantity; see Fig. 5.4. The semi-Markov kernel of $\mathbf{Z}$ is

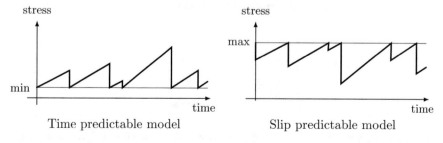

Fig. 5.4 Two semi-Markov models for seismic risk analysis

then

$$Q_{ij}(t) = v(j)F_i(t), \quad t \in \mathbb{R}_+.$$

In both cases, since F_{ij} depends only on one of the two states i and j, necessarily the probability v is the stationary distribution of the embedded chain of the process. If the functions F_i are differentiable for all $i \in E$, the entropy rate of the process can be computed explicitly.

For the slip predictable model,

$$\mathbb{H}(\mathbf{Z}) = \frac{-1}{<v, m>} \sum_{i,j \in E} v(i) \int_{\mathbb{R}_+} v(j) f_i(t) \log[v(j) f_i(t)] \, dt.$$

Since

$$\int_{\mathbb{R}_+} v(j) f_i(t) \log[v(j) f_i(t)] \, dt = v(j) \log[v(j)] + v(j) \int_{\mathbb{R}_+} f_i(t) \log[f_i(t)] \, dt,$$

we get

$$\mathbb{H}(\mathbf{Z}) = \frac{-1}{<v, m>} \sum_{i \in E} v(i) \left[\log v(i) + \int_{\mathbb{R}_+} f_i(t) \log f_i(t) \, dt \right].$$

Similar computation for the time predictable model yields the same formula.
◁

Other applications of semi-Markov processes, linked to reliability, will be studied in Exercises 5.5 and 5.6.

5.3   Exercises

▽ **Exercise 5.1 (Birth-and-death Process)**   The Markov process $(X_t)_{t \geq 0}$ with state space $E = \mathbb{N}$ and generator $A = (a_{ij})$ defined by

$$a_{ij} = \begin{cases} \lambda_i & \text{if } j = i + 1, \ i \geq 0, \\ \mu_i & \text{if } j = i - 1, \ i \geq 1, \\ -(\lambda_i + \mu_i) & \text{if } j = i, \ i \geq 0, \\ 0 & \text{otherwise}, \end{cases}$$

with $\mu_0 = 0$, is called a birth-and-death process. If $\mu_i = 0$ for all i, the process is a birth process; if $\lambda_i = 0$ for all i, it is a death process.

Assume that $\lambda_i \mu_{i+1} > 0$ for all $i \geq 0$. Give a necessary and sufficient condition for the process to have a reversible distribution and give then its stationary distribution.

Solution A reversible distribution π satisfies $\pi(i-1)\lambda_{i-1} = \pi(i)\mu_i$ for $i \geq 1$, that is

$$\pi(i) = \frac{\lambda_{i-1}}{\mu_i}\pi(i-1) = \frac{\lambda_{i-1}\lambda_{i-2}}{\mu_i \mu_{i-1}}\pi(i-2) = \cdots = \frac{\lambda_{i-1}\ldots\lambda_0}{\mu_i \ldots \mu_1}\pi(0) = \gamma_i \pi(0),$$

where $\gamma_0 = 1$ and

$$\gamma_i = \frac{\lambda_0 \ldots \lambda_{i-1}}{\mu_1 \ldots \mu_i}, \quad i \geq 1.$$

Summing on $i \geq 0$ yields

$$1 = \sum_{i\geq 0} \pi(i) = \sum_{i\geq 0} \gamma_i \pi(0).$$

Therefore, the convergence of the sum $\sum_{i\geq 0} \gamma_i$ is a necessary and sufficient condition for the process to have a reversible distribution. We compute then

$$\pi(0) = \left(\sum_{i\geq 0} \gamma_i \right)^{-1},$$

and hence

$$\pi(i) = \frac{\gamma_i}{\sum_{k\geq 0} \gamma_k}, \quad i \geq 0.$$

Thus, according to Proposition 5.41, the reversible distribution is also a stationary distribution. △

▽ **Exercise 5.2 (*M/M/1* Queueing Systems)** At the post office, only one customer can be served at a time. The time of service has an exponential distribution $\mathcal{E}(\mu)$. The times of arrivals of the customers form a homogeneous Poisson process with intensity λ. When a customer arrives, either he is immediately served if the server is available, or he joins the (possibly infinite) queue. Such a system is called an $M/M/1$ queueing system (Fig. 5.5). Let X_t be the random variable equal to the number of customers present in the post office at time t, for $t \in \mathbb{R}_+$.

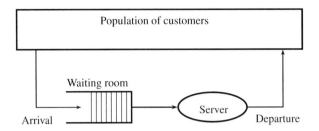

Fig. 5.5 $M/M/1$ queueing system

1. Show that $\mathbf{X} = (X_t)$ is a Markov process. Give its generator.
2. Determine the stationary distribution π of $\mathbf{X}$, when it exists.
3. The initial distribution of $\mathbf{X}$ is assumed to be π.
   a. Compute the average number of customers in the post office at a fixed time t.
   b. Determine the distribution of the time spent in the post office by a customer.
4. Compute the average time during which the post office is empty, for $\lambda/\mu = 1/2$.

Solution

1. The process $\mathbf{X}$ is a birth-death Markov process with state space $E = \mathbb{N}$ and generator determined by $\lambda_i = \lambda$, for $i \geq 0$ and $\mu_i = \mu$, for $i \geq 1$; see Exercise 5.1 for notation.
2. With the same notation,

$$\gamma_0 = 1 \quad \text{and} \quad \gamma_i = \left(\frac{\lambda}{\mu}\right)^i, \quad i \geq 1.$$

Set $a = \lambda/\mu$. For $a < 1$, we have $\sum_{k\geq 0} \gamma_k = 1/(1-a)$, and then—and only then—$\mathbf{X}$ has a stationary distribution π, given by $\pi_i = a^i(1-a)$, for $i \geq 0$. In other words, π is a geometric distribution on $\mathbb{N}$ with parameter a.

3. a. When the initial distribution is π, the process is stationary and the expectation of X_t is the expectation of the geometric distribution $\mathcal{G}(a)$, that is $\mathbb{E} X_t = a/(1-a)$.
   b. Let W be the total time passed in the system by some given customer, arriving at time T_0. Clearly, the (exit) process $(M(t))$ of other customers' exit times, after T_0 and until the customer's exit, is a homogeneous Poisson process with intensity μ. We compute

$$\mathbb{P}(W > t) = \sum_{n\geq 0} \mathbb{P}(W > t, X_{T_0^-} = n) = \sum_{n\geq 0} \mathbb{P}(W > t \mid X_{T_0^-} = n)\mathbb{P}(X_{T_0^-} = n)$$

$$= \sum_{n\geq 0} \mathbb{P}(M_t \leq n)\pi(n) = \sum_{n\geq 0} \left[\sum_{k=0}^{n} e^{-\mu t}\frac{(\mu t)^k}{k!}\right] a^n(1-a) = e^{-\mu(1-a)t},$$

meaning that W has an exponential distribution $\mathcal{E}(\mu(1-a))$.

c. We have $\mathbb{E}\,X_t = \lambda \mathbb{E}\,W$. Note that this formula is a particular case of Little's formula, that characterizes ergodic queueing systems.

4. If $\lambda/\mu = 1/2$, then $a = 1/2$, and hence $\pi_0 = 1/2$. Therefore, the post office will be empty half the time.

See Problem 5.7 for a feed back queue. $\qquad\qquad\qquad\qquad\qquad\qquad\qquad\qquad \triangle$

∇ **Exercise 5.3 (Epidemiological Models)** Consider a population of m individuals. Suppose that exactly one individual is contaminated at time $t = 0$. The others can then be contaminated and the affection is incurable. Suppose that in any time length h, one infected individual can infect a healthy individual with probability $\alpha h + o(h)$ for $h \to 0^+$, where $\alpha > 0$. Let X_t be the number of individuals contaminated at time $t \geq 0$, and let T_i be the time necessary to pass from i contaminated individuals to $i + 1$, for $1 \leq i \leq m - 1$.

1. a. Suppose $X_t = i$. Compute the probability that only one individual is contaminated in the time interval $[t, t + h]$. Show that the probability that two or more individuals are contaminated in the same time interval is $o(h)$.

   b. Show that $\mathbf{X} = (X_t)$ is a Markov process; give its state space and generator.

2. Show that T_i has an exponential distribution; give its parameter.

3. Let T be the time necessary for the whole population to be contaminated; compute its mean and variance.

4. Numerical application: compute an approximate value of the mean of T for $m = 6 \times 10^7$, $\alpha = 6 \times 10^{-8}$ per day, and $h = 1$ day.

Solution

1. a. If i individuals are infected, then each individual among the $m - i$ healthy ones can be contaminated in $]t, t + h]$ with probability $i\alpha h + o(h)$. Thus, the probability that one individual among the $m - i$ will be contaminated in $]t, t + h]$ is

$$\binom{m-i}{1} [i\alpha h + o(h)]^1 [1 - i\alpha h + o(h)]^{m-i-1} = (m - i)i\alpha h + o(h).$$

Similarly, for $k \geq 2$,

$$\binom{m-i}{k} [i\alpha h + o(h)]^k [1 - i\alpha h + o(h)]^{m-i-k} = o(h).$$

b. Therefore, (X_t) is a Markov process with state space $E = [\![1, m]\!]$ and generator $A = (a_{ij})_{(i,j) \in E \times E}$, where

$$a_{ij} = \begin{cases} (m - i)i\alpha, & \text{if } j = i + 1, \\ -(m - i)i\alpha, & \text{if } j = i, \\ 0, & \text{otherwise.} \end{cases}$$

This process is a birth process, also called Yule process.

2. Proposition 5.22 and Corollary 5.27 together yield that $T_i \sim \mathcal{E}(-a_{ii})$.

3. Since $T = T_1 + \cdots + T_{m-1}$,

$$\mathbb{E}\, T = \sum_{i=1}^{m-1} \frac{1}{(m-i)i\alpha}.$$

The variables T_i, for $1 \le i \le m-1$, are independent, so

$$\mathbb{V}\text{ar}\, T = \sum_{i=1}^{m-1} \frac{1}{[(m-i)i\alpha]^2}.$$

4. We compute

$$\mathbb{E}\, T = \frac{1}{m\alpha} \sum_{i=1}^{m-1} \left(\frac{1}{m-i} + \frac{1}{i} \right) \approx \frac{1}{m\alpha} \int_1^{m-1} \left(\frac{1}{m-t} + \frac{1}{t} \right) dt,$$

and $\int_1^{m-1} \left(\frac{1}{m-t} + \frac{1}{t} \right) dt = 2\log(m-1)$, so $\mathbb{E}\, T = 2\log(m-1)/m\alpha$. For the data, $\mathbb{E}\, T \approx 358$ days, around 1 year.   △

∇ **Exercise 5.4 (Reliability of a Markov System)** Consider a system whose stochastic behavior is modelled by a Markov process, $\mathbf{X} = (X_t)_{t \in \mathbb{R}_+}$, with finite state space $E = [\![1, e]\!]$, generator A, transition function $P_t(i, j)$, and initial distribution α. Let $U = [\![1, m]\!]$ be the set of functioning states and $D = [\![m+1, e]\!]$ the set of failed states, for some $m \in [\![2, e-1]\!]$.

1. a. Compute the instantaneous availability $A(t)$ of the system for $t > 0$; see Exercise 4.4 for definition.

   b. Use a. to compute the limit availability when $\mathbf{X}$ is ergodic.

2. Let $T_D = \inf\{t \ge 0 : X_t \in D\}$ be the hitting time of the set D of failed states of $\mathbf{X}$, with the convention $\inf \phi = +\infty$. Consider the process $\mathbf{Y}$ with state space $U \cup \{\Delta\}$—where Δ is an absorbing state, defined by

$$Y_t = \begin{cases} X_t & \text{if } t < T_D, \\ \Delta & \text{if } t \ge T_D. \end{cases}$$

   a. Give the initial distribution and the generator of $\mathbf{Y}$, which is a Markov process.

   b. Use $\mathbf{Y}$ to compute the reliability function of the system, defined by $R(t) = \mathbb{P}(T_D > t)$ for $t \ge 0$.

Solution

1. a. The instantaneous availability is

$$A(t) = \mathbb{P}(X_t \in U) = \sum_{j \in U} \mathbb{P}(X_t = j) = \sum_{j \in U} \sum_{i \in E} \mathbb{P}(X_t = j, X_0 = i)$$

$$= \sum_{j \in U} \sum_{i \in E} \mathbb{P}(X_t = j \mid X_0 = i) \mathbb{P}(X_0 = i)$$

$$= \sum_{j \in U} \sum_{i \in E} \alpha(i) P_t(i, j) = \alpha P_t \mathbf{1}_{e,m} = \alpha e^{tA} \mathbf{1}_{e,m},$$

where $\mathbf{1}_{e,m} = (1, \ldots, 1, 0, \ldots, 0)'$ is the e-dimensional column vector of which the first m components are equal to 1 and the $e - m$ others are equal to 0.

b. Therefore, the limit availability is

$$A = \lim_{t \to \infty} A(t) = \alpha \Pi \mathbf{1}_{e,m} = \sum_{k \in U} \pi(k).$$

2. a. For computing the reliability, it is necessary to consider the partition of A and α between U and D, that is, $\alpha = (\alpha_1, \alpha_2)$ and

$$A = \begin{pmatrix} A_{11} & A_{12} \\ A_{21} & A_{22} \end{pmatrix}.$$

We can write $Y_t = X_{t \wedge T_D}$, and $\mathbf{Y}$ is indeed a Markov process with generator

$$B = \begin{pmatrix} A_{11} & A_{12}\mathbf{1} \\ \mathbf{0} & 0 \end{pmatrix}.$$

Its initial distribution is $\beta = (\alpha_1, b)$ with $b = \alpha_2 \mathbf{1}$.

b. Let Q_t be the transition function of $\mathbf{Y}$. We have

$$R(t) = \mathbb{P}(\forall u \in [0, t], X_u \in U) = \mathbb{P}(Y_t \in U) = \sum_{j \in U} \mathbb{P}(Y_t = j).$$

We compute for all $j \in U$

$$\mathbb{P}(Y_t = j) = \sum_{i \in U} \mathbb{P}(Y_t = j, Y_0 = i)$$

$$= \sum_{i \in U} \mathbb{P}(Y_t = j \mid Y_0 = i) \mathbb{P}(Y_0 = i) = \sum_{i \in U} \alpha(i) Q_t(i, j),$$

that is $R(t) = (\alpha_1, 0) Q_t \mathbf{1}_{s,m} = \alpha_1 e^{tA_{11}} \mathbf{1}_m$. △

∇ **Exercise 5.5 (A Binary Semi-Markov System and Its Entropy Rate)**

1. Model by a semi-Markov process the system of Exercise 4.4.
2. Write the Markov renewal equation and determine the transition function of the system.
3. Show again the availability results.
4. Assume that X_1 (Y_1) has a density f (g) and a finite expected value a (b). Compute the entropy rate of the process. Determine the sojourn times distributions maximizing this rate.

Solution With the notation of Sect. 5.2.

1. If the functioning states are represented by 0 and the failed states by 1, the system can be modelled by a semi-Markov process **Z** defined by

$$Z_t = \sum_{n \geq 0} \mathbf{1}_{(S_n \leq t < S_n + X_{n+1})}, \quad t \geq 0,$$

with semi-Markov kernel Q given by

$$Q(t) = \begin{pmatrix} 0 & F(t) \\ G(t) & 0 \end{pmatrix}.$$

2. The transition function P_t satisfies the Markov renewal equation $P = I - H + Q * P$, or $[I - Q] * P = H$, where

$$H(t) = \begin{pmatrix} F(t) & 0 \\ 0 & G(t) \end{pmatrix}.$$

Its solution is $P(t) = [I - Q(t)]^{*(-1)} * [I - H(t)]$.
On the one hand,

$$[I - Q(t)]^{*(-1)} = \begin{pmatrix} 1 & -F(t) \\ -G(t) & 1 \end{pmatrix}^{*(-1)}$$

$$= [1 - F * G(t)]^{*(-1)} * \begin{pmatrix} 1 & F(t) \\ G(t) & 1 \end{pmatrix}$$

and the renewal function of the alternated renewal process of Exercise 4.4 is

$$m(t) = [1 - F * G(t)]^{*(-1)} = \sum_{n \geq 0} (F * G)^{*(n)}(t).$$

On the other hand,

$$I - H(t) = \begin{pmatrix} 1 - F(t) & 0 \\ 0 & 1 - G(t) \end{pmatrix}.$$

Finally,

$$P(t) = m * \begin{pmatrix} 1 & F(t) \\ G(t) & 1 \end{pmatrix} * \begin{pmatrix} 1 - F(t) & 0 \\ 0 & 1 - G(t) \end{pmatrix}$$

$$= m * \begin{pmatrix} 1 - F(t) & F * (1 - G)(t) \\ G * (1 - F)(t) & 1 - G(t) \end{pmatrix}. \tag{5.17}$$

3. Taking as initial distribution $(1, 0)$, we obtain the availability $A(t) = P_{00}(t) = m * (1 - F)(t)$.

   Note that this approach is much more general than the one obtained by using the alternated renewal process, because all results are given by (5.17). For example,

$$A(t) = P_{01}(t) = m * G * (1 - F)(t),$$

if the system is assumed to be failed at time 0.

4. We compute

$$< v, m > = \int_0^{+\infty} \frac{t}{2} [f(t) + g(t)] \, dt = \frac{a + b}{2},$$

and

$$\sum_{i,j=1}^{2} v(i) \int_0^{+\infty} q_{ij}(t) \log q_{ij}(t) \, dt =$$

$$= \frac{1}{2} \left[\int_0^{+\infty} f(t) \log f(t) \, dt + \int_0^{+\infty} g(t) \log g(t) \, dt \right],$$

so the entropy rate of **Z** is

$$\mathbb{H}(\mathbf{Z}) = \frac{1}{a + b} [\mathcal{I}(X_1) + \mathcal{I}(Y_1)].$$

If f and g are exponential distributions with respective parameters λ and μ, the entropy rate is

$$\mathbb{H}(\mathbf{Z}) = \frac{\lambda \mu}{\lambda + \mu} [2 - \log(\lambda \mu)],$$

and is clearly maximum. $\triangle$

▽ **Exercise 5.6 (A Treatment Station)**   A factory discharges polluting waste at a known flow rate. A treatment station is constructed for sparing environment. A tank is provided for stocking the waste during the failures of the station; this avoids to stop the factory if the repairing is finished before the tank is full. Both the time for emptying the tank and the time necessary for the treatment of its content by the station are assumed to be negligible. The random variable τ equal to the time for filling the tank is called the delay.

This system is modelled by a semi-Markov process with three states.

1. Give the states and the semi-Markov kernel of the process.
2. Determine the transition matrix and the stationary distribution of its embedded chain.
3. Give the limit distribution of the process.
4. Assuming that the system works perfectly at time $t = 0$, determine its reliability R; see Exercise 5.4 for definition.

Solution

1. The states of the process are the following:

   state 1 : the factory is functioning;
   state 2 : the factory has been failed shorter than τ;
   state 3 : the factory has been failed longer than τ.

   The semi-Markov kernel is

   $$Q = \begin{pmatrix} 0 & F & 0 \\ Q_{21} & 0 & Q_{23} \\ B & 0 & 0 \end{pmatrix},$$

   with

   $$Q_{21}(t) = \int_0^t [1 - C(x)]dA(x) \quad \text{and} \quad Q_{23}(t) = \int_0^t [1 - A(x)]dC(x),$$

   where F is the distribution function of the life time of the station, A that of the repairing time of the station, C that of the delay, and B that of the repairing time of the factory.
2. The transition matrix of the embedded chain (J_n) is

   $$P = \begin{pmatrix} 0 & 1 & 0 \\ q & 0 & p \\ 1 & 0 & 0 \end{pmatrix}$$

where $p + q = 1$, with

$$p = \int_0^{+\infty} [1 - A(x)]dC(x).$$

The stationary distribution of (J_n) is $\nu = (1/(2 + p), 1/(2 + p), p/(2 + p))$.
3. The limit distribution of the process is

$$\pi = \frac{1}{<\nu, m>}\mathrm{diag}(\nu(i))m = \frac{1}{m_1 + m_2 + pm_3}(m_1, m_2, pm_3),$$

where $m_1 = \int_0^{+\infty}[1 - F(x)]dx$, $m_2 = \int_0^{+\infty}[1 - Q_{21}(x) - Q_{23}(x)]dx$, and $m_3 = \int_0^{+\infty}[1 - B(x)]dx$.
4. We compute

$$R(t) = M * [1 - F + F * (1 - F - Q_{23})](t),$$

where $M(t) = \sum_{n \geq 0}(F * Q_{21})^{(n)}(t)$. △

▽ **Exercise 5.7 (A Feed Back Queue)** Let us consider again the queue of Exercise 5.2. When a customer arrives, if more than N customers already queue, then he leaves the system. Moreover, once served, either he comes back queueing, with probability $p \in]0, 1[$, or he leaves the system, with probability $1 - p$; see Fig. 5.6. Let X_t be the random variable equal to the number of customers present in the post office at time t.

1. Of which type is the semi-Markov process $\mathbf{X} = (X_t)$? Give its semi-Markov kernel and determine the transition matrix of its embedded chain.
2. Compute the stationary distribution of this chain, the average sojourn times in each state and the limit distribution of the process.

Solution The number $\mathbf{X} = (X_t)$ of customers present in the system is a birth-death semi-Markov process with state space $E = [\![0, N]\!]$ and exponentially distributed sojourn times, that is a Markov process again. The only non-zero entries of its

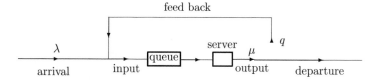

Fig. 5.6 A feed back queue—Exercise 5.7

semi-Markov kernel are

$$Q_{i,i-1}(t) = \frac{\mu_i q_i}{\lambda_i + \mu_i}[1 - e^{-(\lambda_i + \mu_i)t}], \quad 1 \leq i \leq N,$$

$$Q_{i,i}(t) = \frac{\mu_i p_i}{\lambda_i + \mu_i}[1 - e^{-(\lambda_i + \mu_i)t}], \quad 1 \leq i \leq N,$$

$$Q_{i,i+1}(t) = \frac{\lambda_i}{\lambda_i + \mu_i}[1 - e^{-(\lambda_i + \mu_i)t}], \quad 0 \leq i \leq N - 1,$$

2. Its mean sojourn times are given by $m_i = 1/(\lambda_i + \mu_i)$. The stationary distribution v of its embedded chain is given by

$$v_i = \lambda_1 \ldots \lambda_{i-1}(1 + \lambda_i/\mu_i + 1/q_i)v_0/\mu_1 \ldots \mu_{i-1}, \quad 2 \leq i \leq N,$$

$$v_1 = (1 + \lambda_i/\mu_i + 1/q_i)v_0,$$

$$v_0 = \left[\sum_{i=1}^{N} \lambda_1 \ldots \lambda_{i-1}(1 + 1/q_i)/\mu_1 \ldots \mu_{i-1}\right]^{-1}.$$

Note that for $p = 0$, this system amounts to the system described in Exercise 5.2, for a finite queue, that is an M/M/1/(N + 1) queue.   △

Further Reading

Analysis

Bobrowski, A. (2005). *Functional analysis for probability and stochastic processes*. Cambridge: Cambridge University Press.

Rudin, W. (1987). *Complex and real analysis*, 3rd edn. New York: McGraw-Hill.

Probability Theory

Billingsley, P. (1986). *Probability and measure*, 2nd edn. New York: Wiley.

Borkar, V. (1995). *Probability theory, an advanced course*. New York: Springer.

Brémaud, P. (1988). *An introduction to probabilistic modeling*. New York: Springer.

Chamond, L., & Yor, M. (2003). *Exercises in probability*. Cambridge: Cambridge University Press.

Çinlar, E. (2011). *Probability and stochastics*. New York: Springer.

Evans, M. J., & Rosenthal, J. S. (2004). *Probability and statistics*. New York: Freeman and Company.

Feller, W. (1971). *An introduction to probability theory and its applications* (Vols. 1 and 2). New York: Wiley.

Fristedt, B., & Gray, L. (1997). *A modern approach to probability theory*. Boston: Birkhäuser.

Girardin, V., & Limnios, N. (2014). *Probabilités, avec une introduction à la statistique*, 3rd edn. Paris: Vuibert.

Gut, A. (2013). *Probability: A graduate course*, 2nd edn. New York: Springer.

Ibe, O. (2005). *Fundamentals of applied probability and random processes*. Amsterdam: Elsevier Academic Press.

Lamperti, J. W. (1996). *Probability - A survey of the mathematical theory*, 2nd edn. New York: Wiley.

Roussas, G. G. (2014). *An introduction to measure-theoretic probability*, 2nd edn. Boston: Academic.

Shiryaev, A. N. (1996). *Probability*, 2nd edn. New York: Springer.

Resnick, S. I. (1999). *A probability path*. Boston: Birkhäuser.

Stochastic Processes and Applications

Baldi, P., Mazliak, L., & Priouret, P. (2002). *Martingales and markov chains*. Boca Raton: Chapman & Hall.

Barlow, R., & Proschan, F. (1975). *Statistical theory of reliability and life testing.* New York: Holt Rinehart Winston.

Brémaud, P. (1999). *Markov chains. Gibbs fields. Monte Carlo simulation, and queues.* New York: Springer.

Billingsley, P. (1978), *Ergodic theory and information.* Huntington/New York: R. E. Krieger Publishing Company.

Cover, L., & Thomas, J. (1991). *Elements of information theory.* Wiley Series in Telecommunications. New York: Wiley.

Girardin V., & Limnios N. (2014). *Probabilités, processus stochastiques et applications*, 3rd edn. Paris: Vuibert.

Grimmett, G. R., & Stirzaker, D. R. (1992). *Probability and random processes.* New York: Oxford Sciences.

Gut, A. (1988). *Stopped random walks, limit theorems and applications.* New York: Springer.

Haccou, P., Jagers P., & Vatutin, V. A. (2005). *Branching processes. Variation, growth, and extinction of population.* Cambridge: Cambridge University Press.

Higham, D. J. (2004). *An introduction to financial option valuation.* Cambridge: Cambridge University Press.

Iosifescu, M., Limnios, N., & Oprişan, G. (2007). *Modèles stochastiques.* Paris: Hermès Lavoisier.

Janssen, J., & Manca, R. (2006). *Applied semi-Markov processes.* New York: Springer.

Karlin S., Taylor H. (1984) *An introduction to stochastic modeling.* Orlando: Academic.

Kemeny, J. G., & Snell, J. L. (1976). *Finite Markov chains.* New York: Springer.

Kijima, M. (1997). *Markov processes for stochastic modeling.* London: Chapman and Hall.

Koroliuk, V. S., & Limnios, N. (2005). *Stochastic systems in merging phase space.* Singapore: World Scientific.

Limnios, N., & Oprişan, G. (2001). *Semi-Markov processes and reliability.* Boston: Birkhäuser.

Mishura, Y., & Shevchenko, G. (2017). *Theory and statistical applications of stochastic processes.* Hoboken: Iste, Wiley.

Norris, J. R. (1997). *Markov chains.* Cambridge: Cambridge University Press.

Pinsky, M. A., & Karlin, S. (2011). *An introduction to stochastic modeling*, 4th edn. Amsterdam: Elsevier.

Port, S. C. (1994). *Theoretical probability for applications.* New York: Wiley.

Resnick, S. I. (1992). *Adventure in stochastic processes.* Boston: Birkhäuser.

Ross, S. M. (1996). *Stochastic processes*, 2nd edn. New York: Wiley.

Williams, D. (1991). *Probability with martingales.* Cambridge: Cambridge University Press.

Index

absolutely continuous
  measure, 3
  random variable, 3
absorbing state, 120, 222
adapted
  process, 178
  sequence, 12, 85
σ-algebra, 2
almost sure, 2
  convergence, 34
  event, 11
alternated renewal process, 198, 212, 237
Anscombe's theorem, 182
aperiodic
  Markov chain, 120
  semi-Markov process, 240
  state, 240
AR process, 177
arithmetic distribution function, 205
ARMA process, 177, 188
associativity principle, 9
asymptotic
  σ-algebra, 10
  event, 10
attainable state, 129
auto-correlation function, 187
auto-regressive
  -moving-average process, 177
  process, 177, 189, 207, 209

ballot theorem, 51
Bernoulli
  distribution, 13, 15
  random walk, 33
binary unit, 152
binomial distribution, 15
bin packing, 108
Birkhoff theorem, 185
birth-and-death Markov

chain, 118, 120, 133, 164
  process, 242
birth Markov process, 220, 222, 228, 230
Blackwell's renewal theorem, 205
Borel
  σ-algebra, 2
  -Cantelli lemma, 11
  set, 2
bounded random sequence, 8
branching process, 154, 170
Brownian motion, 191

cadlag process, 216
canonical
  Markov chain, 119
  space, 120, 176
Cauchy
  distribution, 30
  sequence, 36
centered random variable, 5
central limit theorem, 182, 204
change of variable formula, 6
Chapman-Kolmogorov equation, 115, 159, 219
characteristic function, 14
Chernoff's
  inequality, 20
  theorem, 43
chi-squared distribution, 27
closed set, 131
cold stand-by system, 228, 231
communicating states, 129, 230
compensator, 89
complementary sets, 9
completely monotonous function, 19
compound Poisson
  distribution, 32, 103
  process, 202
conditional
  density, 65

© Springer Nature Switzerland AG 2018
V. Girardin, N. Limnios, *Applied Probability*,
https://doi.org/10.1007/978-3-319-97412-5

distribution, 64, 72
    function, 65
dominated convergence theorem, 69
expectation, 66
Fatou's lemma, 69
independence, 75
Jensen's inequality, 70
monotone convergence theorem, 69
probability, 60, 61, 64
variance, 71
condition of explosion, 229
conservative
    Markov process, 222
    state, 222
continuous
    process, 176
    time indexed process, 175
convergence
    almost sure, 34
    in distribution, 34
    of martingales, 97
    in mean, 35
    in L^p-norm, 34
    in probability, 34
    with probability one, 34
    in quadratic mean, 35
    of a random sequence, 34, 35
convolution, 25
covariance, 28
    function, 181
Cramér
    -Lundberg process, 202, 204
    transform, 19
critical case, 156
crossing time theorem, 92
cumulant generating function, 17
cyclic class, 141

degenerated distribution, 4
delayed
    renewal process, 198
    system, 250
density, 3
departure time, 84
diffeomorphism, 6
diffusion parameter, 191
Dirac distribution, 4
direct Riemann integrable, 205
discrete
    state space, 113
    time indexed process, 1, 175
distribution, 2
    function, 2

of a Markov process, 220
of a point process, 195
of a random process, 178
of a random sequence, 3
Doob's
    convergence theorem, 98
    decomposition theorem, 89
    L^p inequality, 92
    martingale inequality, 96
    maximal inequality, 90
doubly stochastic matrix, 147
drift parameter, 191

Ehrenfest Markov chain, 118, 143
elementary renewal theorem, 203
embedded chain of a
    Markov chain, 120, 163
    Markov process, 224, 226
    semi-Markov process, 237
entropy, 21
    of a random variable, 23
    of a random vector, 24
    rate, 24
        of a Markov chain, 139
        of a Markov process, 232
        of a random process, 186
        of a random sequence, 24, 187
        of a semi-Markov process, 240
epidemiological model, 245
equi-integrable sequence, 8, 55
ergodic, 189
    Markov chain, 138
    Markov process, 232
    process, 184, 186
    state, 138
    theorem, 184, 234
        of information theory, 187
    theory, 184
    transformation, 184
ergodicity, 184
Erlang distribution, 27
events, 2
exchangeable variables, 52
exit time, 83
expectation, 4
explosive Markov process, 219
exponential distribution, 19
extinction problem, 156

factorial moment, 14
failure rate, 151
Fatou's lemma, 5, 55

feed back queue, 251
filtered space, 82
filtration, 12, 82, 177
final set, 131
finite
    dimensional distribution, 178
        function, 178
    Markov chain, 145
first
    jump time, 224
    passage time, 84
Fubini's theorem, 6
fundamental matrix, 148, 150

Galton-Watson process, 154
gambler's ruin problem, 118, 126
gamma distribution, 18
Gaussian
    distribution, 4, 16
    linear model, 79
    process, 177
    sequence, 4
    vector, 49
    white noise, 176
generated σ-algebra, 2
generating function, 13
geometric distribution, 23

harmonic function, 114, 123, 161
Harris Markov chain, 118
history, 12
hitting time, 83, 124
Hoeffding's inequality, 90
homogeneous
    Markov chain, 114
    Markov process, 216
    Poisson process, 200
    process, 191

i.i.d. random sequence, 12
indefinitely divisible distribution, 26, 46
independence of sequences, 9
    of σ-algebras, 9
    of events, 10
    of random variables, 12
independent random sequence, 12
indisguishable processes, 180
inferior limit, 9
infinitesimal generator, 221
information theory, 21
initial distribution of a

Markov chain, 114
    Markov process, 216
instantaneous
    availability, 212
    power, 181
    state, 222
    variance, 180
insurance, 103, 200, 206
integrable random sequence, 8
integral representation, 188
intensity of a
    point process, 194
    Poisson process, 200
internal history, 12
invariant
    measure, 132, 184, 231
    set, 184
irreducible
    Markov chain, 131
    Markov process, 230
    semi-Markov process, 240

Jacobian, 6
jump
    Markov process, 215, 218, 238
    time of a
        Markov chain, 115
        Markov process, 218

key
    Markov renewal theorem, 240
    renewal theorem, 205
Kolmogorov
    criterion, 38
    equations, 222
    inequality, 37
    integral equation, 226
    L^2-martingale inequality, 100
    theorem, 2, 179
    zero-one law, 10
König formula, 5, 71
Krickeberg's decomposition theorem, 90

Laplace
    functional, 195
    transform, 17
large deviations, 43
law
    of iterated logarithm, 43
    of large numbers, 37, 41, 182
        strong, 39
        weak, 40

least-squares approximation, 79
Lebesgue
    dominated convergence theorem, 5
    monotone convergence theorem, 5
Lévy
    continuity theorem, 36
    theorem, 220
lifetime, 151
liminf, 8
limit
    availability, 212
    distribution of a
        Markov chain, 138
        semi-Markov process, 240
    of a sequence of events, 9
limsup, 8
linear
    approximation, 79
    model, 79
Little's formula, 245

MA process, 177
marginal distribution, 3, 178
Markov
    chain, 113
    kernel, 64
    process, 215
    property, 114, 126, 215
        strong, 124
    renewal
        equation, 238
        process, 236, 237
        theorem, 240
martingale, 59, 85, 122
    at the casino, 86
    L^p-convergence theorem, 98
    difference, 85
    strong convergence theorem, 101
maximum
    of entropy, 49
    entropy method, 24
    likelihood estimator, 171
mean
    measure, 194, 197
    of a random variable, 4
    recurrence time, 230
    return time, 240
    sojourn time, 233, 237
measurable
    function, 2
    space, 1
minimum mean square error predictor, 79
modified renewal process, 198

moment, 4
    generating function, 19
monochromatic wave, 176
Moran's reservoir, 116
moving-average process, 177, 188, 209
multidimensional process, 176
multiplicity, 193
multivariate process, 176

natural filtration, 12, 83, 177
negative binomial distribution, 13
norm, 7
normal
    distribution, 15
    equations, 80
null
    event, 11
    recurrent state, 126, 230, 240

offspring distribution, 154
operational research, 108
orbit, 185

periodic
    Markov chain, 140
    state, 120
periodicity, 240
persistent state, 126
point
    measure, 193
    process, 193, 194
pointwise ergodic theorem, 185
Poisson
    distribution, 13
    process, 196, 200
positive
    measure, 2
    recurrent state, 126, 230, 240
potential of a Markov chain, 127
predictable
    quadratic variation, 100
    sequence, 12
probability, 2
    space, 1
product
    Markov chain, 121
    probability space, 2
production system, 160

quadratic
    approximation, 68, 79

characteristic, 100
quantity of information, 21
queue, 243, 251

Radon-Nikodym
    density, 3
    theorem, 3
random
    distribution, 65
    point measure, 193
    probability, 65
    process, 175
    sequence, 1, 3, 175
    sum, 31, 102
    variable, 2
    vector, 3
    walk, 33, 52, 85, 117, 131
randomly indexed variable, 30
recurrent
    Markov chain, 131
    state, 126, 230, 240
regression, 79
    line, 81
regular
    conditional
        distribution, 66
        distribution function, 66
        probability, 63
    Markov process, 218, 229
    martingale, 87
    matrix, 145
reliability, 48, 151, 152, 169, 196, 228, 246
    function, 246, 250
renewal
    density, 197
    distribution function, 197
    equation, 197, 198
    function, 197, 238
    process, 195, 203, 237, 239
    rate, 197
return time, 124
Reuter's condition of explosion, 230
reversible
    distribution, 231
    measure, 133
risk process, 200, 206

sampling
    of a sequence, 95
    theorem, 95

second order random process, 176
seismic risk, 241
semi-Markov
    kernel, 236
    matrix, 236
    process, 236, 237
sequence of
    events, 8
    random variables, 1
Shannon entropy, 23
shift operator, 125, 176, 178
signal, 175, 177
simple
    point measure, 193
    random walk, 33
simulation of a
    Markov chain, 121
    Markov process, 227
sinusoidal signal, 176, 187, 210
sojourn time, 120, 224, 226, 237
space average, 181
spectral
    density, 188
    measure, 188
square integrable
    martingale, 99
    random sequence, 8
stable
    distribution, 30
    Markov process, 222
    state, 222
standard, variable, 5
state, 175
    space, 113, 175
stationarity, 184
stationary
    distribution, 132
    measure, 132, 231
    process, 183, 186, 187
    transformation, 184
stochastic
    basis, 82, 178
    continuity, 180
    convergence, 34
    element, 175
    integral, 88
    Kronecker lemma, 38
    matrix, 145
    process
        with independent increments, 190
        with stationary increments, 191
stochastically equivalent processes, 180

stopped
    renewal process, 199
    sequence, 84
stopping
    theorem, 94, 109
    time, 82, 93, 178
strictly stationary process, 210
strong
    law of large numbers, 39
    Markov property, 124, 126, 217
    martingale convergence theorem, 101
subcritical case, 156
submartingale, 85
sum of random variables, 25
supercritical case, 156
superior limit, 9, 46
supermartingale, 85
symmetric
    exponential distribution, 50
    simple random walk, 33

tail σ-algebra, 12
time
    of the first success, 106
    series, 176
trajectory, 3
    random sequence, 3
    stochastic process, 175
transfer theorem, 4
transformation, 184
transient

Markov chain, 131
    renewal process, 199
    state, 126, 230, 240
transition
    function, 113, 115, 218, 238
    kernel, 64
    matrix, 114
    probability, 64, 114
travelling salesman problem, 107

uncertainty, 21
uncorelated, 28
uniformisable Markov process, 223
uniformisation method, 235
uniformly integrable sequence, 55

variance, 4
version of a process, 180
visit time, 158

Wald's identity, 84
weak law of large numbers, 40
weakly stationary process, 210
white noise, 188
Wiener process, 191

Yule process, 246

Printed in the United States
By Bookmasters